T0134425

# Studies in Systems, Decision and Control

Volume 378

The series "Studies in Systems, Decision and Control" (SSDC) covers both new developments and advances, as well as the state of the art, in the various areas of broadly perceived systems, decision making and control–quickly, up to date and with a high quality. The intent is to cover the theory, applications, and perspectives on the state of the art and future developments relevant to systems, decision making, control, complex processes and related areas, as embedded in the fields of engineering, computer science, physics, economics, social and life sciences, as well as the paradigms and methodologies behind them. The series contains monographs, textbooks, lecture notes and edited volumes in systems, decision making and control spanning the areas of Cyber-Physical Systems, Autonomous Systems, Sensor Networks, Control Systems, Energy Systems, Automotive Systems, Biological Systems, Vehicular Networking and Connected Vehicles, Aerospace Systems, Automation, Manufacturing, Smart Grids, Nonlinear Systems, Power Systems, Robotics, Social Systems, Economic Systems and other. Of particular value to both the contributors and the readership are the short publication timeframe and the world-wide distribution and exposure which enable both a wide and rapid dissemination of research output.

Indexed by SCOPUS, DBLP, WTI Frankfurt eG, zbMATH, SCImago.

All books published in the series are submitted for consideration in Web of Science.

More information about this series at http://www.springer.com/series/13304

Aboul-Ella Hassanien · Sally M. Elghamrawy ·
Ivan Zelinka
Editors

# Advances in Data Science and Intelligent Data Communication Technologies for COVID-19

## Innovative Solutions Against COVID-19

*Editors*
Aboul-Ella Hassanien
Faculty of Computers and Artificial Intelligence
Cairo University
Giza, Egypt

Sally M. Elghamrawy
Department of Computer Engineering
MISR Higher Institute for Engineering and Technology
Mansoura, Egypt

Ivan Zelinka
Faculty of Electrical Engineering and Computer Science
VŠB-TU Ostrava
Ostrava, Czech Republic

ISSN 2198-4182 ISSN 2198-4190 (electronic)
Studies in Systems, Decision and Control
ISBN 978-3-030-77304-5 ISBN 978-3-030-77302-1 (eBook)
https://doi.org/10.1007/978-3-030-77302-1



This Springer imprint is published by the registered company Springer Nature Switzerland AG
The registered company address is: Gewerbestrasse 11, 6330 Cham, Switzerland

# Preface

This book presents different insights into the role of IoT against COVID-19 and its potential applications. It gives the state-of-the-art improvements in the cloud and edge computing, addressing data security and privacy for securing COVID-19 data. This will offer perceptions for developing intelligent data communication technologies. The main issues for big data analysis are investigated for tackling the huge COVID-19's data produced. This book presents different approaches for fighting against COVID-19 using different methodologies addressing data classification, data prediction, data visualization, and various data analysis techniques. In brief, this book collects original and superior studies that provide innovative solutions for data science and intelligent data communication technologies against the COVID-19 pandemic.

## Data Science Against COVID-19

### *Intelligent Computing, Machine Learning, and Data Mining*

In the chapter "Content-based Retrieval of COVID-19 Affected Chest X-rays with Siamese CNN," an automated diagnostic system is developed to separate COVID-19-affected images from the bulk image obtained from a mass screening process. Image retrieval has been carried out based on the absolute difference between the encoded features of twin images obtained from the Siamese convolutional neural network (SCNN).

Chapter "A Machine Learning System for Awareness, Diagnosing and Predicting COVID-19" proposes a system to help individual and national healthcare systems in curtailing the COVID-19 pandemic by offering chatbot, initial diagnosis for COVID-19 using chest X-ray, about symptoms, precautions, and safety measures, in early detection for COVID-19 cases.

In the chapter "Social Distancing Model Utilizing Machine Learning Techniques," the authors use artificial intelligence algorithms to keep an eye on the distances between people to reduce exposure to the COVID-19 virus. Genetic neural network (GNN) is utilized as the pioneer algorithm in dealing with features.

## *Big Data Analysis*

In the chapter "The Applications of Artificial Intelligence to Control COVID-19," the authors elaborate on the usefulness of collecting and analyzing big data obtained from different resources to forecast the spread of COVID-19. The chapter presents the numerous technologies of artificial intelligence used in the fight of COVID-19 in terms of their applications, impact, and future implications.

In the chapter "System of Systems as a Solution to Mitigate the Spread of Covid-19," the authors provide system of systems as a solution where different applications can be combined to operate in real time to mitigate the spread of COVID-19 and assist the government agencies and medical staff in working effectively.

## *Data Classification and Prediction*

In the chapter "Data Classification Model for COVID-19 Pandemic," a CXR data classification based on feature extraction, selection, and ensemble modeling is presented. The authors discuss and implement data classification tasks for early diagnosis and prognosis of the COVID-19 pandemic using CXR image. The histogram of oriented gradient (HOG) is used for the feature descriptor stage to extract useful features from the CXR image. Simultaneously, the principal component analysis (PCA) is applied for the feature selector stage to select the most salient features to help the model produce an optimal classification.

In the chapter "A Hybrid Automated Intelligent COVID-19 Classification System Based on Neutrosophic Logic and Machine Learning Techniques Using Chest X-Ray Images," a robust and intelligent hybrid framework for automatic detection of COVID-19 utilizing available resources from digital chest X-rays is proposed. The framework presented is based on the fusion of two techniques, neutrosophic techniques (NTs) and ML. Classification features are extracted from X-ray images using morphological features (MFs) and principal component analysis (PCA).

Chapter "COVID 19 Prediction Model Using Prophet Forecasting with Solution for Controlling Cases and Economy" presents a prediction model that helps plan and organize things as precautionary measures and presents the analysis of COVID-19 data. The research and prediction are performed using two methods, viz. random forest and time series. The idea behind analyzing the available dataset and the comparison of two prediction models is to supply some solutions to control the spreading of COVID-19. Gender-specific detailed analysis is also presented in this chapter with different age groups in India.

### *Data Visualization*

In the chapter "Artificial Intelligence for Strengthening Administrative and Support Services in Public Sector Amid COVID-19: Challenges and Opportunities in Pakistan," the authors aim to highlight artificial intelligence's use in governance and support services through the online survey from the local departments and administrative service providers. The scope addressed in the study included the key state divisions, trailed by the National Health and Services, Science and Technology, and Relief and Welfare departments, along with the key stakeholders in localities. The study examines the facilitative governance, shared objectives, imparting information, communication, socializing, AI expertise, and decision making over coronavirus governance. The results reflect a comprehensive strategy through artificial intelligence application to minimize the devastation caused by the event of coronavirus.

In the chapter "Artificial Intelligence in Healthcare and Medical Imaging: Role in Fighting the Spread of COVID-19," the authors aim to define the need and importance of artificial intelligence in the healthcare sector in general and the medical imaging and radiology procedures in specific. The chapter applies the AI in the radiology department and highlights its role to help the radiologists in better diagnosis, and how it could help in increasing the efficiency of the operations by using the database of information that is being gathered by the modern techniques.

## Intelligent Data Communication Technologies Against COVID-19

### *Cloud and Edge Computing*

In the chapter "An Intelligent Cloud Computing Context-Aware Model for Remote Monitoring COVID-19 Patients Using IoT Technology," the authors propose an intelligent context-aware model that adopts a hybrid architecture with both local and cloud-based components used to monitor patients, particularly COVID-19 patients with mild symptoms while they are in their homes. The cloud-based part of the system makes storing and processing easier, especially that the data generated by ambient assisted living systems are huge, particularly with patients suffering from chronic diseases, and require more frequent readings. The proposed model uses context-aware techniques by monitoring different physiological signals, surrounding ambient conditions, and the patient activities simultaneously to better understand the COVID-19 patient's health status in real time.

## Communications and Networking Technologies

Chapter "The Relationship Between the Government's official Facebook Pages and Healthcare Awareness During COVID-19 in Jordan" aims to explore the role of Facebook in health awareness concerning COVID-19 in Jordan. It has adopted the descriptive approach, using a content analysis form that consists of main categories and sub-topics carefully selected to analyze three Facebook pages for the official Jordanian institutions. These Facebook pages are directly linked to COVID-19, namely: the Prime Ministry, the Ministry of Health, and the Crisis Management Center (CMC).

In the chapter "The Influence of YouTube Videos on the Learning Experience of Disabled People During the COVID-19 Outbreak," the authors investigate the influence of YouTube usage as an educational tool on the learning process of disabled people during the COVID-19 pandemic. The researchers selected individuals working as disability specialists and used structural equation modeling to examine the proposed study model. The results revealed that there is a positive relationship between YouTube videos and e-learning among disabled individuals.

## Internet of Things (IoT)

In the chapter "IoT-Based Wearable Body Sensor Network for COVID-19 Pandemic," the authors review the role of IoT and wearable body sensor technologies in fighting COVID-19 and presents an IoT-based wearable body sensor architecture to combat the COVID-19 outbreak. IoT-based wearable body sensor is used to control and track patient conditions in both towns and cities using an internal network, thus minimizing pressure and tension on healthcare professionals, eliminating medical faults, reducing workload and medical staff productivity, reducing long-term healthcare costs, and enhancing patient satisfaction during COVID-19 pandemic.

## Data Security and Privacy

Finally, the chapter "The Dark Side of Social Media: Spreading Misleading Information During COVID-19 Crisis" aims to underscore social media's existence and how it paves the way to smooth the flow of information and communication worldwide. This chapter highlights social media's disadvantages in increasing false information that drives more uncertainty, sadness, anger, and lack of confidence among the public. The chapter identifies how governments deal with rumors in crisis, as the COVID-19 outbreak, and deliver the best practices that manage and control the fake information in social media, and determines the best ways to spread reliable and sufficient facts.

Giza, Egypt
Mansoura, Egypt
Ostrava, Czech Republic
July 2021

Aboul-Ella Hassanien
Sally M. Elghamrawy
Ivan Zelinka

# Contents

# About the Editors

**Aboul-Ella Hassanien** is Founder and Head of the Egyptian Scientific Research Group (SRGE) and Professor of information technology at the Faculty of Computer and Artificial Intelligence, Cairo University. He has more than 1000 scientific research papers published in prestigious international journals and over 50 books covering diverse topics such as data mining, medical images, intelligent systems, social networks, and smart environment. He won several awards including the Best Researcher of the Youth Award of Astronomy and Geophysics of the National Research Institute, Academy of Scientific Research (Egypt, 1990). He was also granted a scientific excellence award in humanities from the University of Kuwait for the 2004 Award and received the scientific in technology—University Award (Cairo University, 2013). Also, he was honored in Egypt as the best researcher in Cairo University in 2013. He was also received the Islamic Educational, Scientific and Cultural Organization (ISESCO) prize on Technology (2014) and received the state award of excellence in engineering sciences 2015. He held the Medal of Sciences and Arts from the first class from President of Egypt in 2017.

**Sally M. Elghamrawy** is Head of Communications and Computer Engineering Department at MISR Higher Institute for Engineering and Technology and Part-Time Associate Professor in Electrical and Computers Engineering Department, Faculty of Engineering, British University in Egypt (BUE) and in Computers Engineering Department—Faculty of Engineering, Mansoura University in Egypt. She received a Ph.D. degree in 2012 in distributed decision support systems based on multi-intelligent agents from Computers Engineering Department, Faculty of Engineering, Mansoura University, received a M.Sc. degree in automatic control systems engineering in 2006 from the same department, and received B.Sc. in computers engineering and systems in 2003. She delivered lectures and supervised graduation projects, master's thesis, and doctoral dissertations. She was a practical trainer in the grants from the Ministry of Communications and Information Technology in collaboration with IBM. She certified in Cloud Application Developer 2018 Mastery Award for Educators, Mobile Application Developer Mastery Award for Educators 2017, and A+ International Inc. CompTIA. She is Member in Scientific Research Group in Egypt. Her research focuses on big data analysis, artificial intelligence, information retrievals, and software engineering. She is Author of number of peer-reviewed publications, receiving best paper awards. She is also IEEE Member. She is Editor and Reviewer in number of international journals and Judge on IEEE Young Professionals' competitions.

**Ivan Zelinka** is currently working at the Technical University of Ostrava (VŠB-TU), Faculty of Electrical Engineering and Computer Science and National Supercomputing Center IT4Innovations. He graduated consecutively from the Technical University in Brno (1995—M.Sc.), the UTB in Zlin (2001—Ph.D.), and again the Technical University in Brno (2004—Assoc. Prof.) and VSB-TU (2010—Professor). Before his academic career, he held positions of TELECOM Technician, Computer Specialist (HW+SW), and Commercial Bank Supervisor (computer and LAN operations). During his academic career, he proposed and opened more than 10 different lectures. He also has been invited for lectures at universities in various EU countries as well as for keynote and tutorial presentations at various conferences and symposia. He is responsible for supervising the research grant from the Czech grant agency GAČR named Soft computing methods in control and was Co-supervisor of the grant FRVŠ—laboratory of parallel computing. He has also participated in numerous grants and two EU projects as Member of the team (FP5—RESTORM) and as Supervisor (FP7—PROMOEVO) of the Czech team. Currently, he is Head of the Department of Applied Informatics and throughout his career he has supervised numerous M.Sc. and B.Sc. diploma theses in addition to his role of supervising doctoral students, including students from abroad. He was awarded the Siemens Award for his Ph.D. thesis and received an award from the journal Software news for his book about artificial intelligence. He is Member of the British Computer Society, IEEE (a committee of Czech section of Computational Intelligence), and serves on international program committees of various conferences and three international journals (Soft Computing, SWEVO, Editorial Council of Security Revue). He is Author of numerous journal articles as well as books in Czech and the English language.

# Data Science Against COVID-19

# Content-Based Retrieval of COVID-19 Affected Chest X-rays with Siamese CNN

**Shuvankar Roy, Mahua Nandy Pal, Srirup Lahiri, and N. C. Pal**

**Abstract** In December 2019, the first reported case of COVID-19 was brought to notice in Wuhan, China. The virus has novel characteristics, its harshness is unpredictable, its transmission ability is extremely powerful, and its incubation period is comparatively larger. Thus the outbreak emerged as a pandemic worldwide. World health and socio-economy is getting continually affected by COVID-19 since its outbreak. It will be easier to handle the situation if an automated diagnostic system is developed, capable of separating COVID-19 affected images from bulk images obtained from a mass screening process. Kaggle's online chest X-Ray image dataset has been considered for this work evaluation. Healthy and COVID-19 affected chest X-Ray images were used for evaluating the performance of content-based image retrieval. Image retrieval has been carried out based on the absolute difference between the encoded features of twin images obtained from the Siamese Convolutional Neural Network (SCNN). The retrieval performance is awe-inspiring as the Siamese network used for retrieval is a relatively shallow network. SCNN does not require resource-hungry training with huge samples as part of its underlying implementation characteristics. The execution time is also very encouraging as the simplicity of the method is concerned. The method achieves 94% average precision and 100% average reciprocal rank while rank = 5 has been considered. Till now, no work has been reported on content-based retrieval of COVID-19 chest X-Ray images. Thus, a comparative study of evaluation metrics and execution time requirements of similar work could not be provided.

S. Roy
Tata Consultancy Services, Kolkata, West Bengal, India

M. Nandy Pal (✉)
Computer Science and Engineering Department, MCKV Institute of Engineering, 243, G.T.Road(N), Howrah 711204, India

S. Lahiri
Computer Science and Engineering Department, JIS College of Engineering, Kalyani, Nadia, India

N. C. Pal
Former Radiologist, Dr. B.C. Roy Polio Clinic and Hospital for Crippled Children, Kolkata, India



A.-E. Hassanien et al. (eds.), *Advances in Data Science and Intelligent Data Communication Technologies for COVID-19*, Studies in Systems, Decision and Control 378, https://doi.org/10.1007/978-3-030-77302-1_1

**Keywords** Chest X-ray images · COVID-19 · Siamese convolutional neural network (SCNN) · Siamese distance measurement · Mean average precision (MAP) · Mean reciprocal rank (MRR)

## 1 Introduction

COVID-19 is a representative of SARS, i.e., Severe Acute Respiratory Syndrome virus. COVID-19 outbroke at the end of 2019 in Wuhan province of China. People thought this outbreak was an epidemic within China, but later, the whole world got affected by it. In February 2020, COVID-19 was declared a pandemic all over the World by WHO. The condition was out of control and devastating. The countries like the USA, Britain, Italy, France, etc., with standard and scientific remedial amenities, face an exponential increase in the death rate. In an immensely populous country like India, the number of affected cases is adding up almost 75,000–95,000 per day to the cumulative data at the time of article composition, placing India at the top of the affected country list worldwide. However, daily deceased cases and recovery cases against daily COVID-19 infection cases in India are quite optimistic in terms of a recovery trend. The coronavirus causes a variable range of symptoms such as pneumonia, fever, difficulty in breathing, coughing and lung contamination, diarrhea, vomiting, etc. This virus is common in animals worldwide, but very few cases had been reported to affect humankind before its outburst in late December of 2019. WHO coined the term as a novel coronavirus that infected the lower respiratory tract of the people. WHO named the disease coronavirus disease (COVID-19). Avoiding contact with affected people, washing hands in regular intervals, not touching the face frequently, covering nose and mouth while coughing and sneezing, wearing masks etc. are different Infection Prevention and Control (IPC) measures.

Researchers are currently replenished with adequate data and are quite efficient in predicting this pandemic's rapid growth trend using learning and prediction tools. This type of prediction enables us to understand possible consequences that may affect enormously the socio-economical growth. Chest X-Ray is the first imaging modality and has a significance contribution to the detection of COVID-19. Content-based image retrieval (CBIR) is retrieving similar images based on the visual content of data. It is an active and current research field. In this era of digital imaging, ever-growing visual data are handled through content-based image retrieval. Retrieval system helps in developing automated image-based diagnostic system.

Further, ample availability of chest X-Ray images compared to other modalities and limited availability of properly annotated data makes the retrieval process of COVID-19 affected Chest X-Ray images more significant. The process retrieves similar images from a large database of collected chest X-Ray images based on similarity measurement. Manual retrieval of similar images from the extensive database is very time-consuming and requires many human resources. This may be considered as the motivation of this work.

## 2 Related Works

An emergency meeting was called off on 30th January, 2020 by WHO following the International Health Regulations (IHR) (2005) [1] about the epidemic of novel coronavirus 2019 in China. In their opinion, China took rapid action to help other nations to combat COVID-19. According to a research paper in The Lancet [2], the mathematical model proposed that the epidemic growth can be controlled if the transmission rate decreases to 0.25. Santosh [3] suggests AI-driven tools to forecast the nature of the spread of COVID-19 outbreaks. Currently, deep learning is applied to the medical image domain extensively. Fu et al. [4] and Liskowski and Krawice [5] are instances of deep learning applications in the medical imaging domain. Deep learning network architectures learn from a large volume of training data. The training is usually having high computational complexity and is very time-consuming with expensive resources. In [6], Pereira et al. classified COVID-19 affected chest X-ray images using multi-class and hierarchical classification approach. They used a resampling algorithm to remove the problem of data imbalance. They extracted features using some texture descriptors and passed through a pre-trained CNN model. They composed a dataset, namely RYDLS-20, for work evaluation. They achieved an F1-score of 0.65 and 0.89 in a multi-class and hierarchical classification approach. ELGhamrawy and Hassanien [7] discussed Covid-19 patient response to treatment based on CNN and Whale Optimization technique. Detection of affected images from lung CT scan images is presented in [8]. Li et al. [9] proposed a Siamese neural network-based severity score measurement system. The system automatically detects COVID-19 pulmonary disease severity in chest X-Ray images. Abbas et al. [10], used a Deep CNN—Decompose, Transfer, and Compose (DeTrac) to classify COVID-19 chest X-Ray images with an accuracy of 95.12%. They made use of the AlexNet pre-trained model for transferring knowledge gained from training on ImageNet. They evaluated work efficiency on datasets collected from several hospitals. Two works by Koch et al. [11] and by Vinyals et al. [12] tried to overcome the challenge of learning from few samples. Both the works are regarding character recognition. Koch et al. [11] exploited powerful discriminative features to optimize the predictive power of the network. They used Omniglot for their work evaluation. Vinyals et al. [12] implemented one-shot learning on ImageNet and Omniglot. They reported an accuracy of 93.2% on ImageNet and 93.8% on Omniglot. Ramachandra et al. [13], Zhang and Peng [14] and Yin et al. [15] are other different types of applications of the Siamese network.

As per our survey, no research work has been published to date, which deals with content-based retrieval of COVID-19 Chest X-Ray images. The convolution network itself takes care of the feature generation required for characterizing images. Siamese Convolutional Neural Network (SCNN) has been used for retrieval tasks in this work. The absolute difference between the encoded features of images is considered as the similarity metric. The advantage of using the Siamese network is that it does not require large training data for the network's successful training. Thus execution time requirement has also become very impressive.

The contribution of this work may be cited as follows. Remarkable retrieval performance has been achieved within a concise period of execution time with minimal effort for both feature extraction and network training. It helps categorize image samples from huge medical image screening dataset and generate a collection of labeled medical image samples. Furthermore, the outcome of the work has been verified by a registered radiologist for evaluation of its applicability in the relevant field.

Sections of this chapter are arranged as follows. Section 3 discusses deep neural networks and Siamese CNN. Section 4 is dataset description, Sect. 5 is implementation requirements. Sections 6 and 7 are the representations of similarity measurement and evaluation metrics. Section 8 is description about experimental results. Section 9 presents the discussion. Section 10 is the conclusion of the chapter.

## 3 Preliminaries

### *3.1 Convolutional Neural Network (CNN)—Its Advantages and Disadvantages*

A convolutional neural network (CNN) is an artificial neural network with one or more convolution layer/s in between input and output layers. The neurons of the fully connected layers are interconnected with weighted connections. The network computes and adjusts the correct mathematical weights to convert the input into the output. The impulse or signal passes through the layers finding the probability distribution at the output layer. Network hyperparameters are some values that are required to be decided before starting the training process of the network, such as learning algorithm, learning rate, number of hidden layers, batch size, dropout value, etc.

The advantage of using deep CNN is that the network itself can extract features of the input patterns. Thus, the programmer does not require establishing efficient handcrafted features to represent input patterns.

The disadvantage of using CNN is that they require a huge number of labeled samples for efficient and successful training. Particularly in medical applications, collecting huge labeled samples is quite difficult and, in many cases, impossible. A lot of trained samples are required to generate the output probability. One-shot learning helps to remove this disadvantage of a large number of sample requirements.

## 3.2 *Siamese CNN*

In this application, we successfully utilized the benefit of one-shot learning in the retrieval of COVID-19 affected images using Siamese Convolutional Neural Network (SCNN). Siamese means twins. Thus twin network is used to evaluate reference image and test image using the same network initialization. Hence the output vectors become comparable to each other. At this end, the absolute difference between the two feature encodings is available. Based on the similarity score generated, images are retrieved from the database where both normal and COVID-19 affected images are available simultaneously. Figure 1 shows the schematic diagram for the network used and Table 1 represents the corresponding architecture.

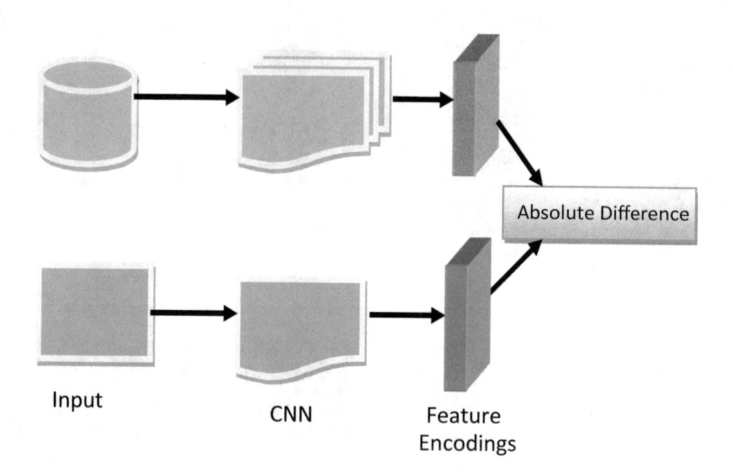


**Fig. 1** Siamese CNN model

**Table 1** Siamese CNN architecture

| Layer | (Type) | Output shape | Param# |
|---|---|---|---|
| Input_1 | (Input Layer) | [(None,1,48,48)] | 0 |
| Input_2 | (Input Layer) | [(None,1,48,48)] | 0 |
| Sequential_1 | Sequential | (None, 4096) | 302,594,880 |
| Lambda_1 | (Lambda) | (None, 4096) | 0 |
| Dense_1 | (Dense) | (None, 1) | 4097 |
| Total Params: 302,599, 977 | | | |
| Trainable Params: 302, 598, 977 | | | |
| Non-Trainable Params: 0 | | | |

## 4 Dataset Description

The dataset [16] is provided by Kaggle community. It is Covid-19 Radiography Dataset. COVID-19 affected, Normal and Viral Pneumonia affected images are kept in this dataset to help research about this pandemic. Number of COVID-19, Normal and viral pneumonia images are 219, 1341 and 1345 in the dataset. Image resolution is 1024 × 1024. Figures 2 and 3 provide some healthy and some COVID-19 samples.

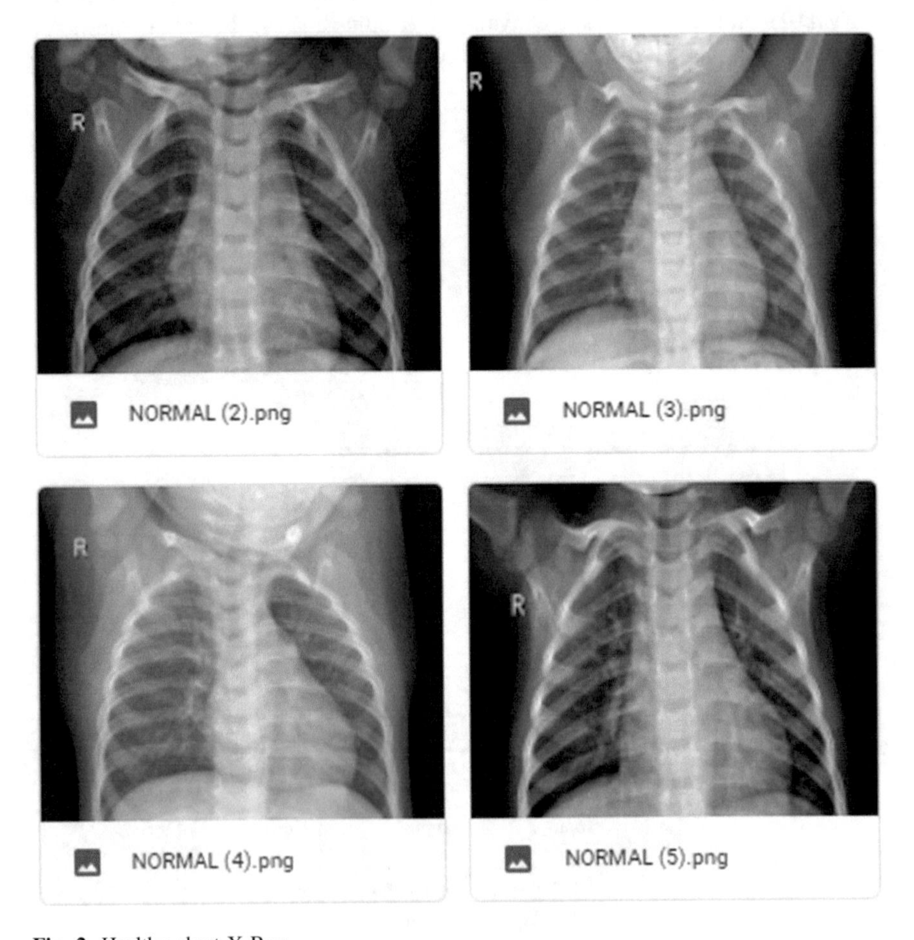


**Fig. 2** Healthy chest X-Ray

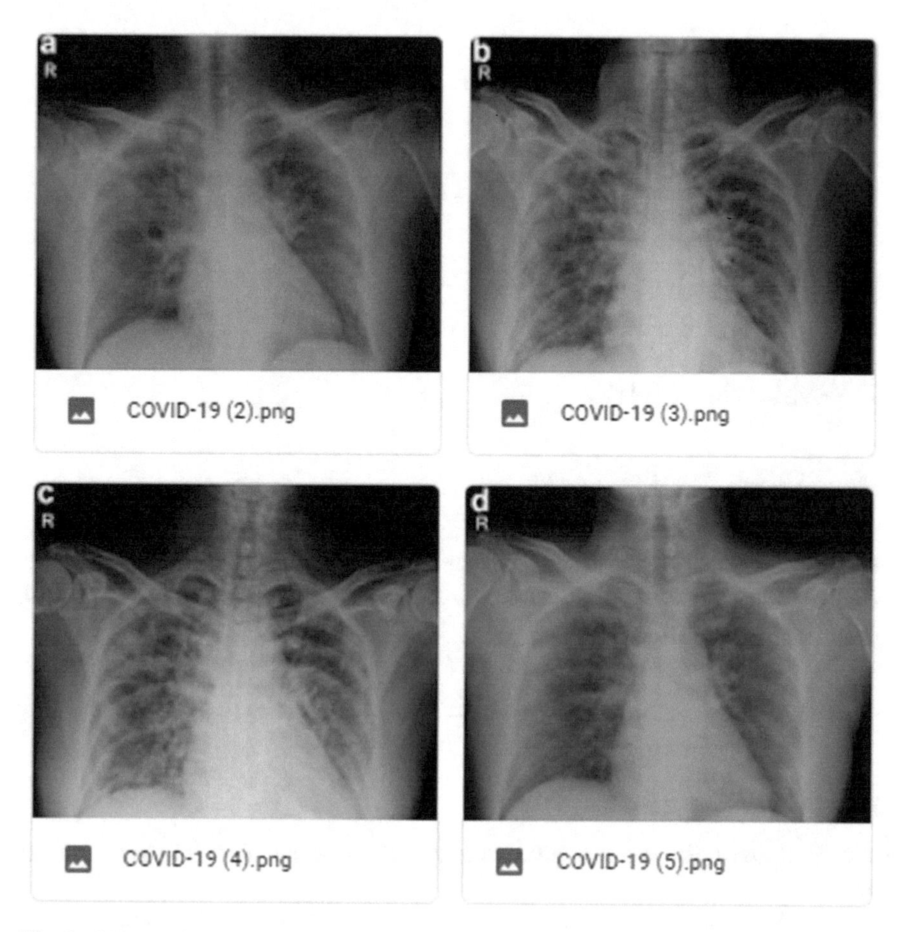


**Fig. 3** COVID-19 affected chest X-Ray

## 5 Implementation Requirements

Python 3.6.8 and Google's deep learning library TensorFlow have been used for the implementation of the work. Code execution took place in Google online cloud computing environment—Colab (Tesla K80 GPU, 12 GB VRAM).

## 6 Similarity Measurement

The similarity between two images in one shot learning by SCNN is measured as follows. X and Y are two feature encodings available from the network itself.

$$X = \begin{bmatrix} x_{11} & \cdots & x_{1n} \\ \vdots & \ddots & \vdots \\ x_{m1} & \cdots & x_{mn} \end{bmatrix} \quad Y = \begin{bmatrix} y_{11} & \cdots & y_{1n} \\ \vdots & \ddots & \vdots \\ y_{m1} & \cdots & y_{mn} \end{bmatrix}$$

$$|X - Y| = |x_{ij} - y_{ij}| \tag{1}$$

where, $x = 1, 2, 3 \ldots n$ and $y = 1, 2, 3, \ldots m$.

## 7 Evaluation Metrics

**Mean Average Precision—MAP**

MAP is the most commonly used metric for content-based image retrieval. Average Precision, AP(q) is defined by an average of precisions for a single query considering multiple ranks (k). The Mean Average Precision [21] (MAP) of a set of queries, Q is defined by,

$$\mathrm{MAP} = \frac{\sum_{\mathrm{q}=1}^{\mathrm{Q}} \mathrm{AP(q)}}{\mathrm{Q}} \tag{2}$$

**Mean Reciprocal Rank—MRR**

The Reciprocal Rank for a given query, RR(q), is the first relevant retrieval position.

The Mean Reciprocal Rank [21] (MRR) of a set of queries Q is defined by,

$$MRR = \frac{\sum_{q=1}^{Q} RR(q)}{Q} \tag{3}$$

## 8 Experimental Results

We considered 40 COVID-19 affected chest X-Ray images and 40 normal chest X-Ray images for carrying out our experiment. Images are converted to grayscale and passed through Siamese CNN. When a query image is represented to the network, it compares with all the other 79 images present.

The system retrieves images according to sorted distance. As the images are passed through CNN, representative convolution features are encoded automatically inside the network. The absolute difference between the encoded features is considered as distance measurement between two images. Two sample CSV files representing retrieved images according to distance areas in Tables 2 and 3.

### Performance Evaluation

The system performance has been estimated on 79 database images and a query image at rank k = 5 and 10. Average precisions of 80% at k = 10 and 94% at k = 5 have been computed on experimentation. Thus mean average precision of the system is 87%, considering both the ranks. In each of the retrieval results, the most similar image, retrieved is of similar grade always. Therefore, the Mean Reciprocal Rank (MRR) of our system is 100%. Image wise average metric values and mean average metric of some of the images are presented in Tables 4 and 5.

### Execution Time Requirement

To find the retrieval result for all 80 images of the resolution, 1024 $\times$ 1024 requires a retrieval time of 385.3332 s. Thus it is considered that the single image retrieval time is 4.8167 s. The rest part of the work, like importing packages and implementing functions, takes around 2.179 s. Thus creating a CSV file by comparing absolute distances of the tensors is very time efficient.

**Table 2** Sample retrieval result for the query image COVID-19(1).png according to Siamese distance

| Query image | Retrieved images | Siamese distance |
|---|---|---|
| COVID-19 (1).png | COVID-19 (7).png | 276.0009 |
| | COVID-19 (5).png | 276.2203 |
| | COVID-19 (6).png | 276.3576 |
| | COVID 19 (9).png | 276.4625 |
| | COVID-19 (2).png | 276.5063 |
| | NORMAL (9).png | 276.7117 |
| | COVID-19 (4).png | 276.7484 |
| | COVID-19 (3).png | 276.8206 |
| | NORMAL (7).png | 277.0488 |
| | NORMAL (6).png | 277.1277 |
| | NORMAL (2).png | 277.1604 |
| | COVID-19 (8).png | 277.1806 |
| | NORMAL (10).png | 277.2783 |
| | NORMAL (1).png | 277.3585 |
| | NORMAL (8).png | 277.4481 |
| | NORMAL (3).png | 277.5428 |
| | NORMAL (4).png | 277.5961 |
| | COVID-19 (10).png | 277.6339 |

**Table 3** Sample retrieval result for the query image COVID-19(8).png according to Siamese distance

| Query image | Retrieved images | Siamese distance |
|---|---|---|
| COVID-19 (8).png | COVID-19 (4).png | 276.3051147 |
| | COVID-19 (3).png | 276.326355 |
| | COVID-19 (7).png | 276.5588074 |
| | COVID-19 (9).png | 276.8031616 |
| | COVID-19 (2).png | 276.8279114 |
| | COVID-19 (5).png | 276.9436035 |
| | COVID-19 (6).png | 277.0921021 |
| | NORMAL (3).png | 277.1562195 |
| | NORMAL (4).png | 277.1602173 |
| | COVID-19 (1).png | 277.1805725 |
| | NORMAL (2).png | 277.184082 |
| | NORMAL (1).png | 277.2429504 |
| | COVID-19 (10).png | 277.2723999 |
| | NORMAL (6).png | 277.3723145 |
| | NORMAL (9).png | 277.4696045 |
| | NORMAL (10).png | 277.4779968 |
| | NORMAL (8).png | 277.500061 |
| | NORMAL (7).png | 277.6011658 |
| | NORMAL (5).png | 277.6700439 |

**Table 4** Mean average precision

| Query image | Precision | | Metric |
|---|---|---|---|
| | K = 5 | K = 10 | Avg Precision over different ranks (AP) |
| COVID-19(1) | 100 | 70 | 85 |
| COVID-19(2) | 100 | 80 | 90 |
| COVID-19(3) | 100 | 90 | 95 |
| COVID-19(4) | 100 | 90 | 95 |
| COVID-19(5) | 100 | 80 | 90 |
| COVID-19(6) | 100 | 80 | 90 |
| COVID-19(7) | 100 | 80 | 90 |
| COVID-19(8) | 100 | 80 | 90 |
| COVID-19(9) | 80 | 80 | 80 |
| COVID-19(10) | 60 | 70 | 65 |
| Mean average precision (MAP) | | | 87 |

**Table 5** Mean reciprocal rank

| Query image | Reciprocal rank | | Metric |
|---|---|---|---|
| | K = 5 | K = 10 | Reciprocal Rank (RR) |
| COVID-19(1) | 100 | 100 | 100 |
| COVID-19(2) | 100 | 100 | 100 |
| COVID-19(3) | 100 | 100 | 100 |
| COVID-19(4) | 100 | 100 | 100 |
| COVID-19(5) | 100 | 100 | 100 |
| COVID-19(6) | 100 | 100 | 100 |
| COVID-19(7) | 100 | 100 | 100 |
| COVID-19(8) | 100 | 100 | 100 |
| COVID-19(9) | 100 | 100 | 100 |
| COVID-19(10) | 100 | 100 | 100 |
| Mean reciprocal rank (MRR) | | | 100 |

## 9 Discussion

Chest X-Ray is a popular imaging modality in detecting COVID-19 infection in patients. The characteristic variations found in COVID-19 affected chest X-Rays compared with normal chest X-Ray may be represented as follows. A normal Chest X-Ray shows normal air translucency of whole lung fields with normal lobes and fissures. Tracheal transparency is in the midline. While the chest X-Ray of a patient suffering from COVID-19 infection shows multiple rounded or patchy areas of non-homogeneous consolidation throughout both lung fields and streaky opacities at both hilum. These characterizations are tried to be featured and compared automatically by CNN application to the images. CNNs have achieved outstanding performances in different research applications as they can imitate non-linear learning successfully. When deep learning capability is restricted with fewer supervised examples available or with no provision for exhaustive and expensive training, One-shot learning with Siamese CNN comes into the rescue. This variation of machine learning is still less explored than other machine learning algorithms and is mostly used in the domain of character recognition. Siamese CNN automatically acquires features through convolution layers and the model can evaluate images in pair-wise fashion. Depending upon the modulus distance between the feature representations of a pair of query and database image, content-based retrieval of images occurs. Thus Siamese distance has been used for similarity measurement between the images.

The retrieval system's performances, without any pre-processing adopted, are very impressive with a mean average precision of 87% and a mean reciprocal rank of 100%. Graphical representation of precision data has also been presented for better visualization in Fig. 4. Execution time is also very encouraging, with 4.8167 s of retrieval time per image in Google's freely available cloud execution

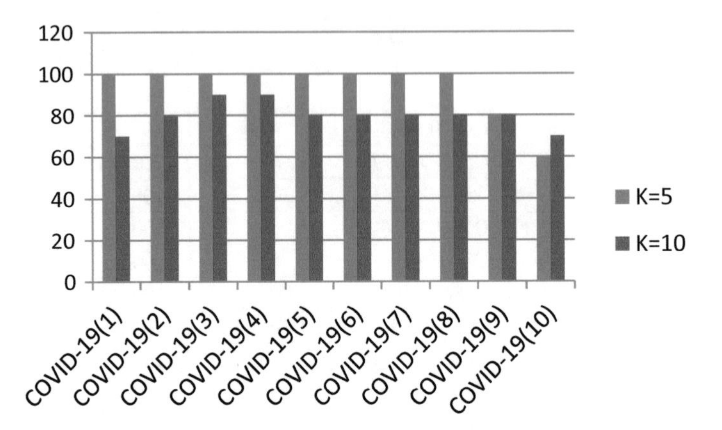


**Fig. 4** Precision graph of 10 arbitrary images for rank = 5 and 10

environment. It is worth mentioning that the CBIR results have been further verified by a registered medical practitioner (Registration number—29,631 of West Bengal Medical Council, India) for their applicability and acceptability.

Comparison with state of the art is not provided as no CBIR work has been published yet on COVID-19 chest X-Ray image dataset. Only one instance of Siamese CNN on COVID-19 chest X-ray images is available [9] though the application domain is different. In [9] it is used for pulmonary disease severity detection. We applied Siamese CNN to a completely different domain of content based image retrieval based on Siamese distances. Other Siamese CNN related works available in literature are not only from different research domain but also from different application area.

## 10 Conclusion and Future Work

X-Ray is the most primitive medical imaging technique. Chest X-Ray of COVID-19 patients plays an essential role in the diagnosis of the disease. The retrieval performance of a content-based image retrieval system crucially depends on the feature representation and similarity measurement.

Initially, database images are not annotated with class values. To capture the semantics of query image and database images and annotate database images successfully, we implemented a Siamese CNN application. CNN is supposed to extract high-level human perceived content descriptors, which help reduce the semantic gap between the actual image and its encoded features. Moreover, for workflow optimization, one-shot learning has been adopted. Thus, arbitrarily collected chest X-Ray images are automatically annotated with proper class values.

The work's future direction is to apply the proposed method of content-based chest X-Ray image retrieval to a larger volume of data and evaluate the retrieval results using the conventional metrics. Experimentation on model architecture is another scope of further work progression. The model training turns out to be more efficient with less available data and can frame the retrieval task more correctly. We aim in the future to extend the experiment with CBIR application to bacterial pneumonia chest X-Ray images also.

## References


1. Statement on the second meeting of the International Health Regulations (2005) Emergency committee regarding the outbreak of novel coronavirus (2019-nCoV). World Health Organization (WHO). 30 Jan 2020 Switzerland Geneva
2. Wu, J.T., Leung, K., Leung, G.M.: Nowcasting and forecasting the potential domestic and international spread of the 2019-nCoV outbreak originating in Wuhan, China: a modelling study. The Lancet 1–9 (2020) https://doi.org/10.1016/S0140-6736(20)30260-9
3. Santosh, K.C.: AI-driven tools for coronavirus outbreak: need of active learning and cross-population train/test models on multitudinal/multimodal data. J. Med. Syst. **44**, 93 (2020)
4. Fu, H., Xu, Y., Wong, D.W.K.: Retinal vessel segmentation via deep learning networkand fully connected conditional random fields. In: 2016 IEEE Symposium on Biomedical Imaging (2016)
5. Liskowski, P., Krawice, K.: Segmenting retinal blood vessels with deep neural networks. IEEE Trans. Med. Imaging **35**(11), 2369–2380 (2016)
6. Pereira, R.M., Bertolini, D., Teixeira, L.O., Silla, Jr., C.N., Costa, Y.M.: COVID-19 identification in chest X-ray images on flat and hierarchical classification scenarios. Comput. Methods Prog. Biomed. 105532 (2020)
7. ELGhamrawy, S.M., Hassanien, A.E.: Diagnosis and Prediction Model for COVID19 Patients Response to Treatment based on Convolutional Neural Networks and Whale Optimization Algorithm Using CT Image (2020). medRxiv 2020.04.16.20063990, https://doi.org/10.1101/2020.04.16.20063990
8. Rajinikanth, V., Dey, N., Joseph Raj, A.N., Hassanien, A.E., Santosh, K.C., Sri Madhava Raja, N.: Harmony-Search and Otsu based System for Coronavirus Disease (COVID-19) Detection using Lung CT Scan Images. (2020). arXiv:2004.03431
9. Li, M.D., Arun, N.T., Gidwani, M., Chang, K., Deng, F., Little, B.P., ... Parakh, A.: Automated assessment of COVID-19 pulmonary disease severity on chest radiographs using convolutional Siamese neural networks. medRxiv (2020)
10. Abbas, A., Abdelsamea, M.M., Gaber, M.M.: Classification of COVID-19 in chest X-ray images using DeTraC deep convolutional neural network. (2020). arXiv:2003.13815
11. Koch, G., Zemel, R., Salakhutdinov, R.: Siamese neural networks for one-shot image recognition. In: ICML Deep Learning Workshop, vol. 2 (2015)
12. Vinyals, O., Blundell, C., Lillicrap, T., Wierstra, D.: Matching networks for one shot learning. In:Advances in Neural Information Processing Systems, pp. 3630–3638 (2016)
13. Ramachandra, B., Jones, M., Vatsavai, R.: Learning a distance function with a Siamese network to localize anomalies in videos. In: The IEEE Winter Conference on Applications of Computer Vision, pp. 2598–2607 (2020)
14. Zhang, Z., Peng, H.: Deeper and wider siamese networks for real-time visual tracking. In: Proceedings of the IEEE Conference on Computer Vision and Pattern Recognition, pp. 4591–4600 (2019)

15. Yin, Z., Wen, C., Huang, Z., Yang, F., Yang, Z.: SiamVGG-LLC: Visual tracking using LLC and deeper siamese networks. In: 2019 IEEE 19th International Conference on Communication Technology (ICCT), pp. 1683–1687. IEEE (2019)
16. COVID-19 Radiography Database. https://www.kaggle.com/tawsifurrahman/covid19-radiography-database/data?select=COVID-19+Radiography+Database. Accessed on 19 Sept 2020

# A Machine Learning System for Awareness, Diagnosing and Predicting COVID-19

**Rania ElGohary, Ahmed Hisham, Mohamed Salama, Yousef A. Yousef Selim, and M. S. Abdelwahab**

**Abstract** Technology plays a vital role in our lives, and its role magnifies in crises like the COVID-19 pandemic. Technology reduced the effects of lockdown by helping in education, healthcare, industry sectors. This book chapter introduces an innovative system that uses contemporary machine learning techniques to stop the COVID-19 virus outbreak. This system provides guidance and awareness for individuals through chatbot, initial diagnosis for COVID-19 using chest X-ray. Moreover, it gives predictions for COVID-19 new cases. The proposed system can help individual and national healthcare systems curtailing the COVID-19 pandemic by offering chatbot about symptoms, precautions, and safety measures in early detection for COVID-19 cases. The developed system Predict chest X-ray for new coronavirus new case and the similar diagnosis symptoms to support governments by automatically reports for the future of the pandemic and helping the decision-makers make better decisions in quarantine lockdown.

**Keywords** COVID-19 · Artificial intelligence · Deep learning · Recurrent neural networks · LSTMs · Medical image processing · Time series forecasting

---

R. ElGohary (✉) · M. S. Abdelwahab
Misr University for Science and Technology, Giza, Egypt
e-mail: Rania_elgohary@must.edu.eg

M. S. Abdelwahab
e-mail: mswahab@must.edu.eg

A. Hisham
Artificial Deep Intelligence Technology Center, Misr University for Science and Technology, Giza, Egypt
e-mail: Ahmed_mesbah@must.edu.eg

M. Salama
Artificial Intelligence Technology Center, Misr University for Science and Technology, Giza, Egypt
e-mail: Mohamed_salama@must.edu.eg

Y. A. Y. Selim
Souad Kafafi Hospital, Giza, Egypt
e-mail: youssef.selim@must.edu.eg



A.-E. Hassanien et al. (eds.), *Advances in Data Science and Intelligent Data Communication Technologies for COVID-19*, Studies in Systems, Decision and Control 378, https://doi.org/10.1007/978-3-030-77302-1_2

## 1 Introduction

Coronaviruses are considered an expansive family of infections known to cause sickness varying from the common cold to more serious illnesses such as Middle East Respiratory Syndrome (MERS) and Severe Acute Respiratory Syndrome (SARS). The novel coronavirus (COVID-19) was investigated in 2019 in Wuhan, China. This new coronavirus has not previously been diagnosed in humans [1]. The first infection is suspected to be a live animal market, but the Virus is now transferring from one person to another. It is essential to understand that person-to-person spread can happen on a continuum. Some viruses are highly contagious (like measles), while other viruses might not be the same criteria [2]. The Virus that causes COVID-19 seems to be spreading easily and sustainably in the community ("community spread") in so many geographic areas. Community spread implies individuals have been contaminated with the infection in a region, counting a few who are not, beyond any doubt, how or where they have to be contaminated [3]. The infection that causes COVID-19 is transferring from an individual to another. Someone who is effectively wiped out with COVID-19 can spread the ailment to others. That is why CDC suggests that these patients be isolated either within the hospital or at domestic (depending on how sick they are) until they feel better and not subjected to the chance of infecting others. How long somebody is effectively sick can change when to release somebody from isolation made on a case-by-case premise in meeting with specialists. Disease avoidance and control specialists and open health authorities are considering each circumstance's specifics, counting illness severity, sickness signs, and side effects, and it comes about research facility testing for that understanding [4, 5]. Then moving to symptoms of this new Virus, these symptoms may appear 2–14 days after exposure. Many countries started the stage of lockdown to control pandemic growth. However, the pandemic did not descend, and the pandemic kept rising in a trending way. Some countries using current technology systems such as big data, machine learning, and data science to control the pandemic in a faster smarter way, to highlight one of the experiences. The South Korean use of AI and big data to fight the COVID-19, using big-data analysis, intelligent-powered super alarm systems, and intensive observation methodology, South Korea has already managed to bring the country's virus situation under spot during a short time. The government-run big-data platform stores all citizens and resident foreign nationals' data and integrates all government organizations, hospitals, financial services, mobile operators, and other services into it. South Korea is using the analysis, data and references given by this integrated data all different real-time responses and knowledge produced by the platform promptly conveyed to people with different AI-based applications. Whenever someone tested positive for COVID-19, all the people within the vicinity given the infected person's travel details, activities, and commute maps for the previous fortnight through mobile notifications sent as a push system. South Korea also introduced drive-through coronavirus testing, during which an individual drives his car inside a quantized testing lab, gets his samples collected while staying inside the car, and gets test results within a couple of minutes. If found to be infected, they're immediately isolated and brought to specialized treatment facilities. Not only but also If any infected person lived or worked at an outsized building, temporary medical centers are found out there to supply medical tests to all or any residents.AI data analysis recommends to the government about possible clusters of the Virus, or areas with the most risk, thus enabling prompt medical services and mobilizing awareness initiatives in those areas [6].

## 2 Literature Review

Background and cardinal clinical features Fudan University, Shanghai, China, a research group proposed a time forecasting model using a Modified autoencoder to predict the recorded cases per day. They used the new case by case, as a day function, and their model predicted the results from Jan 2020, until 20 April 2020, by using the feature clustering. The system able to predict the time series for thirty-one states in the Chinese lands, the features were grouped from nine clusters in three main spots (including the Outbreak land Hubei), where each group of regions in the three divisions formed a cluster, while Hubei had its own cluster. The architecture of the modified auto encoder as shown in Fig. 1.

The MAE architecture consists of the following: an input layer: first latent layer, second latent layer, and the output layers, that are consisted of 8,32,2,1 Nodes respectively, then the team segmented the data into eight days with 128 segments as the training data, then normalizing the data and later taking the average mean element of the time forecasting element per day, the team was able to predict the following result [7] (Table 1).

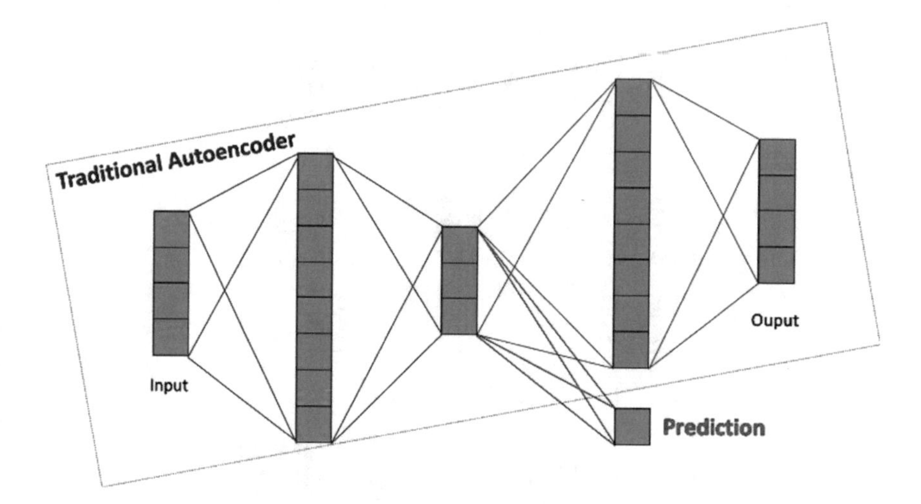


**Fig. 1** Mean auto-encoder

**Table 1** MAE predicted time series

| Date | Actual | 1-Step prediction | 1-Step error | 2-Step error | 3-Step error | 4-Step error | 5-Step error | 6-Step error | 7-Step error | 8-Step error | 9-Step error | 10-Step error |
|---|---|---|---|---|---|---|---|---|---|---|---|---|
| 18/2/2020 | 72,528 | 71,757 | −1.06% | | | | | | | | | |
| 19/2/2020 | 74,280 | 74,005 | −0.37% | −1.34% | | | | | | | | |
| 20/2/2020 | 74,675 | 75,564 | 1.19% | 1.31% | −0.06% | | | | | | | |
| 21/2/2020 | 75,569 | 76,685 | 1.48% | 1.36% | 1.49% | 0.12% | | | | | | |
| 22/2/2020 | 76,392 | 76,305 | −0.11% | 1.49% | 1.41% | 1.81% | 0.15% | | | | | |
| 23/2/2020 | 77,042 | 77,827 | 1.02% | 0.02% | 1.58% | 1.39% | 2.30% | 0.13% | | | | |
| 24/2/2020 | 77,262 | 79,837 | 3.33% | 2.18% | 0.83% | 2.53% | 2.46% | 3.30% | 1.08% | | | |
| 25/2/2020 | 77,780 | 79,155 | 1.77% | 3.01% | 2.14% | 0.47% | 2.28% | 2.36% | 3.15% | 0.76% | | |
| 26/2/2020 | 78,191 | 77,957 | −0.30% | 1.61% | 2.70% | 2.15% | 0.28% | 2.20% | 2.49% | 3.21% | 0.80% | |
| 27/2/2020 | 78,630 | 78,646 | 0.02% | −0.53% | 1.49% | 2.47% | 2.22% | 0.21% | 2.37% | 2.46% | 3.37% | 0.73% |
| Average absolute error | 1.07% | 1.43% | 1.46% | 1.56% | 1.62% | 1.64% | 2.27% | 2.14% | 2.08% | 0.73% | | |

The model was tested for ten days giving excellent results, where the predicted result was compared to that of what it was expected 10 days later, as a result, the step error was noticed to have a variable average absolute error along the time series, where its unstable error rate was very noticeable [7].

Another approach was hybrid AI time series system for prediction, and this system differs from the previous system as it uses the LSTM instead of auto encoder, the hybrid model, the model is based on infection rate and NLP (natural language processing) where it's based on the following (MLP: multi-layer perceptron, LSTM, NLP, CDC data).

The hybrid model in Fig. 2 shows the hybridization of the LSTM and the NLP. where the NLP is used to extract the data from the news, where the output prediction is fed as the input label for the next recurrent unit then hybridized with NLP, at time step t + 1 to produce the prediction (t + 1), thus this architecture could handle more stabilized model in the short term prediction better than that of the MAE (modified autoencoder). As shown in Table 2 [8–10].

Vinay Kumar Reddy Chimmula, Lei Zhang from Faculty of Engineering and Applied Science, University of Regina, Regina, Saskatchewan, S4S0A2 Canada proposed a Time series forecasting of COVID-19 transmission method and applied this method in Canada and Italy. The proposed method used the LSTM network to forecast the future COVID-19 cases. They predicted the possible end of COVID-19 outbreak will be around June 2020 using the results of their models. Data provided from Hopkins University and Canadian Health authority used in training and testing their models. This data contains a number of COVID-19 positive cases, fatalities and recovered patients for each day until March 31. The dataset divided 80% for training and 20% for testing.The team built two LSTM models with RMSE loss function to predict short-term and long-term prediction in Canada. First model trained and tested on Canadian dataset, error was 34.83 with an accuracy of 93.4 for short-term predictions and 45.70 with an accuracy of 92.67% for long-term predictions. Second LSTM model trained on Italian dataset to predict short-term and long-term. For short-term predictions, the error is about 51.46%.

In Fig. 3 The red dotted lines represent the sudden changes from where the number of infections started following exponential trends. The black dotted lines in the figure represent the training data or available confirmed cases. However, this study was not completed where this study targets a specific period where the long-term prediction ended in 28–04–2020, while the pandemic is still rising [11].

## 3 Stage One (Awareness Stage)

This section proposes the first stage, an awareness chatbot responsible for answering different queries about the novel Virus (COVID-19). Those queries are answered using NLP (natural language processing), where the CDC (center of disease control and prevention) trains the bot on some FAQs and the WHO (Worldwide health organization). The trusted answers about the awareness and

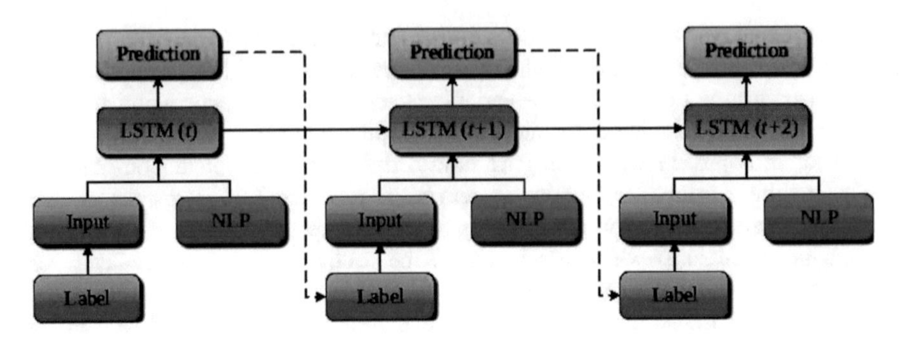


**Fig. 2** The hybrid model

**Table 2** The hybrid model performance over the MAE model and MAPE model

| | SI | ISI | ISI + LSTM | ISI + NLP + LSTM | GT |
|---|---|---|---|---|---|
| Feb. 19 | 45,794 | 45,260 | 45,175 | 44,970 | 45,027 |
| Feb. 20 | 47,030 | 45,997 | 46,307 | 45,504 | 45,346 |
| Feb. 21 | 48,130 | 46,628 | 46,918 | 45,872 | 45,660 |
| Feb. 22 | 49,106 | 47,138 | 47,665 | 46,163 | 46,201 |
| Feb. 23 | 49,970 | 47,532 | 48,045 | 46,265 | 46,607 |
| Feb. 24 | 50,732 | 47,842 | 48,538 | 46,439 | 47,071 |
| MAE | 2475 | 747.5 | 1122.67 | 239.83 | 0 |
| MAPE | 0.0535 | 0.0162 | 0.0243 | 0.0052 | 0 |

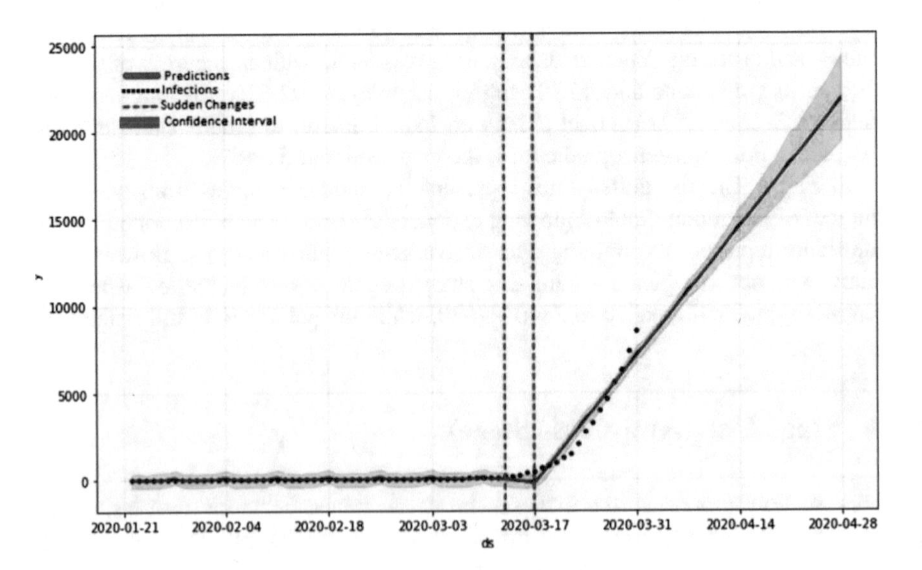


**Fig. 3** Predictions of the LSTM model on exposed and infectious cases (Red solid line)

definitions of some of the Arabic language's virus terminologies are represented. This Corpus contains the following: Coronavirus definition Coronaviruses are a family of viruses that can cause illnesses such as the common cold, severe acute respiratory syndrome (SARS) and Middle East respiratory syndrome (MERS), severe common diagnostic symptoms, video tutorials on washing hands in the right way and putting on facemasks, common interactive questions [12].

The Chabot built through three main stages: the first one is the corpus where the Chabot training data collected from the gathered data in CDC. WHO and Egyptian ministry of health and these data provided in a main corpus, the Arabic language, where the third corpus is the useful YouTube tutorial links for washing hand, explanation of the right way to put the facial mask and more health care issues. With the Use of Google Firebase (for storing conversation interactions to evaluate the bot at the backend and modify it), the Chabot will be able to answer most of the queries from the end users about COVID-19, sustaining the target of the awareness. DialogFlow also enabled the idea of implementing and providing more corpus, as much as adding the bot to webhook, twitter, telegram and other messaging platforms that people usually use in their interaction.

As mentioned in Fig. 4, the Chabot uses the three main databases corpora. the three main corpora are the English language queries of the COVID-19, Arabic language queries of the COVID-19 and the useful video links for self-care,before the machine learning algorithms used the preprocessing part where the cleaning of the text occurred as the system uses Stop-Words and Stemming, where the stage of using the Stop-Words refers to some words in the English language which are not high effectiveness. Words such as "a", "the", etc. are common that they are generally not effective when doing intent mapping. While the stage of Stemming refers to Words, which have a common root, such as connect, Dialogflow as well as other NLU bot frameworks as being similar or even identical treat connecting, connected, and connection. This root is called the "stem" of the word and stemming is what helps Dialogflow manage multiple variants of the same basic "word concept" so it can do better intent matching. The searching and the matching algorithms used are a set hybrid machine learning algorithms such as reinforcement learning where each intent from the mentioned list like (COVID-19 symptoms, tests, latency period.) are trained with at least 15 phrases to force the agent to focus on the main keywords to become more intelligent.

## 4 Stage Two (Chest X-Ray-Diagnosis)

In this section, a solution is introduced that provides an initial diagnosis for COVID-19 Virus using medical chest X-ray images based on deep learning [13–16]. In COVID-19, CXR shows patchy or diffuse reticular–nodular opacities and consolidation, with basal, peripheral, and bilateral predominance [17] in Fig. 5 shows the effects of COVID-19 on chest X-ray. This solution could be used in health care facilities to accelerate checking suspects of COVID-19 and isolate

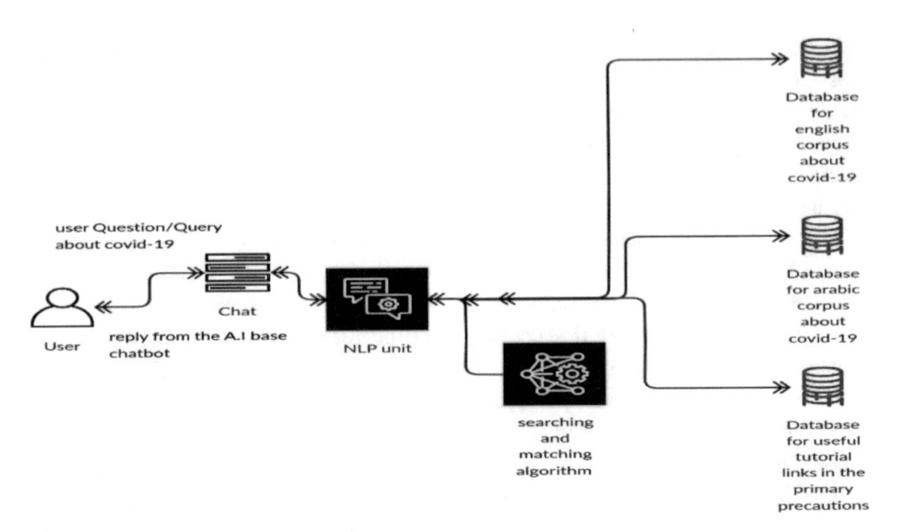


**Fig. 4** Chatbot architecture

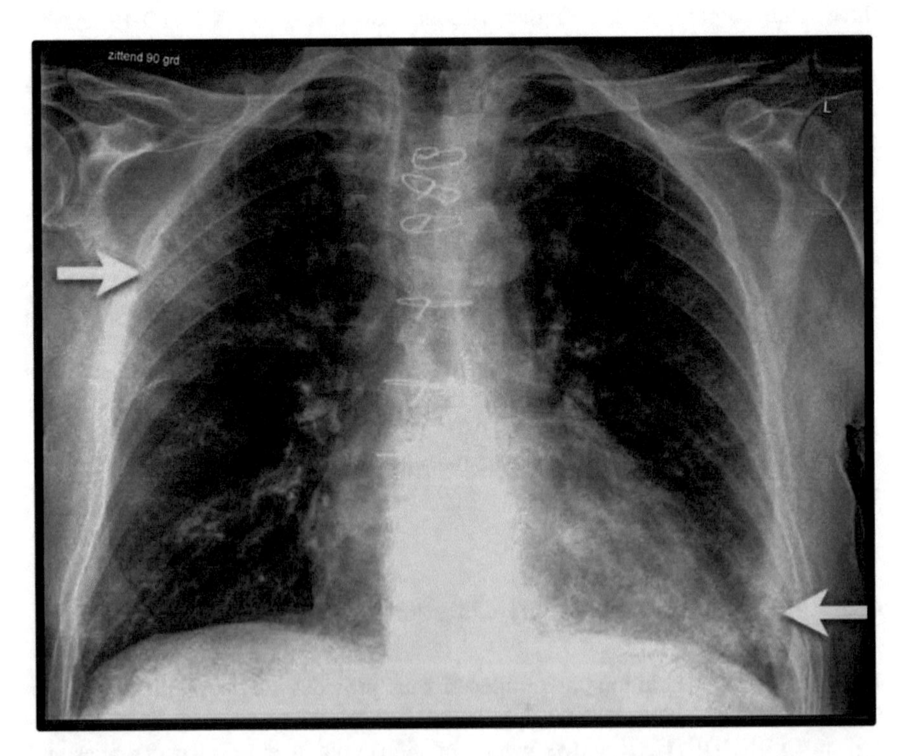


**Fig. 5** Bilateral patchy areas of peripheral consolidation. Right peripheral consolidation. No pleural effusion. The mediastinum is unremarkable

positive cases in a short time. Isolating positive cases is an essential step to stop spreading COVID-19. Also, this solution could be used by individuals to give them initial diagnosis. As this solution is a part of a mobile app, users' errors are considered in this solution by taking care of the possibility of uploading a non-chest X-ray image. This solution consists of two convolutional neural networks, each image passing through these models in series. Uploaded images pass through a model to check whether this image is a chest X-ray or regular image. Secondly, if the image is not a chest X-ray image, the system outputs a not valid image message and if the image is a chest X-ray, it passes through the second model. This model detects whether it is an image for a COVID-19 patient or not. In Fig. 6 a high-level flowchart for this solution.

a. **Dataset**

One of the difficulties of using deep learning is that deep learning models need a huge amount of data to train and test. Two datasets are needed, one for the model that checks whether an image is a chest X-ray or not and the other dataset for the model that detects whether it is a positive case or a negative. Since, COVID-19 is a new disease; there is no well-structured dataset for these problems and datasets used in training and testing our models collected from multiple resources.

First Dataset is a combination of images from three different datasets. A public dataset of Covid-19 chest X-ray images is available at IEEE 8023 [18]. This dataset collected from multiple public sources and through indirect collection from hospitals and physicians. This database is updated with new images regularly. At the time of implementing this model, this database contains 458 novel coronavirus chest X-ray. A dataset provided on Kaggle for Pneumonia detection [19]. This dataset contains 1583 normal chest X-ray and 4262 Pneumonia chest X-ray. Finally, A dataset for scene recognition problems that contain 3907 images for ten classes: airport, bakery, bedroom, greenhouse, gym, kitchen, operating room, pool, restaurant and toy store. So, a dataset is created containing 6303 chest X-ray images and 3907 non-chest X-ray images. The second dataset was created using the COVID-19 chest X-ray public dataset containing 458 COVID-19 chest X-ray images and 458 randomly chosen images from the Pneumonia detection dataset. All images resized to be 224 * 224 pixels using Pillow library, an open-source Python Imaging Library. Datasets split randomly to 80% training data and 20 validation data.

b. **Model architecture and results**

Vgg16 [20] model architecture pre-trained on ImageNet dataset is used in the two models, followed by three layers, a flatten layer, 128 neuron feed forward layer with Relu activation function and output layer with sigmoid activation function. In Figs. 7 and 8 Vgg16 architecture and model architecture are shown respectively.

Models are trained with Adam optimizer [21, 22] with 32 batch size and binary cross-entropy loss function. First model that checks images whether chest X-ray or not reached 100% accuracy for training and validation data after 5 epochs of

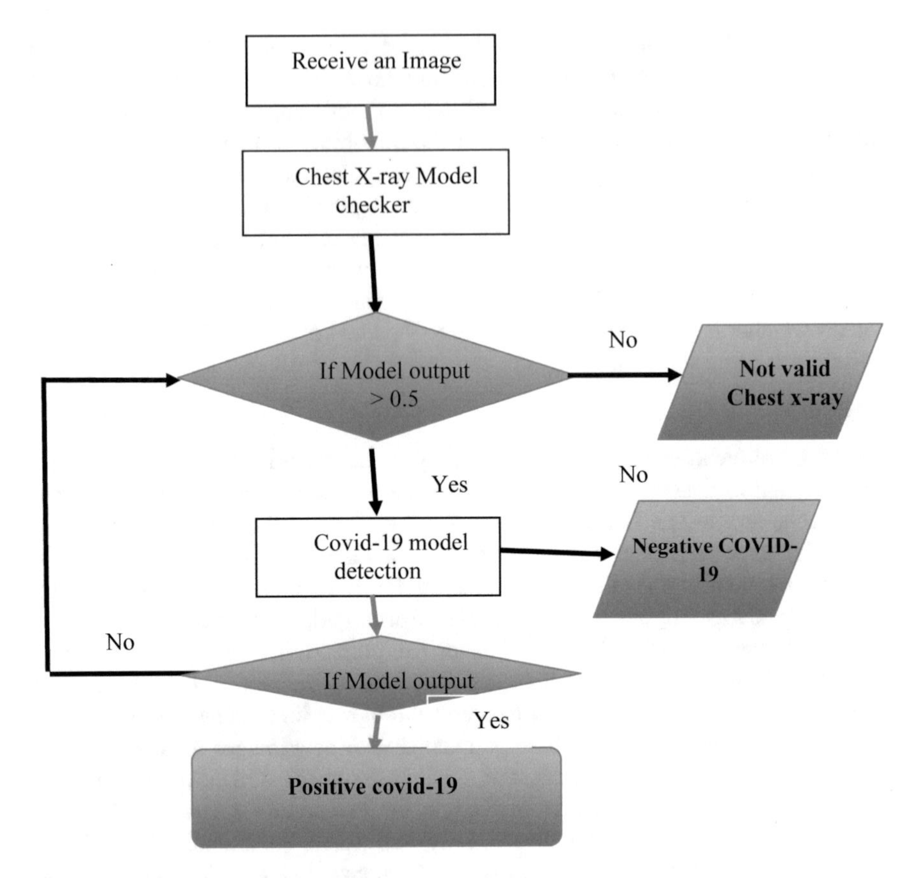


**Fig. 6** Flowchart describing our solution

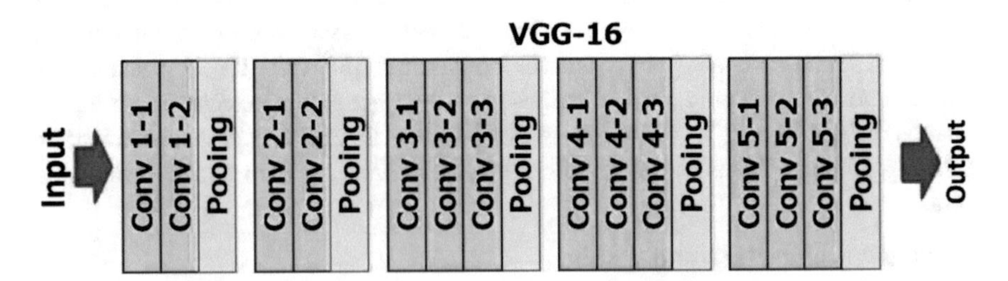


**Fig. 7** Vgg16 architecture

training as there is enough amount of training data and differences between chest X-ray and other images are well defined and easy to detect. Second model that detected COVID-19 from a chest X-ray Image reached 100% accuracy for testing and validation data after 25 epochs.

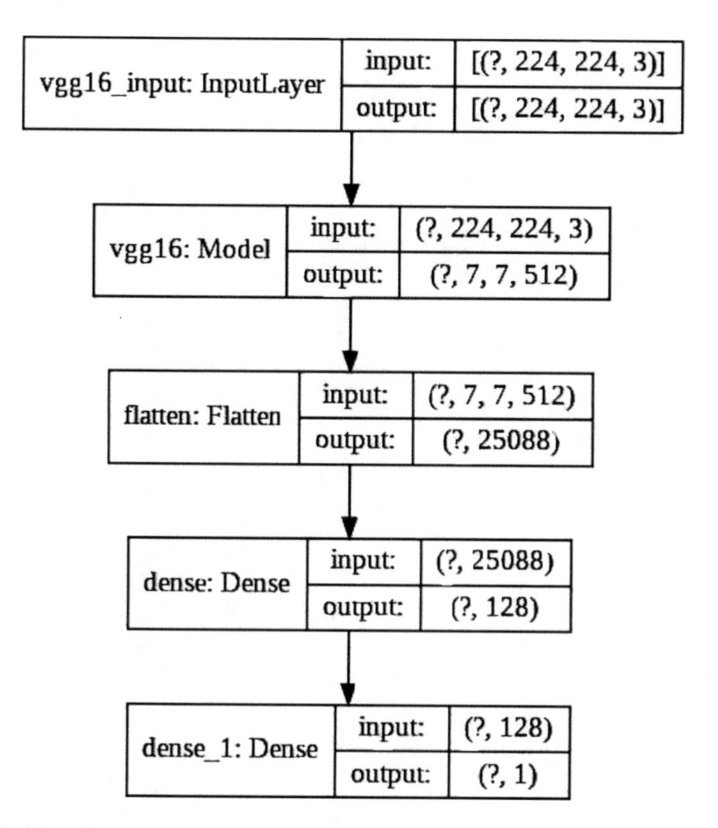


**Fig. 8** Model architecture

Tensor Flow and Keras frameworks used to build and train these models. Figures 9 and 10 loss and accuracy curves shown for both models and Table 1. Figure 1 shows precision, recall, f1-score and support results on validation data and ROC curve (Table 3).

## 5 Stage Three (COVID-19 Predictor Forecast Model)

The live Predictor using time series forecasting prediction, data for the infected and non-infected cases worldwide coming from Kaggle and Johns Hopkins University. The methodology in developing the live Predictor inherits the features of the weather forecasting model. The operation is following the weather forecasting model in the historic numerical values storing. Furthermore, these histories are used as its training weights. The variation is stored and treated as the training data, where the model can understand the changes occurring in the days/records, so the model can adapt and update itself no matter how the variations changed.

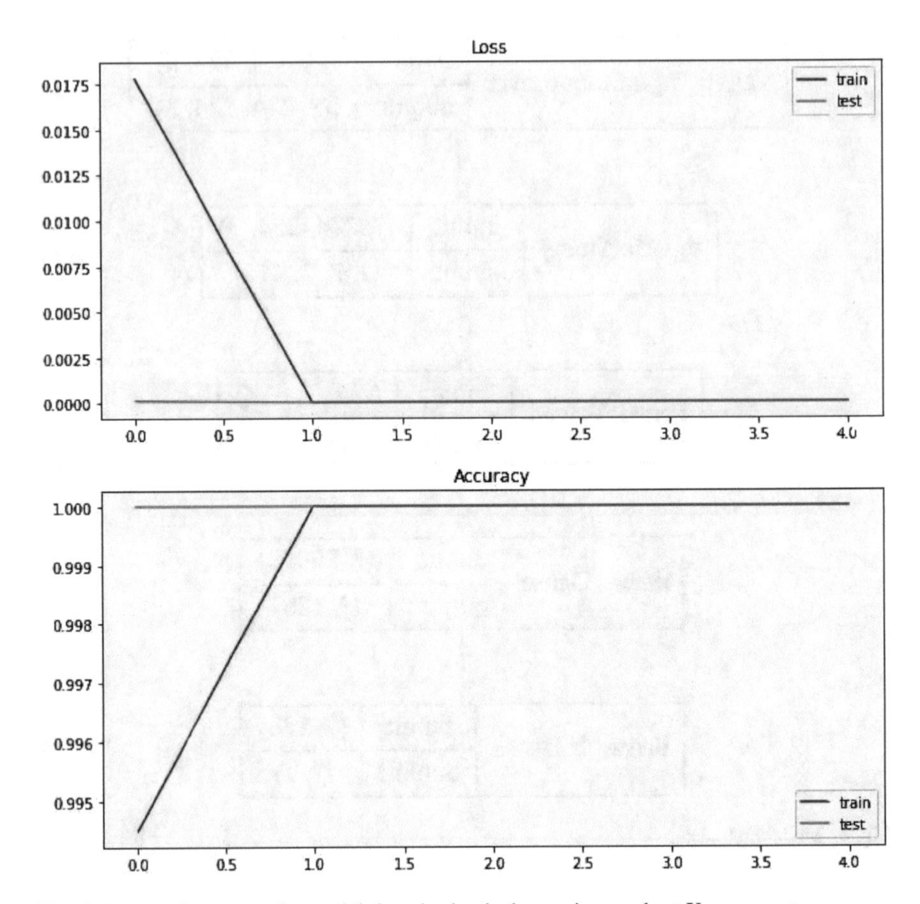


**Fig. 9** Loss and accuracy for model that check whether an image chest X-ray or not

The selected model is Bidirectional LSTM (long short term memory), BiLSTMs [23, 24] are commonly used in sequence analysis and prediction, referred to the introduced problem definition. This data type is a time-series dataset that is considered to be a sort of sequence data-type. The proposed design is illustrated in Figs. 11, 12 and 13.

LSTMs structure is shown in Fig. 14 where each LSTM cell consists of four gates update gate Γu that calculate how much past should matter now, relevance gate Γr where responsible for dropping previous information, forget gate Γf that erase a cell or not and output gate Γo that calculate how much to reveal of a cell. Values of c(t), c ∼ (t) and a(t) are calculated using this gates values as follow

$$c\sim(t) = \tanh\,(Wc[\Gamma r{*}a(t-1),\ x(t)] + bc) \tag{1}$$

$$c(t) = \Gamma u{*}c\sim(t) + \Gamma f{*}c(t-1) \tag{2}$$

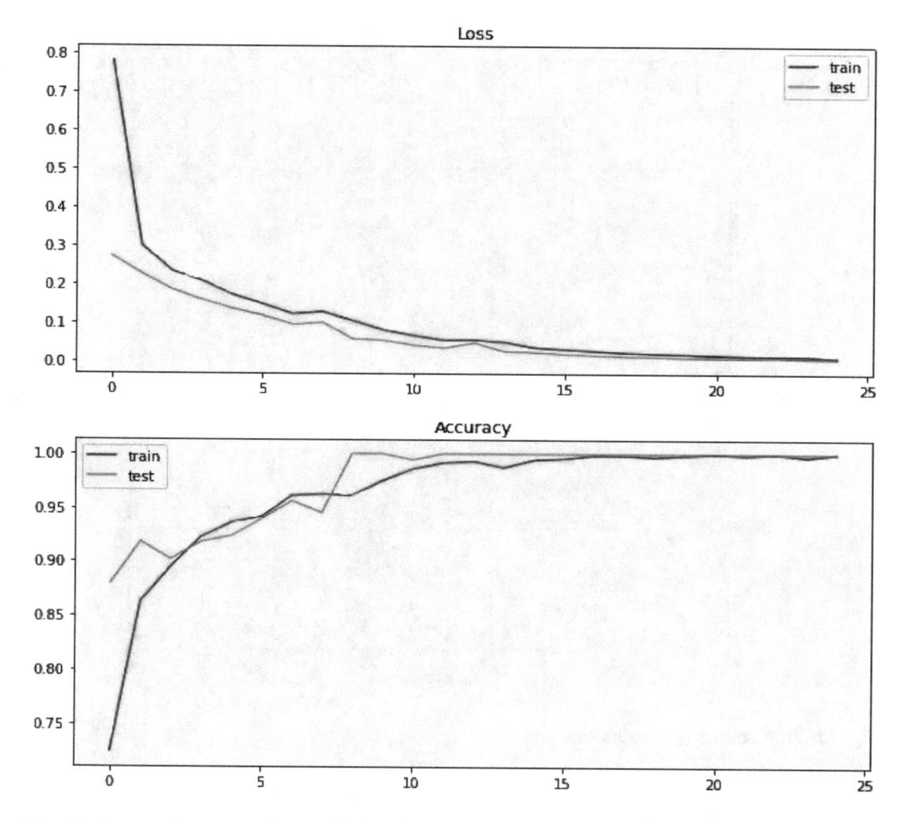


**Fig. 10** Loss and accuracy for model that check whether an image is a positive COVID-19 case or not

**Table 3** Precision, recall, F1-score and support on validation data

| | Precision | Recall | F1-Score | Support |
|---|---|---|---|---|
| 0 | 0.98 | 0.94 | 0.96 | 99 |
| 1 | 0.94 | 0.98 | 0.96 | 99 |
| Accuracy | | | 0.96 | 198 |
| Macro average | 0.96 | 0.96 | 0.96 | 198 |
| Weighted average | 0.96 | 0.96 | 0.96 | 198 |

$$a(t) = \Gamma o{*}c(t) \tag{3}$$

The LSTM and traditional Bi directional RNN as a ground truth represented in Fig. 15. It will lead to bi-directional LSTM,which act as more powerful network that can understand better the variance through the data, where in the bidirectional methodology each LSTM, RNN, GRU units are multiplied by 2 one in the forward direction and the other in the backward direction [25–27].

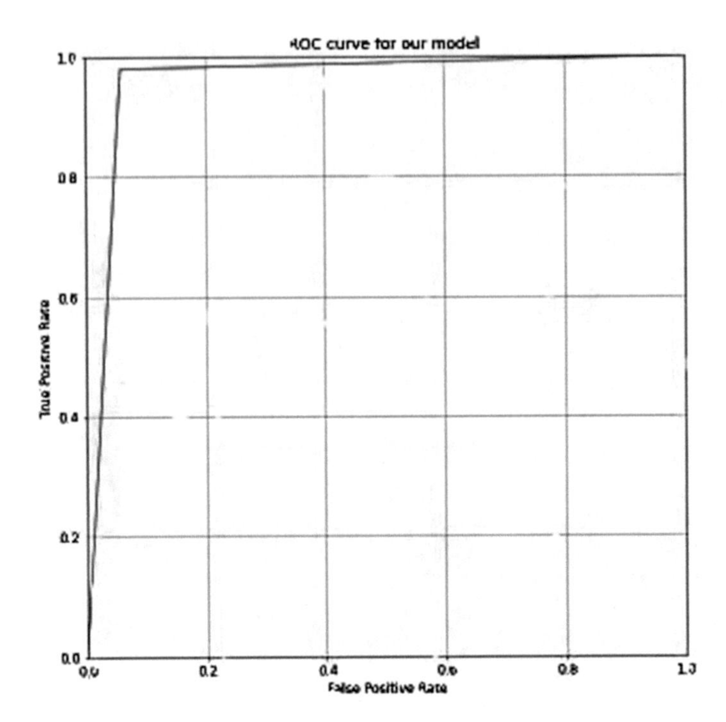


**Fig. 11** ROC curve for our model

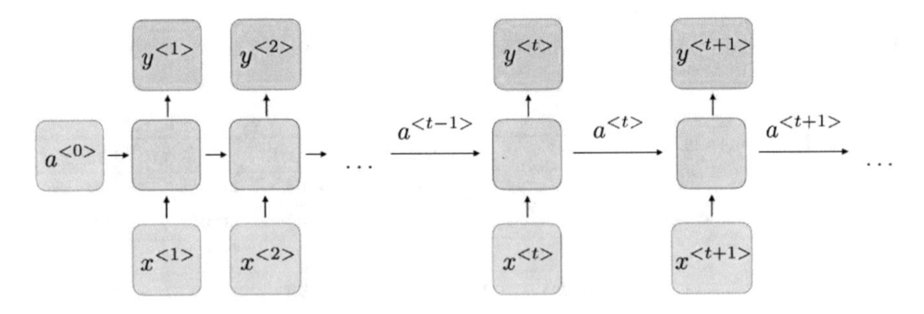


**Fig. 12** Traditional RNN diagram

Using the same concept, to generalize the bidirectional methodology which can be applied to RNN, LSTM, GRUs, since they all follow the same structure, the idea of bi-directionality can be applied to LSTMs Fig. 16.

c. **Model flow and architecture**

The model states the following, the input takes 5 previous days as an input in order to calculate the prediction accurate to the nearest 5 previous days, and then in the middle layer which is the bidirectional LSTM layer, it contains 8000 LSTM units due to the explained bi-directional methodology. Model summary shown in Fig. 17.

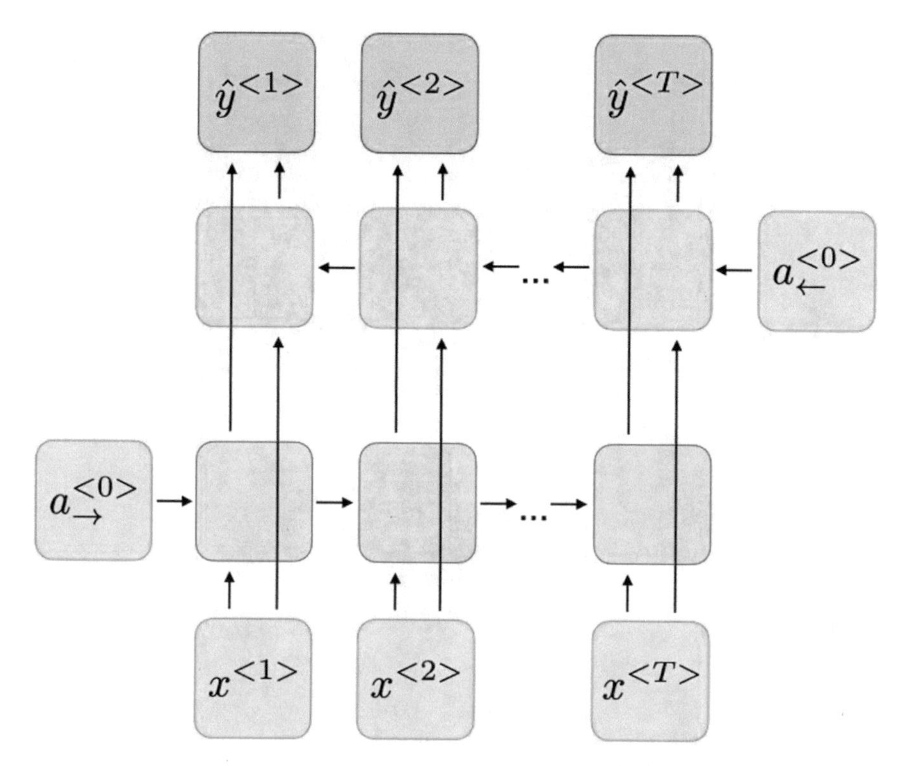


**Fig. 13** Traditional bidirectional RNN

The problem of weight update; in order for the model to give close accurate results of the predictions close to the real world results announced by the Governments. The weights need updated daily with the latest announcements in order to adjust the latest input weights, which give better results compared to the ground truth-values announced by the governments. Moreover, the network applies the mathematical formulas explained in the bi-directional LSTM section where the input should be updated daily after the announcement to be working as the feedback signal to the network as shown in Fig. 18.

d. **Data preprocessing**

The LSTM model must learn a function that maps a sequence of previous observations as input to an output observation. As such, the sequence of observations must be transformed into multiple examples from which the LSTM can be trained from and learn.

e. **The network hyperparameters**

The hyperparameters are selected as the following:

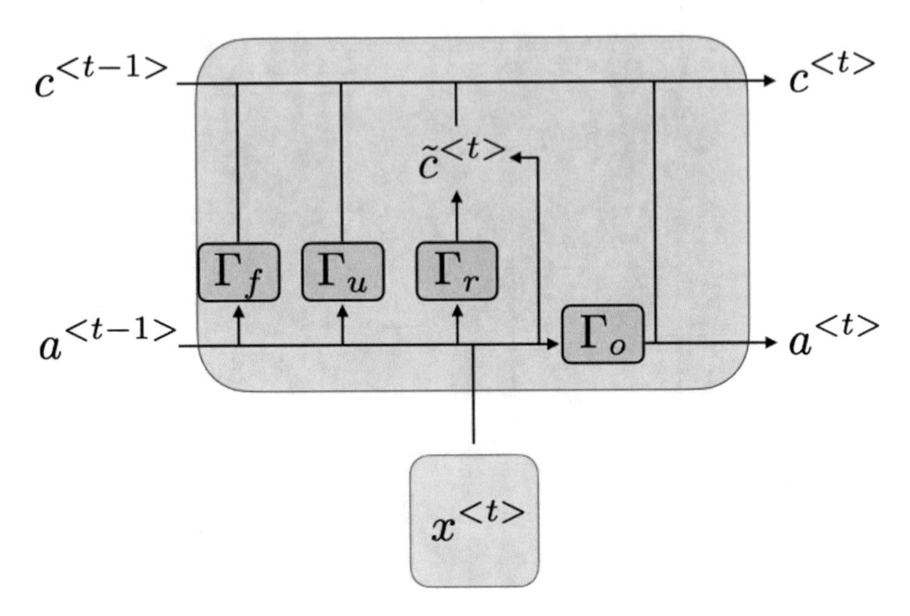


**Fig. 14** LSTM Cell

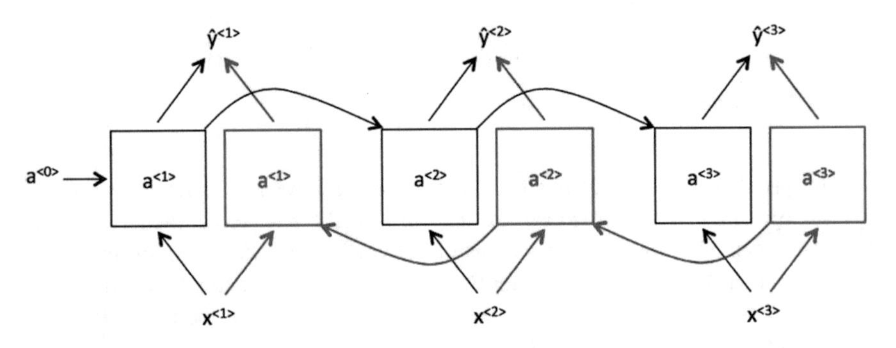


**Fig. 15** Bi-directional methodology diagram

1. A number of steps: The depending days in the network are considered equivalent to the number of steps that the network is required to select from the time series data to predict the next day.
2. Sequence splitting: Dividing the sequence into various input/output patterns called samples, where five-time steps are used as input and one-time steps are used as output for the one-step prediction learned.
3. Samples, time steps and features: The model expects the input to be in the form of samples, time steps and features, splitting the univariate series into 120 samples which is considered to be the factorial of the 5 days where each sample

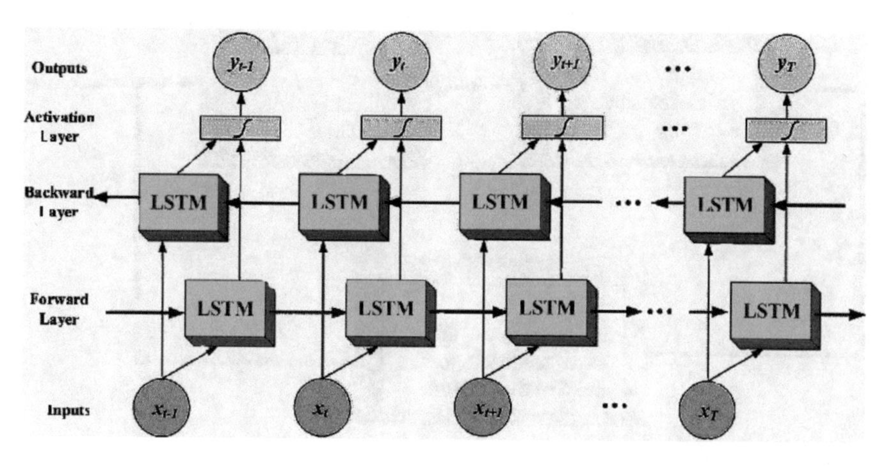


Fig. 16 Bi-directional LSTM

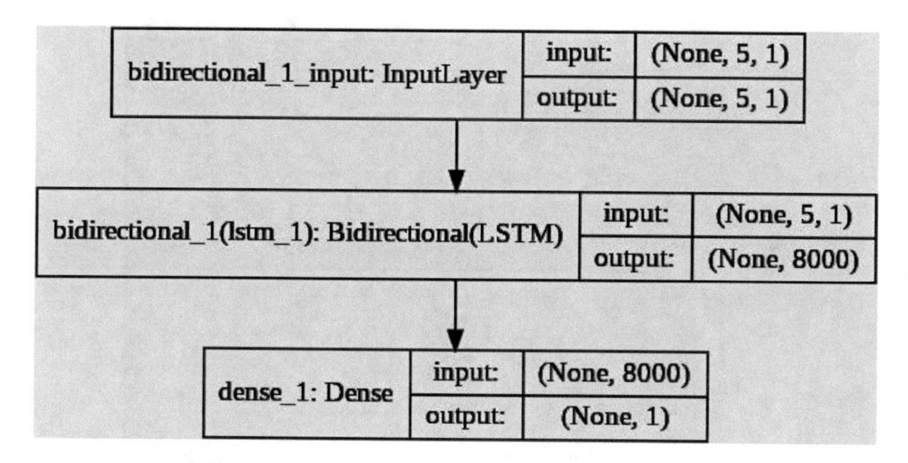


Fig. 17 Network model summary

has five input time steps and one output time step. Number of features is one since the case is univariate.

4. Optimizer: The utilized optimizer in such cases is ADAM optimizer and getting the best results using the mean squared error function.
5. Number of epochs: Number of epochs defines the number of training iterations that the learning algorithm will work through the entire training dataset, in This case the number of epochs is 500.

f. **Countries test and results**

The model tested on Germany in the study period from (7–6–2020) until (16–6–2020), where it predicted the COVID-19 recorded cases. In this period utilizing the

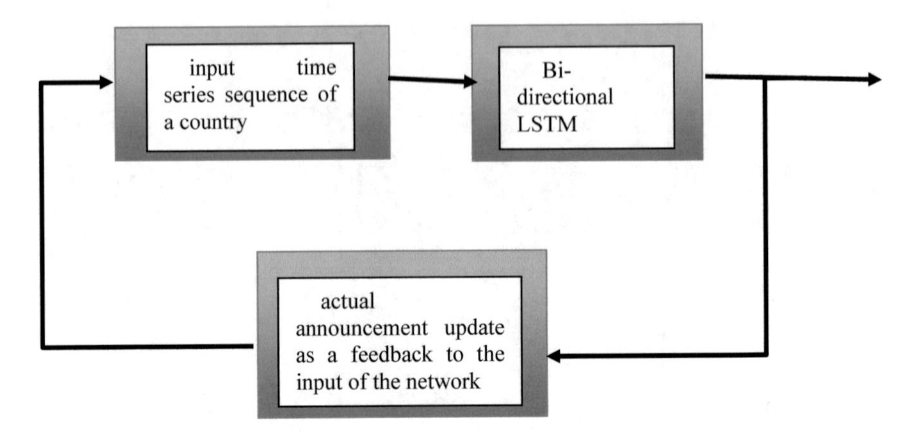


**Fig. 18** Model flow methodology diagram

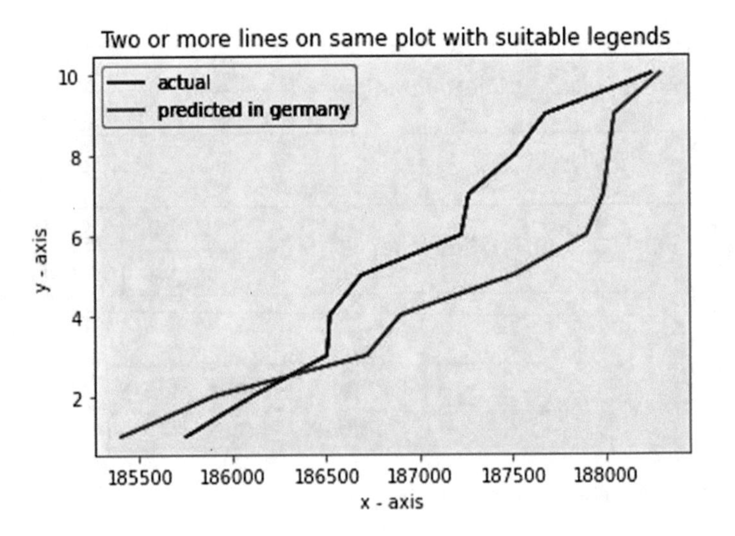


**Fig. 19** Prediction in Germany for 10 days each 5 are separated

benefit of the bi-directional LSTM of the forward learning scheme and backward scheme to provide the following curve in Fig. 19 and Table 4. The same model was tested on Egypt in the period of 17–5–2020 until 1–6–2020 as shown in Fig. 20 and Table 5, not only but also it's tested on the USA which is considered to be the most challenging test due to the very high changes in the records. The test results are shown in Fig. 21 and Table 6, where it is considered to be successful in predicting big numbers in Millions maintaining likelihood symmetric curvature nature.

**Table 4** Germany recorded cases predicted cases versus actual cases on a 10-day study

| Days | Predicted | Actual |
|---|---|---|
| 7/6/2020 | 185,402 | 185,750 |
| 8/6/2020 | 185,900 | 186,109 |
| 9/6/2020 | 186,720 | 186,506 |
| 10/6/2020 | 186,900 | 186,522 |
| 11/6/2020 | 187,509 | 186,691 |
| 12/6/2020 | 187,901 | 187,226 |
| 13/6/2020 | 187,990 | 187,267 |
| 14/6/2020 | 188,020 | 187,518 |
| 15/6/2020 | 188,050 | 187,682 |
| 16/6/2020 | 188,300 | 188,252 |

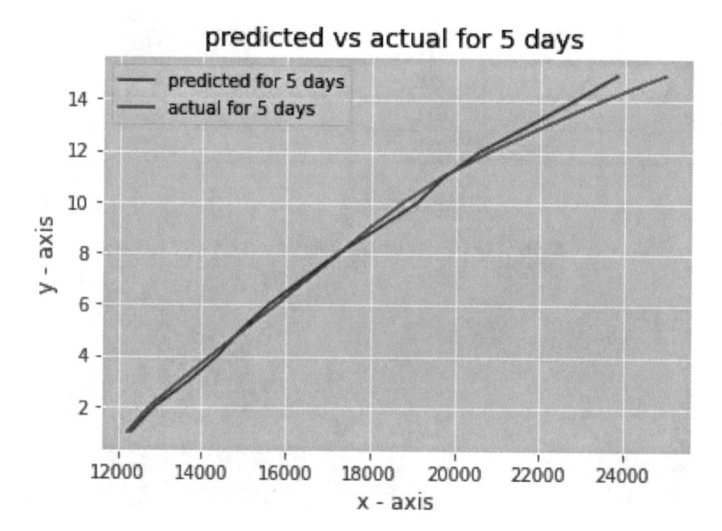


**Fig. 20** Prediction of Egypt in 15 days each 5 days are separated

# 6 System Deployment in Production

The deployment of the chest X-ray detection module and live predictor module was really challenging. For both modules, a Linux server is used for deployment with operating system Centos. Chest X-ray detection module consist of two deep learning models as mentioned before and api that handle users' requests and image preprocessing. TensorFlow serving is used with docker to deploy both models. Flask web framework and pillow library were used to implement the api. Api response is one of three possible responses: positive COVID-19 case, negative COVID-19 case and not valid image. Chest X-ray detection module's architecture is shown in Fig. 22.

Table 5 Egypt recorded cases predicted cases versus actual cases on a 15-day study

| Day | Prediction | Actual |
|---|---|---|
| 17–5–2020 | 12,314 | 12,229 |
| 18–5–2020 | 12,881 | 12,764 |
| 19–5–2020 | 13,697 | 13,484 |
| 20–5–2020 | 14,397 | 14,229 |
| 21–5–2020 | 14,943 | 15,003 |
| 22–5–2020 | 15,607 | 15,786 |
| 23–5–2020 | 16,413 | 16,513 |
| 24–5–2020 | 17,229 | 17,265 |
| 25–5–2020 | 18,187 | 17,967 |
| 26–5–2020 | 19,097 | 18,756 |
| 27–5–2020 | 19,718 | 19,666 |
| 28–5–2020 | 20,564 | 20,793 |
| 29–5–2020 | 21,687 | 22,082 |
| 30–5–2020 | 22,788 | 23,449 |
| 31–5–2020 | 23,821 | 24,985 |
| 1–6–2020 | 25,851 | 26,384 |

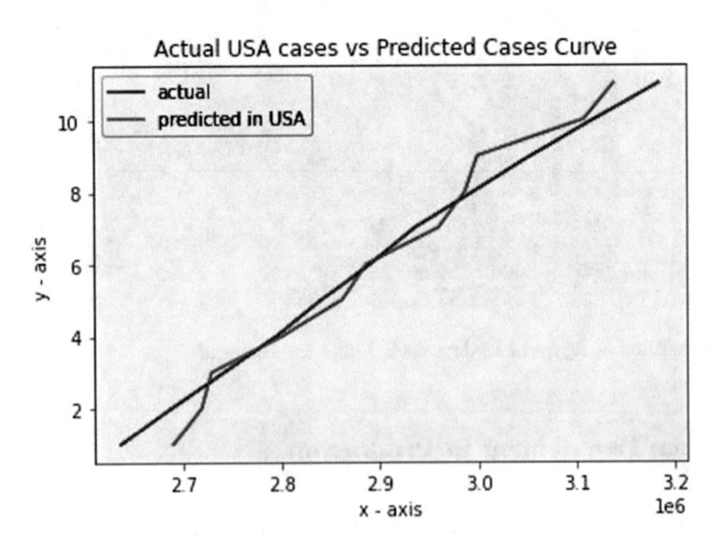


**Fig. 21** USA Curve in 10 days each 5 days are separated

In the live Predictor, this module consists of a data handler, prediction model and api that handle database and users' requests. As mentioned before, the prediction model must be updated daily with new records of COVID-19 cases, so data handler is responsible for tracking any new COVID-19 data updates, in case any update prediction model is called for retraining with the new data and sending new

**Table 6** USA recorded cases predicted cases vs actual cases on a 11-day study

| Day | Predicted | Actual |
|---|---|---|
| 30–6–2020 | 2,689,464 | 2,636,414 |
| 1–7–2020 | 2,718,626 | 2,687,588 |
| 2–7–2020 | 2,728,884 | 2,742,049 |
| 3–7–2020 | 2,801,422 | 2,795,361 |
| 4–7–2020 | 2,862,129 | 2,841,241 |
| 5–7–2020 | 2,887,189 | 2,891,124 |
| 6–7–2020 | 2,959,966 | 2,936,077 |
| 7–7–2020 | 2,987,196 | 2,996,098 |
| 8–7–2020 | 2,999,899 | 3,054,699 |
| 9–7–2020 | 3,107,548 | 3,117,946 |
| 10–7–2020 | 3,137,905 | 3,184,573 |

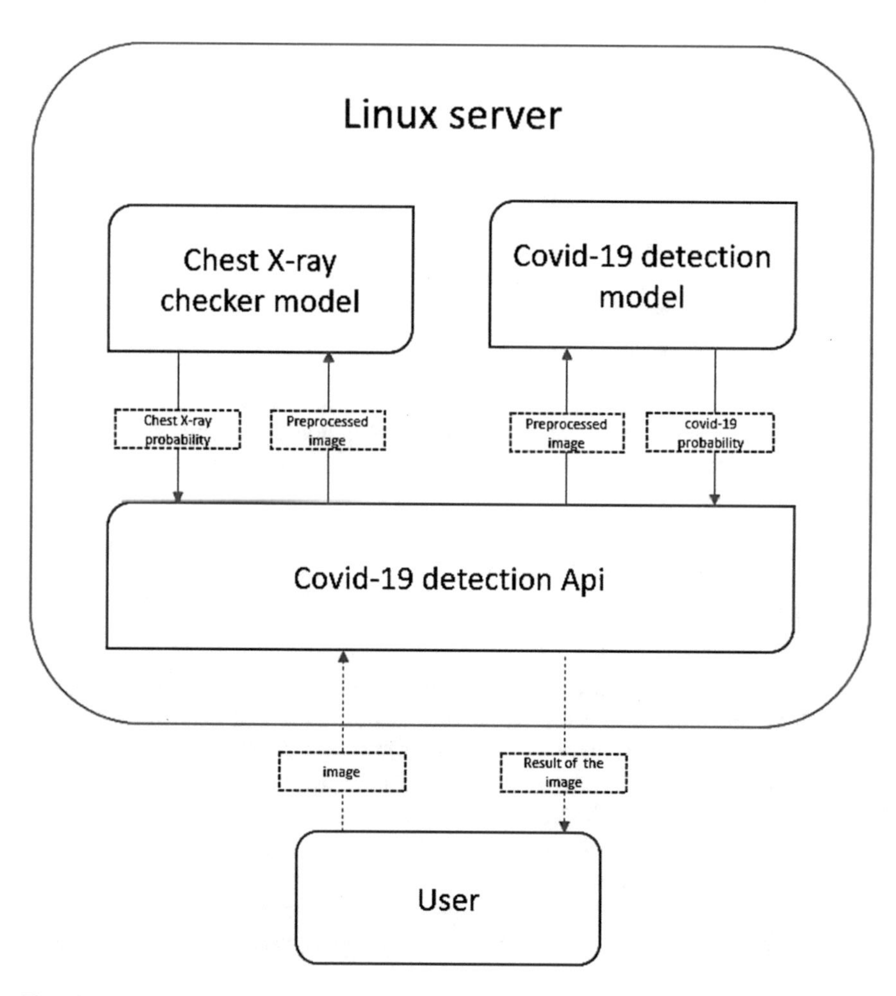


Fig. 22 Chest X-ray module deployment architecture

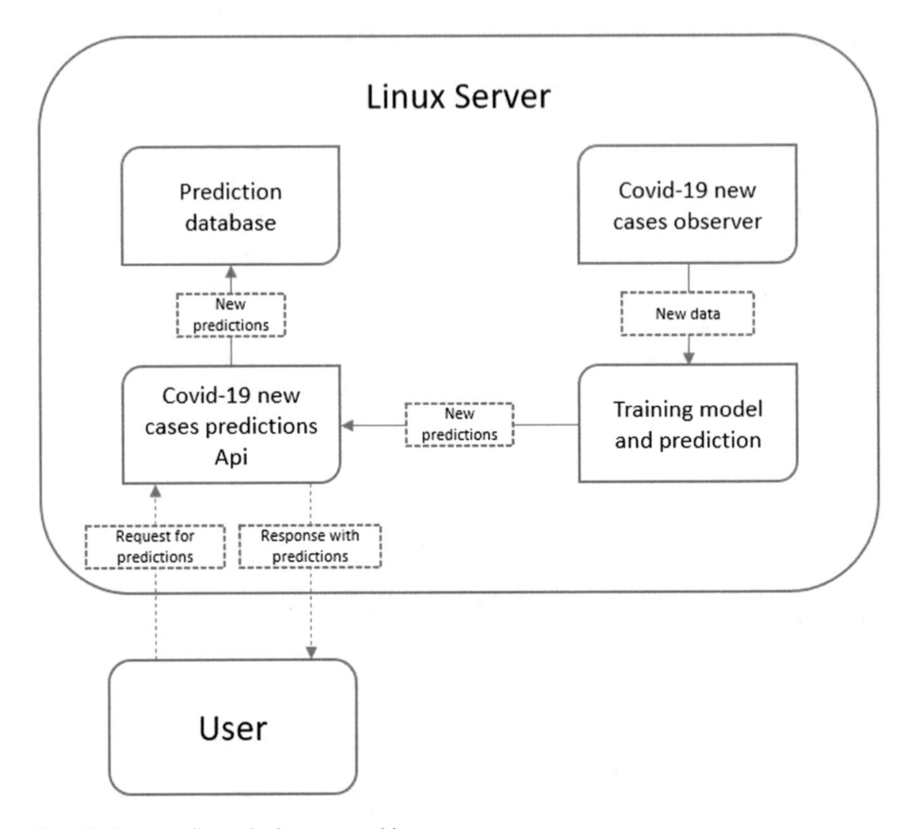


**Fig. 23** Live predictor deployment architecture

predictions for the api. This api in turn saves prediction in the database and sends prediction to users upon request. Live predictor module's architecture is shown in Fig. 23.

## 7 Conclusion

Finally, to sum up, after reviewing the critical issues that are accompanied by the COVID-19 and how it affected worldwide health, as it had been reviewed, the Chinese crisis, also the south Korean solution to fight COVID-19, which was very obvious to the world that using A.I and Big data analytics in the next year is not a choice, it's an obligatory field that could help a nation to survive or wait for the other nations to develop a solution and ask for this solution. South Korea and China survived due to the Big-Data and analysis, which led them to understand the causes and the regions of interest where this Virus is very active. Thus they controlled the spread issue and began to retake control. Our proposed solution will work mainly through AI in Increasing the awareness of this Novel Virus first, then adding Tips about the current statistics, tutorials from experts in the right way of curing in video material, then in this stage, we are collecting the new data concerning the chest X-rays about COVID-19 infection, which later on will be linked with the PCR test results as a future enhancement.

**Acknowledgements** We want to record our appreciation and gratitude for the academy of scientific research and technology to cooperate and provide all possible resources to help us complete this research and build our innovative AI system.

## References

1. Kimberlin, D.W., Brady, M.D., Jackson, M.A., Long, S.S.: American Academy of Pediatrics. FAAP: Committee on Infectious Diseases
2. https://www.who.int/health-topics/coronavirus
3. https://www.cdc.gov/coronavirus/2019/ncov/prepare/transmission.html?CDC_AA_refVal=https%3A%2F%2Fwww.cdc.gov%2Fcoronavirus%2F2019ncov%2Fabout%2Ftransmission.html
4. Al-Abdely, H.M., Midgley, C.M., Alkhamis, A.M.: Middle east respiratory syndrome coronavirus infection dynamics and antibody responses among clinically diverse patients, Saudi Arabia. Emerg Infect. Dis. **25**(4), 753–66 (2019)
5. Chan, J.F., Yuan, S., Kok, K.H.: A familial cluster of pneumonia associated with the 2019 novel coronavirus indicating person-to-person transmission: a study of a family cluster. Lancet **395**(10223), 514–23 (2020)
6. https://spectrum.ieee.org/the-human-os/artificial-intelligence/medical-ai/companies-ai-coronavirus
7. Hu, Z., Ge, Q., Jin, L., Xiong, M.: Artificial Intelligence Forecasting of COVID-19 in China
8. Randhawa, G.S., Soltysiak, M.P., El Roz, H., de Souza, C.P., Hill, K.A., Kari, L.: Machine learning using intrinsic genomic signatures for rapid classification of novel pathogens: COVID-19 case study (2020). https://doi.org/10.1101/2020.02.03.932350
9. Gers, F.A., Schraudolph, N.N., Schmidhuber, J.: Learning precise timing with lstm recurrent networks. J. Mach. Learn. Res. **3**, 115–143 (2002)
10. Palangi, H., Deng, L., Shen, Y., Gao, J., He, X., Chen, J., Song, X., Ward, R.: Deep sentence embedding using long short-term memory networks: analysis and application to information retrieval. IEEE/ACM Trans. Audio, Speech Language Process. **24**(4), 694–707 (2016)
11. Chimmula, V.K.R., Zhang, L.: Time series forecasting of COVID-19 transmission in Canada using LSTM networks. Chaos Soliton Fract **135**, 109864 (2020). https://doi.org/10.1016/j.chaos.2020.109864
12. https://www.who.int/ar/emergencies/diseases/novel-coronavirus-2019/advice-for-public/q-a-coronaviruses
13. Horry, M., Chakraborty, S., Paul, M., Ulhaq, A., Pradhan, B., Saha, M., Shukla, N.: X-ray image based COVID-19 detection using pre-trained deep learning models (2020) https://doi.org/10.31224/osf.io/wx89s
14. Makris, A., Kontopoulos, I.,Tserpes, K.: COVID-19 detection from chest X-ray images using deep learning and convolutional neural networks (2020) https://doi.org/10.1101/2020.05.22.20110817
15. Apostolopoulos, I.D., Mpesiana, T.A.: COVID-19: automatic detection from X-ray images utilizing transfer learning with convolutional neural networks. Phys. Eng. Sci. Med. 1 (2020)
16. Abbas, A., Abdelsamea, M.M., Gaber, M.M.: Detrac: transfer learning of class decomposed medical images in convolutional neural networks. IEEE Access **8**, 74 901–74 913 (2020)

17. Cozzi, D.,Albanesi, M., Cavigli, E., Moroni, C., Bindi, A., Luvarà, S., Lucarini, S., Busoni, S., Mazzoni, L., Miele, V.: Chest X-ray in new Coronavirus Disease 2019 (COVID-19) infection: findings and correlation with clinical outcome. La Radiologia Medica. **125** (2020) https://doi.org/10.1007/s11547-020-01232-9
18. Cohen, J.P., Morrison, P., Dao, L.: COVID-19 image data collection. arXiv:2003.11597 (2020)
19. https://www.kaggle.com/paultimothymooney/chest-xray-pneumonia
20. Karen, S., Andrew, Z.: Very deep convolutional networks for largescale image recognition. ICLR (2015)
21. Kingma, D.P., Adam, J.B.: A method for stochastic optimization. In: The 3rd International Conference for Learning Representations, San Diego (2015)
22. Elghamrawy, S., Hassanien, A.E.: Diagnosis and prediction model for COVID-19 patient's response to treatment based on convolutional neural networks and whale optimization algorithm using CT images (2020) medRxiv2020.04.16.20063990. https://doi.org/10.1101/2020.04.16.20063990
23. Chung, J., Gulcehre, C.,Cho, K.H., Bengio, Y.: Empirical evaluation of gated recurrent neural networks on sequence modeling. arXiv:1412.3555 (2014)
24. Zhou, P., Qi, Z., Zheng, S., Xu, J., Bao, H. Xu, B.: Text classification improved by integrating bidirectional lstm with two-dimensional max pooling. arX-iv:1611.06639 (2016)
25. Greff, K., Srivastava, R.K., Koutnik, J., Steunebrink, B.R., Schmidhuber, J.: Lstm: a search space odyssey. IEEE Trans. Neural Netw Learn Syst (2017)
26. Graves, A., Jaitly, N., Mohamed, A.-R.: Hybrid speech recognition with deep bidirectional lstm. Automatic Speech Recognition and Understanding (ASRU), IEEE Workshop on. IEEE, pp. 273–278 (2013)
27. Wu, Y., Tan, H.: Short-term traffic flow forecasting with spatial temporal correlation in a hybrid deep learning framework. arXiv:1612.01022 (2016)

# Social Distancing Model Utilizing Machine Learning Techniques

**Sherine Khamis Mohamed and Bassem Ezzat Abdel Samee**

**Abstract** Inside the nonappearance of a cure inside the time of a widespread, isolating social measures show up to be the preeminent practical intercession to direct the spreading sickness. Different simulation-based ponders been conducted to explore the adequacy of these measures. Whereas those think about collectively affirm the moderating impact of social removing on illness spread, the detailed viability changes from 10% to more than 90%, lessening the number of diseases. For the most part, this level of instability is due to the complex elements of scourges and their time-variant parameters. In any case, genuine value-based information can diminish vulnerability and give a less loud picture of social removal's viability. In any case, individuals, as a rule, don't take note of their removal, and here comes the part of Artificial Intelligence (AI). This chapter uses AI algorithms to keep an eye on the distances between people to reduce exposure to the virus. Genetic Neural Network (GNN) use utilized as a neural network is the pioneer algorithm in dealing with features. A genetic algorithm is a metaheuristic search algorithm that is good at enhancing the choice of the features, leading to better results.

**Keywords** COVID-19 · Genetic algorithm · Convolutional neural network · Social distancing · People detection · Machine learning

## 1 Introduction

December 2019 was a month that will be remembered all over history. A unique coronavirus (COVID-19) broke call at Wuhan, China, quickly spreading over China and the remainder of the globe. As of 10 May 2020, there are quite 81.2 million

S. K. Mohamed (✉) · B. E. Abdel Samee
Department of Information Technology, Institute of Graduate Studies and Research, University of Alexandria, 163 Horyya Road Elshatby, P.O:832, Alexandria 21526, Egypt
e-mail: sherinekhamis@alexu.edu.eg

B. E. Abdel Samee
e-mail: igsr.bassem.ezzat@alexu.edu.eg



A.-E. Hassanien et al. (eds.), *Advances in Data Science and Intelligent Data Communication Technologies for COVID-19*, Studies in Systems, Decision and Control 378, https://doi.org/10.1007/978-3-030-77302-1_3

affirmed cases of COVID-19 in 218 countries and territories all over the world, with 1.77 million deaths worldwide. As a typical virus, effective pharmaceutical intercessions do not seem to be required to be available for months; meanwhile, population prevention measures such as stay-at-home orders and lock-downs are commonly executed by influenced nations pointing to smooth the bend of the scourge affecting an all-inclusive approximate 3 billion people. Social distancing was one of all the extreme acts to minimize the spread of the infection [1].

Social isolation applies to non-pharmaceutical steps to minimize the recurrence of physical encounters and the separation of contacts between persons in the sense of an irresistible flare-up of illness. It is possible to divide social removal methods into transparent and individual steps. Open interventions include the closure or decreasing of instructive teaching and working environments, the cancellation of mass social occasions, travel restrictions, border protection and building quarantine. Personal interventions consist of confinement, isolation and support to establish physical distinctions between individuals [2]. Despite the fact that these interventions may have a few adverse effects on the economy and the opportunity for individuals, they play a key role in reducing the severity of a widespread [2].

A few systematic methods are routinely used for the evaluation of social separation interventions. The critical generation number is one of the most relevant parameters for assessing social distance measurements $R_0$. Which normally speaks to how a case (i.e., an irresistible person) can contaminate multiple individuals in its entire irresistible period [3]. To be illustrated, $R_0 < 1$. It shows that less than 1 person would be infected in each case, and thus the infection is decreasing within the population considered. Since the recognition of $R_0$ represents how rapidly the malady is spreading, $R_0$ is one of the main key indicators for social elimination initiatives to assess how quickly the epidemic is spreading [4, 5]. Scientifically, $R_0$ can be decided by:

$$R_0 = \int_0^\infty b(a)F(a)da, \tag{1}$$

where $b(a)$ is the average number of new cases an infectious person will infect per unit of time during the infectious period a, and $F(a)$ is the probability that the individual will remain infectious during the period $a$.

A typical approach is to degree the attack rate that is the rate of infected individuals in a very vulnerable population (where no one is protected at the beginning of the disease) at the time of measurement in order to determine the feasibility of social distance. The assault rate represents the magnitude of a disease at a given time, and in the infection flare-up there are different values during this way. Among these principles, in order to assess the ability of this system to cope with the crest number of patients, the crest attack rate is typically considered and compared to this healthcare capability (e.g., severe care unit capacity). After the flare-up is over, data is normally gathered to assess the last word attack rate, which is that the entire number of contaminated incidents is divided by the overall population over the entire duration of the episode [6].

Social separation steps are seen to be convincing when properly revised. Various forms of social elimination interventions may have varying degrees of effectiveness for the spread of the disease. The effect of social separation initiatives on workplaces is assessed through an agent-based reenactment method in [7]. Specifically, six common strategies in the working world that minimize the amount of workdays are re-enacted. It appears that reducing the number of working days will minimize the ultimate attack rate viably (e.g., up to 82% in case three successive workdays are decreased). All things considered, reducing the number of workdays in a pandemic-level flu entails an inherently weaker impact, i.e. 3% (one additional day off) to 21% decrease (three additional continuous days off). A few others are thinking about showing comparable outcomes. It appears in [6] that social removal of the working environment will reduce the ultimate rate of attack by up to 39%. In addition, [3] shows that, depending on the recurrence of contacts among employees, various kinds of interventions will reduce the at-tack rate from 11 to 20%.

Since a virus is taken into account in COVID-19, the entire globe needs the aid of artificial intelligence to reduce its consequences: either in health, in the economy or in society. When countries around the world realized that the lockout took an unnecessary amount of time and impacted the economy, the precautionary lock-down began to decrease. Normal life has started to return again. Sadly, this has contributed to a rise in the number of infected individuals. One of the most successful solutions approved by the Health Organization (Who) of the world is social distancing. Machine Learning (ML), which can be a computer branch, may help to keep an eye on the distance between individuals and warn them when space becomes less than 1.5 m. In our chapter, this is always what we are looking for here. We built a fusion between the genetic algorithm and the neural network to detect and measure the distance between individuals through surveillance cameras' photographs. As a metaheuristic optimization tool, the genetic algorithm is used to search for the most powerful features that can be given to the neural network to detect people [8–10].

That chapter is structured as takes after: numerous of recent related works are put forward Sect. 2. The comprehensive description of the proposed model has been made in Sect. 3. In Sect. 4, a short overview is given for the results and description of the dataset is put forward. As a final point, the conclusion is annotated in Sect. 5.

## 2 Literature Review

As a convincing non-pharmacological strategy and a critical inhibitor for restricting the transmission of infectious diseases such as H1N1, SARS, and COVID-19, numerous research papers have shown social elimination. A more comprehensive Gaussian bend with a shorter spike within the extent of the profit potential of the well-being system makes it less demanding for patients to combat the infection by embracing the well-being treatment organizations' consistent and timely assistance.

Any surprising sharp rise and rapid rate of contamination would lead to the disappointment of profit and hence exponential growth in the number of fatalities [10].

The use of Artificial Intelligence, Computer Vision, and Machine Learning enables us to find a high-level feature relationship. For example, it empowers us to get it and predict individual actions on foot in activity scenes, sporting activities and exercises, clinical imagery, discovery of inconsistency, etc. by examining spatio-temporal visual data and investigating the image arrangements with factual details. In health-related areas, a mixture of visual and geo-location cellular data would be too feasible to foresee the disease slant of specific regions, to estimate the thickness of individuals in open spaces, or to determine to exclude people from prevalent swarms34. Be that as it may, such research ventures face obstacles such as talented manpower or the preparation and upgrading of the infrastructure toll [11].

In AI, Machine Vision, Deep Learning, and Design Recognition, on the other hand, later propels empower the machines to get it and decode the visual knowledge from computerised images or videos. It helps computers to identify and recognise distinctive object types, too. Such skills can also play a critical role in communicating, inspiring, and conducting recognition and estimations that are socially distinct. For instance, computer vision may transform CCTV cameras into "smart" cameras within the current infrastructure capacity, not as screen people, but can also determine whether or not individuals obey the social separation rules. Such systems need extraordinarily accurate algorithms for human detection [12, 13].

One of the most important sub-branches within the field of object detection and computer vision is the detection of people in image sequences. Despite the fact that the field of human location and human activity identification has been exhausted by multiple inquiries regarding works, most of them are either restricted to indoor applications or suffer from accuracy issues under difficult lighting conditions under open air. A number of other inquiries rely on manual tuning techniques to separate the exercises of individuals, be that as it may, constantly restricted functionality was an issue. In order to provide ex-traction and complex object classification, Convolutional Neural Networks (CNN) have played a very vital role. CNNs allow analysts to build accurate and quick locators compared to ordinary models with the advancement of faster CPUs, GPUs, and expanded memory space. In any case, the longtime planning, position pace and accomplishing superior exactness, are still remaining challenges to be unravelled [14].

Narinder et al. used a deep neural arrangement (DNN) based locator to test the separate infringement list at the side of the Profound sort calculation as a query tracker for person discovery. In any event, there is no factual investigation of the outcome of their findings. In addition, no dialogue is given about the legitimacy of the estimates of the removal. In another discussion by Prateek et al., the developers appeared to delete the people in a given factory. MobileNet V2 has been used as a light-weight detector to minimise computing costs, which in turn gives less accuracy compared to a few other popular models [15].

No observable analysis is carried out, compared to the other investigation, on the occurrence of the separate estimate inside. A distinction between two common kinds of DNN models (YOLO and Faster R-CNN) was made by the creators; in any case, the structure was as attempted on a simple dataset. Since the topic is unusually unused, there has not been much research on the accuracy of discoveries, no exploration has been carried out on difficult datasets, no normal comparison has been made on common datasets, and no explanatory considerations or post-processing have been taken into account after the detection step [16].

Taking into account the research gaps described above, a model is introduced that not only performs more reliably and faster than the state-of-the-art, but will also be trained and tested in challenging environments and lighting conditions using broad and detailed datasets. This will ensure that the model can work in real-world environments, particularly in covered shopping centres where the lighting conditions are not as ideal as outdoor lighting. Also, analytical solutions for post-detection and post-processing to minimise the spread of the virus is provided.

## 3 Proposed Methodology

We propose a model that generates a few errands such as counting person identification, person tracing, and measuring distances of separation between each of the two recognised individuals as an add-up to social distance observation solution. The system can be organised and linked with any determination from VGA to Full-HD on all kinds of CCTV reconnaissance cameras, with real-time execution.

### *3.1 People Detection*

A recently proposed crossover operator planned by Mishra and Shukla [17] could be the Ternary crossover operator. Ternary crossover operator, as recommended by the word, involves the crossover between three parents and falls within the specifics of the three new siblings. Including a ternary hybrid administrator, we suggest an adapted GA and call it a ternary hereditary measurement (TGA). We have used the idea of learning domain transfer and used VGG-16 as a pre-trained show. Within the VGG-16 architecture, for a bunch of two or three to make a bit, the convolution layers are stacked together. Afterwards, we plan one block at a time. Because a square consists of the convolution layers in VGG-16 (and no pooling layer is also a bit of the square), we see it as a 'layer' within the remainder of this analysis. Within the VGG-16 architecture, up to 5 parts (named as layers within the rest of the first copy) are added [18].

We apply a technique of layer-wise preparation during which one layer is ready at a time because it is, while the rest of the layers remain unchanged. During this process, we prepare the first and also the moment layer at the same time because it was one exception (Layer-1 and layer-2 are the start layers of the pre-trained show, i.e., VGG-16. These layers are competent of extricating the essential highlights and have less number of trainable parameters; thus we prepare these two layers simultaneously). We plan within the starting point since it was the thick layer, and the first and moment layers are prepared at the moment, while the opposite layers are solidified. The third layer is ready at that moment, and the remaining four layers are then kept settled. The fourth layer is ready from there on, and the remaining four layers are therefore kept settled. Furthermore, the fifth layer is often prepared thus holding the rest of the four layers solidified. This complete handle is rehashed again, i.e., each layer is prepared another time separately (but layer-1 and layer-2 are prepared at the same time), thus keeping the opposite layers resolved.

For each layer, the taking after two hyper-parameters have been optimized utilizing the fundamental GA and the TGA: number of epochs, learning rate parameter [19].

When we use the basic GA to optimize these two hyper-parameters, we refer to the approach as 'our model-1,' and when we use the TGA to optimize these two hyper-parameters, we refer to the approach as 'our-model-2.' We have used both the basic GA and the TGA binary encoding technique. In such a way that it represents the values of both these hyper-parameters for each of the five layers separately, for two distinctive periods, we have designed a chromosome. Consequently, up to twenty optimization parameters are applied (ten diverse values for the number of ages and ten distinctive values for the learning rate parameter). To make it easier to understand, we used a logical classification to talk about each of these parameters.

## 3.2 People Tracking

We use the Simple Online and Real-time (SORT) tracking technique that solves the problem of frame data association $t$. Within a time frame, using already computed tracking results from the past $t - \Delta t_1, \ldots, t - 1$. Together with at-time input detections $t, \ldots, t + \Delta t_2 - 1$. Let $\mathcal{D}$ identify the collection of video sequence detections to be tracked, which decomposes into body detection. $\mathcal{D}^B$ and joint detections $\mathcal{D}^J$. More, $\mathcal{D}_t$ encompasses all detections at frame $t$. A line $T \subset \mathcal{D}$ comprises of all detections belonging to a person. Let $\delta(T)$ designate the newest time imprint for which $T$ encompasses detections and let $\overline{T} := T \cap \mathcal{D}_{\delta(T)}$ encompass the matching detections at the tail of $T$. As a final point, the set $T_{t-\Delta t_1}, t-1$ comprises all lines $T$ where $\delta(T)$ is within $t - \Delta t_1, \ldots, t - 1$. Now the tracking task is to find optimal associations between the previously computed lines $T_{t-\Delta t_1}, t-1$ and detections $\mathcal{D}_{t,t+\Delta t_{2-1}} := \cup_{i=t}^{t+\Delta t_2 - 1}$ within the sliding window, which also integrates to categorize freshly seemed objectives. We note that the skimming window has a size of $\Delta t_1 + \Delta t_2$ and is a result a delay of $\Delta t_2 - 1$ frames. For a skimming window around frame $t$, we resolve the data connotation problem by discovering a min cost graph labeling solution. In particular, we create an undirected weighted graph $\mathcal{G} = (v, \varepsilon, \mathcal{C})$, with the vertex set.

$$\mathcal{V} := T_{t-\Delta t_1, t-1} \sqcup \mathcal{D}_{t, t+\Delta t_{2-1}} \tag{2}$$

The edge set $\varepsilon$ comprises all possible connections between precomputed trajectories $T_{t-\Delta t_1, t-1}$ and detections $\mathcal{D}_{t, t+\Delta t_{2-1}}$ as well as between any two detections of $\mathcal{D}_{t, t+\Delta t_{2-1}}$. Affinity costs $c_e$ for $e \in \varepsilon$ reflect how likely an edge connects inputs belonging to the same person. Accordingly, costs $c_v$ for $v \in \mathcal{V}$ reflect how likely an input is a true positive. The association problem can then be formulated as a min cost graph labeling problem:

$$\underset{\mathfrak{T} \in \{0,1\}^{P \times |\mathcal{V}|}}{\arg\min} \sum_{l=1}^{P} \sum_{v \in \mathcal{V}} c_v \mathfrak{T}_{v'}, l + \sum_{e=\{v,v'\} \in \varepsilon} c_e, \mathfrak{T}_v, l \mathfrak{T}_{v'}, l \tag{3}$$

subject to $\sum_{l=1}^{P} \mathfrak{T}_{v,l} \leq 1$ for all $v \in \mathcal{V}$. Here, $P$ is an upper bound on the number of persons in the sliding window. If for an indicator variable $\mathfrak{T}_{v,l} = 1$ holds, then vertex $v$ is assigned to person $l$. Accordingly, the aim of (3) is to compute the assignment of the indicator variables $\mathfrak{T}_{v,l}$ such that the most plausible data association is selected that is consistent in space and time, with respect to both input detectors then update the trajectories and shift the sliding window one-time step forward.

### 3.3 Distance Between People Measuring

Stereo-vision may be a common distance estimating technique; be that as it may, when we point to incorporation of an effective solution, relevant in all open places using as it was a fundamental CCTV camera, this is often not an achievable approach in our investigation. We then lead to a monocular solution. On the other hand, the projection of a 3-D world scene into a 2-D perspective image plane by the use of a single camera leads to arbitrary pixel distances between the objects. Typically referred to as consequences from the point of view; in which within the whole picture we do not see uniform dissemination of separations. For instance, at the skyline, parallel lines cross and more distant people to the camera seem much shorter than the individuals closer to the camera facilitate the middle. Three parameters ($x$, $y$, $z$) are connected with the middle or reference point of each holding box in the three-dimensional space, while the original 3D space is reduced to two-dimensions ($x$, $y$) within the image obtained from the camera, and the depth parameter ($z$) is not accessible In such a lowered-dimensional space, the Euclidean coordinate uses the removal model to inter-popular degree. In order to apply a calibrated IPM step, we must first have a camera calibration by setting $z = 0$ to dispense with the repercussions of the viewpoint. In addition, the camera area, its stature, point of view and the determination of optics must be understood (i.e. the camera natural parameters).

By applying the IMP, the 2D pixel points (u, v) will be mapped to the corresponding world coordinate points ($X_w$, $Y_w$, $Z_w$):

$$[u \quad v \quad 1]^T = K\, RT[X_w, Y_w, Z_w \quad 1]^T \tag{4}$$

where $R$ is the rotation matrix:

$$R = \begin{bmatrix} 1 & 0 & 0 & 0 \\ 0 & \cos\theta & -\sin\theta & 0 \\ 0 & \sin\theta & \cos\theta & 0 \\ 0 & 0 & 0 & 1 \end{bmatrix}, \tag{5}$$

$T$ is the translation matrix:

$$T = \begin{bmatrix} 1 & 0 & 0 & 0 \\ 0 & 1 & 0 & 0 \\ 0 & 0 & 1 & -\frac{h}{\sin\theta} \\ 0 & 0 & 0 & 1 \end{bmatrix}, \tag{6}$$

and $K$, the intrinsic parameters of the camera are shown by the following matrix:

$$K = \begin{bmatrix} f * ku & s & c_x & 0 \\ 0 & f * kv & c_y & 0 \\ 0 & 0 & 1 & 0 \end{bmatrix} \tag{7}$$

where $h$ is the camera height, $f$ is focal length, and $ku$ and $kv$ are the measured calibration coefficient values in horizontal and vertical pixel units, respectively. ($c_x$, $c_Y$) are the principal point changes that correct the image plane's optical axis. The camera produces a picture that falls on a retina plane with a projection of three-dimensional points in the coordination of the globe. The relationship between three-dimensional points and the corresponding projection image points can be seen as follows, using homogeneous coordinates:

$$\begin{bmatrix} u \\ (v) \\ 1 \end{bmatrix} = \begin{bmatrix} m_{11} & m_{12} & m_{13} & m_{14} \\ m_{21} & m_{22} & m_{23} & m_{24} \\ m_{31} & m_{32} & m_{33} & m_{34} \end{bmatrix} \begin{bmatrix} X_w \\ \begin{bmatrix} Y_w \\ Z_w \end{bmatrix} \\ 1 \end{bmatrix}, \tag{8}$$

where $M \in \mathbb{R}^{3\times4}$ is the transformation matrix with $m_{ij}$ elements in Eq. 8, that provided by the Camera Intrinsic Matrix K (Eq. 7), Rotation Matrix R (Eq. 5) and the Translation Matrix T, maps the world coordinate points into the image points based on the camera position and the reference frame (Eq. 6). The dimensions of the above equation can be reduced to the following shape of the camera image plane perpendicular to the Z access in the world coordination system (i.e. z = 0):

$$\begin{bmatrix} u \\ (v) \\ 1 \end{bmatrix} = \begin{bmatrix} m_{11} & m_{12} & m_{13} \\ m_{21} & m_{22} & m_{23} \\ m_{31} & m_{32} & m_{33} \end{bmatrix} \begin{bmatrix} X_w \\ (Y_w) \\ 1 \end{bmatrix}, \tag{9}$$

in the following scalar form, the transition from perspective space to inverse perspective space (BEV) can also be expressed:

$$(u, v) = \left( \frac{m_{11} \times X_w + m_{12} \times Y_w + m_{13}}{m_{31} \times X_w + m_{32} \times Y_w + m_{33}}, \frac{m_{21} \times X_w + m_{22} \times Y_w + m_{23}}{m_{31} \times X_w + m_{32} \times Y_w + m_{33}} \right). \tag{10}$$

# 4 Implementation and Experimental Results

## *4.1 Implementation*

We first assessed their importance and contribution to predictive accuracy to assess the possible benefits of the proposed model; we continue to equate it to other status-of-the-art systems. In the suggested algorithm implemented in the MATLAB (R2017b) simulator, the machine functions are written. MATLAB is an environment for software development that proposes high-performance numerical computing, data analysis, visualisation capabilities, and tools for application development. It is an object-oriented programming language that can be used to write code to run within the MATLAB environment or in a wide range of programmes published by other MATLAB-operating businesses. MATLAB has several advantages: (1) it enables recompilation of the research algorithms immediately. It allows anything to be typed on the command line or a segment in the editor to be executed and displays the results instantly and greatly facilitates the creation of algorithms. (2) The built-in graphics tools and GUI builders from MATLAB ensure that customizing the data and models allows the user to view their data more easily for faster decision-making. (3) The functionality of MATLAB can be extended by adding toolboxes [20] to run on HP Pavilion—15-cs3008tx—CoreTM i7-8 GB DDR4-2666 SDRAM running Windows 10.

## *4.2 Dataset*

To assess the demonstrated model, Caltech Person on foot Dataset Benchmark was utilized. The Cal-tech Person on foot Dataset comprises of generally 10 h of 640 × 480 30 Hz video taken from a vehicle driving through standard activity in an urban environment. Approximately 250,000 outlines (in 137 around miniature long sections) with a total of 350,000 bounding boxes and 2300 special people on foot were clarified. The comment incorporates transient correspondence between bounding boxes and detailed occlusion labels.

## 4.3 Model Evaluation

The goodness-of-fit and prediction power of the suggested models were evaluated based on statistical measures such as overall success rate, positive predictive value, negative predictive value, specificity, sensitivity (Eqs. 11 and 12)

$$\text{Overallsuccessrate} = \frac{TP + NT}{TP + NT + FP + FN};\ \text{Sensitivity} = \frac{TP}{TP + FN} \tag{11}$$

$$\text{Specificity} = \frac{TN}{FP + TN};\ PPV = \frac{TP}{FP + TP};\ NPV = \frac{TN}{FN + TN} \tag{12}$$

where *TP* (True Positive) and *TN* (True Negative) are samples that are categorised correctly within the training or validation datasets. In the training or validation datasets, *FP* (False Positive) and *FN* (False Negative) samples are misclassified. The total success rate is split into all points by the number of persons and non-people points that are correctly categorised. The Positive Predictive Value (*PPV*) is the probability of points that are classified as individuals, while the probability of points that are classified as non-people is the Negative Predictive Value (*NPV*). Sensitivity is that the percentage-age of right people points, while specificity is that inside the training or validation datasets, the percentage of right non-people points.

## 4.4 Experimental Results

To validate the suggested model in people detection, it was compared by deep learning model suggested in [21]. The suggested model was a little bit better. That's is because the deep learning model dealt with feature directly. In the suggested model, genetic algorithm chooses the best features first then passes it to the convolutional neural network. Table 1 shows the comparison results.

In our second experiment, the proposed model was compared by a model based on the Random Forest (RF) algorithm. The proposed model gave better results, 86.7% accuracy, whereas the RF gave 86% accuracy. That is because the RF can fit over noisy datasets but for data with different values, attributes with more values will have greater repercussions on random forests, so the attribute weights generated by random forests on such data are not credible [22].

**Table 1** Comparison between the suggested model and neural network model based on accuracy

| | Tensorflow (%) | CMSIS-NN (%) | Proposed model |
|---|---|---|---|
| Train | 81.2 | 80.9 | 97.2 |
| Validation | 77.4 | 76.4 | 97.6 |
| Test | 76.9 | 76.7 | 97.7 |

As distant as Following is concerned, the proposed Demonstrate was compared with Range-Doppler location approach employing a machine learning algorithm fed with information from a radar to allow controlling little versatile robots without coordinate human operation. The machine learning usage of motion acknowledgment depends on choice tree. The comparison was based on a cruel victory rate (genuine positives). The comparison result is displayed in Fig. 1.

As for estimating distance between people, the corona virus is still new and no work was done in this point yet. We compared our approach with an obstacle avoidance system for road robots equipped with RGB-D sensor that captures scenes of its way forward. This system is based on RGB-D Semantic Segmentation. This work results were based on SIM10k dataset as the source domain and the Cityscapes dataset as the target domain. So, we used the same dataset in the comparison. The accuracy rate for the proposed system was 97.2 in [23] while our accuracy rate was 97.8.

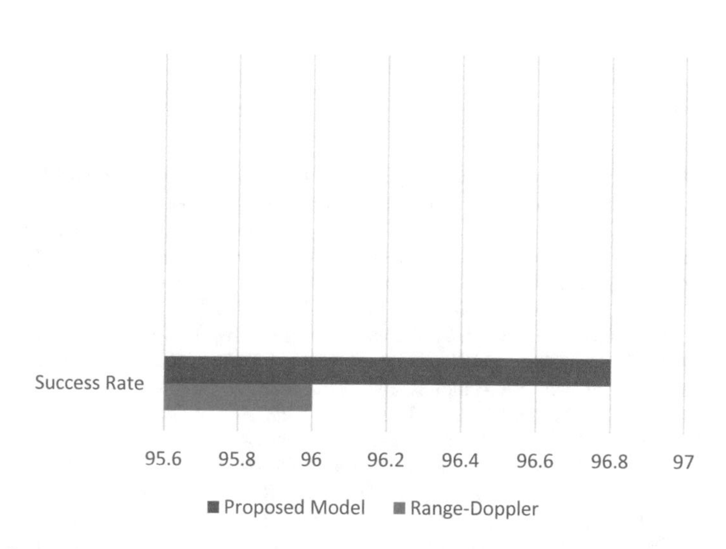


**Fig. 1** Comparison between range-doppler detection approach and the proposed model based on success rate

## 5 Conclusion

This chapter proposes an automated model is used in keeping an eye on social distancing. It uses the images coming from surveillance camera to detect persons, trace them and calculate the distance between them in order to keep social distancing and decrease the spread of the Corona Virus. The model has been able to work in a number of difficulties, including occlusion, lighting variations, shades, and partial visibility, and has demonstrated a substantial increase in accuracy and speed compared to precision and speed compared with state-of-the-art approaches. For this model, the adapted tracking algorithm to estimate the inter-personal distances, and to track people's moving trajectories, evaluation of infection risk and interpretation has a benefit for health authorities and governments.

The model presents a Genetic Convolutional Neural Network point of view-independent algorithm for human identification. The findings of this chapter are therefore directly applicable to a broader group of researchers, regardless of camera angle and location, not only in the fields of computer vision, artificial intelligence and health, but also in other industrial applications, including pedestrian detection for driver assistance systems, autonomous vehicles, public and crowd anomaly activity detection, surveillance protection systems. Results acquired are encouraging. To set a plan for future works, the model may utilize other machine leaning techniques such as using particle swarm optimization to optimize the features that are used to detect people in a better way.

## References

1. Ather, A., Patel, B., Ruparel, N.B., Diogenes, A., Hargreaves, K.M.: Coronavirus disease 19 (COVID-19): implications for clinical dental care. J. Endod. (2020)
2. Nguyen, C.T., Saputra, Y.M., Van Huynh, N., Nguyen, N.T., Khoa, T.V., Tuan, B.M., Nguyen, D.N., Hoang, D.T., Vu, T.X., Dutkiewicz, E., Chatzinotas, S.: A comprehensive survey of enabling and emerging technologies for social distancing—Part I: fundamentals and enabling technologies. IEEE Access **8**, 153479–153507 (2020)
3. Timpka, T., Eriksson, H., Holm, E., Strömgren, M., Ekberg, J., Spreco, A., Dahlström, Ö.: Relevance of workplace social mixing during influenza pandemics: an experimental modelling study of workplace cultures. Epidemiol. Infect. **144**(10), 2031–2042 (2016)
4. Ferguson, N.M., Cummings, D.A., Fraser, C., Cajka, J.C., Cooley, P.C., Burke, D.S.: Strategies for mitigating an influenza pandemic. Nature **442**(7101), 448–452 (2006)
5. Fraser, C., Riley, S., Anderson, R.M., Ferguson, N.M.: Factors that make an infectious disease outbreak controllable. Proc. Natl. Acad. Sci. **101**(16), 6146–6151 (2004)
6. Kumar, S., Grefenstette, J.J., Galloway, D., Albert, S.M., Burke, D.S.: Policies to reduce influenza in the workplace: impact assessments using an agent-based model. Am. J. Public Health **103**(8), 1406–1411 (2013)
7. Mao, L.: Agent-based simulation for weekend-extension strategies to mitigate influenza outbreaks. BMC Public Health **11**(1), 522 (2011)
8. Dardari, D., Closas, P., Djurić, P.M.: Indoor tracking: theory, methods, and technologies. IEEE Trans. Veh. Technol. **64**(4), 1263–1278 (2015)

9. Liu, W., Cheng, Q., Deng, Z., Chen, H., Fu, X., Zheng, X., Zheng, S., Chen, C., Wang, S.: Survey on CSI-based indoor positioning systems and recent advances. In: 2019 International Conference on Indoor Positioning and Indoor Navigation (IPIN), pp. 1–8. IEEE (2019)
10. Mazuelas, S., Bahillo, A., Lorenzo, R.M., Fernandez, P., Lago, F.A., Garcia, E., Blas, J., Abril, E.J.: Robust indoor positioning provided by real-time RSSI values in unmodified WLAN networks. IEEE J. Sel. Top. Sign. Process. **3**(5), 821–831 (2009)
11. Fong, M.W., Gao, H., Wong, J.Y., Xiao, J., Shiu, E.Y., Ryu, S., Cowling, B.J.: Nonpharmaceutical measures for pandemic influenza in nonhealthcare settings—social distancing measures. Emerg. Infect. Dis. **26**(5), 976 (2020)
12. Liu, L., Ouyang, W., Wang, X., Fieguth, P., Chen, J., Liu, X., Pietikäinen, M.: Deep learning for generic object detection: a survey (2019). arXiv 2018, arXiv:1809.02165
13. Brighente, A., Formaggio, F., Di Nunzio, G.M., Tomasin, S.: Machine learning for in-region location verification in wireless networks. IEEE J. Sel. Areas Commun. **37**(11), 2490–2502 (2019)
14. Gawande, U., Hajari, K., Golhar, Y.: Pedestrian Detection and Tracking in Video Surveillance System: Issues, Comprehensive Review, and Challenges. Recent Trends in Computational Intelligence, IntechOpen Publisher ( 2020)
15. Khandelwal, P., Khandelwal, A., Agarwal, S.: Using computer vision to enhance safety of workforce in manufacturing in a post COVID world (2020). arXiv preprint arXiv:2005.05287
16. Yang, D., Yurtsever, E., Renganathan, V., Redmill, K., Özgüner, U.: A vision-based social distancing and critical density detection system for COVID-19. Image Video Process. (2020)
17. Mishra, A., Shukla, A.: Mathematical analysis of the cumulative effect of novel ternary crossover operator and mutation on probability of survival of a schema. Theoret. Comput. Sci. **666**, 1–11 (2017)
18. Simonyan, K., Zisserman, A.: Very deep convolutional networks for large-scale image recognition (2014). arXiv preprint arXiv:1409.1556
19. Protopapadakis, E., Schauer, M., Pierri, E., Doulamis, A.D., Stavroulakis, G.E., Böhrnsen, J. U., Langer, S.: A genetically optimized neural classifier applied to numerical pile integrity tests considering concrete piles. Comput. Struct. **162**, 68–79 (2016)
20. Katsikis, V. ed.: MATLAB: a fundamental tool for scientific computing and engineering applications, vol. 3. BoD–Books on Demand (2012)
21. Cerutti, G., Prasad, R., Farella, E.: Convolutional neural network on embedded platform for people presence detection in low resolution thermal images. In: ICASSP 2019–2019 IEEE International Conference on Acoustics, Speech and Signal Processing (ICASSP), pp. 7610–7614. IEEE (2019)
22. Castanheira, J., Curado, F., Tomé, A., Gonçalves, E.: Machine learning methods for radar-based people detection and tracking. In: EPIA Conference on Artificial Intelligence, pp. 412–423. Springer, Cham (2019)
23. Hua, M., Nan, Y., Lian, S.: Small obstacle avoidance based on RGB-D semantic segmentation. In: Proceedings of the IEEE International Conference on Computer Vision Workshops, pp. 0–0 (2019)

# The Applications of Artificial Intelligence to Control COVID-19

**Mukhtar AL-Hashimi and Allam Hamdan**

**Abstract** The usability of AI technology has often been underestimated before the spread of COVID-19. AI serves as the most appropriate tool in predicting and controlling the spread of the pandemic. The use of AI technologies such as; Machine Learning, Computer Vision applications, and Natural Language Processing (NLP), which uses big data applications, is critical under the spread of COVID-19. BlueDot is an organization that uses AI technology to track and identify the spread of the virus quicker than any other organization. The usability of deep learning models is another example that helps find the minute details that are often unnoticed through the naked eye. Notably, the model is effective as it provides rapid analysis of many chest X-rays, which is generally time-consuming when done by healthcare experts. Upon the literature reviews, it can be concluded presently, AI plays a landmark role in controlling the spread of the virus. This chapter will elaborate on numerous technologies of artificial intelligence used in the fight of coronavirus COVID-19 in terms of their applications, impact, and future implications.

**Keywords** AI technology · Applications · Artificial intelligence · Big data · Coronavirus · COVID-19

## 1 Introduction and Background

The spread of the global pandemic COVID-19, which was first identified in December 2019 in China, has been identified as the most life-threatening disease caused by the SARS-CoV-2 virus [1]. AI's role is critical and serves as the most potentially powerful tool in the fight against the spread of the growing pandemic [2]. AI, in this regard, can be referred as the Natural Language Processing (NLP), Machine Learning (ML), and Computer Vision applications, which use big-data based models for pattern recognition, its explanation along with the prediction [2].

M. AL-Hashimi (✉) · A. Hamdan
Ahlia University, Manama, Bahrain



A.-E. Hassanien et al. (eds.), *Advances in Data Science and Intelligent Data Communication Technologies for COVID-19*, Studies in Systems, Decision and Control 378, https://doi.org/10.1007/978-3-030-77302-1_4

To deal with the deadly disease like COVID-19, natural language processing technology is being used [3]. For instance, Canada's most appropriate use of NLP and machine learning has been incorporated through its BlueDot technology, which is used to identify, track, and report the spread of virus quicker than other organizations such as the US Centre for Disease Control and Prevention (CDC) and the WHO. In addition, in the future prospects, the technology will be further used to predict zoonotic infections risks by considering factors such as human activity and climate change. Virtual Healthcare Assistants such as Chatbots is another such example. Stallion AI, widely used in Canada, is the virtual healthcare agent that answers the questions and provides reliable and necessary information regarding the COVID-19. The technology is further affective as it monitors and checks for important symptoms and thus advises people for necessary actions required to see the doctor or patient's screening [4]. Researchers are now utilizing the Machine Learning to add certain improvements in the existing virus nucleic detection test. Metsky et al. [5] in their study indicated that ML in combination with CRISPR (refers to a tool that employs the use of enzymes to change genomes through cleaving the specific strands of genetic code) to formulate assay designs that will be used to detect the 67 respiratory viruses such as; SARS-COV-2. The technology is further affective in understanding the infectivity and severity of infection. Presently, Support Vector Machines (SVM) on genomes are being used to identify which part pf the coronavirus protein sequences cause high fatality rate in comparison to those which cause low fatality rate [5]. Besides, several computer vision models are now being used to conduct X-ray of COVID-19 patients. Another successful contribution of the AI technology is the use of model COVID-Net. Human developed principles network design prototyping is combined with the machine-driven design to develop a network that may easily identify patients with COVID-19 diseases through Chest X-rays. Farooq and Hafeez [6] added that the proposed model provided effective results with 80% sensitivity and 92.4% accuracy. These AI technology functions can be of greater interest in recognizing, predicting and developing a treatment plan for the COVID-19 infection, while managing its socio-economic impact [7–10]

However, for the effective management of the global pandemic, many doctors and medical experts have incorporated AI and other data tools to fight against the prevailing impact of the disease [11, 12]. Following this, six areas have been identified where AI may contribute in effective management of the pandemic. This includes; early warnings and timely alerts, predicting and tracking the prevalence of the disease, formation of data dashboards, diagnosis and prognosis, treatment and cure, resulting in the social control. AI is further useful in tracking, forecasting and predicting the spread of COVID-19 disease by the passage of time [4].

Consider an example of the last pandemic which was spread in 2015 and is known as the Zikra virus. The spread of this virus was tracked through the development of the dynamic neural network. Such models can play a huge role in dealing with the pandemic situation, like that happening in the present world due to the spread of COVID-19 [13]. One of such examples regarding the use of AI technology is BlueDot which is identified as the low-cost AI tool and showed greater accuracy in predicting the humans infected with the outbreak of the developing disease. The efficiency of the BlueDot was observed when it successfully predicted the outbreak of infection in 2019 and triggered warnings to its clients before WHO. Bogoch et al. [14] added that researchers through the use of BlueDot were able to spot 20 destinations where passengers of Wuhan might be affected due to the virus. It further raised a warning against cities which might be affected in future [15, 16].

Hao [17] explained that at Carnegie Mellon University, researchers are now working on training algorithms to forecast and predict the spread of seasonal flu. Besides, several initiatives are now under process in collecting the training data regarding the pandemic's current spread. The usefulness of AI in healthcare can be understood because several measures have been undertaken for accurate and timely diagnosis of the COVI-19, which can save numeral lives while limiting the spread of the deadly disease. It further generates data which helps provide training to the AI models. Contributions of AI can be meaningful in this regard as it provides information regarding the proper diagnosis, based on the chest radiography. Zhang et al. [18] added that the use of AI technology such as deep learning is highly effective to assist radiologists in rapidly screening the affected patients. On experimenting with the effectiveness of deep learning in providing accurate chest X-rays to detect patients with COVID 19, findings of the study indicated that the method was reliable for 96% as it was able to detect most of the patients affected by the disease [18].

Bullock et al. [2] in his recent review related to the role of AI in fighting against the spread of COVID-19 stated that the usability of AI can be as accurate as humans and is helpful in saving time of the radiologists, while providing a faster and cheaper diagnosis of the disease in comparison to the present standard tests held for COVID-19. Other important advantages of AI in recent times are its support in providing the clinical diagnoses of the disease through medical imaging and developing alternative ways to track disease evolution through non invasive devices. Besides, following the societal perspective AI is further used in different areas of the epidemiological research that includes empirical data and thus remained helpful in forecasting the number of cases prevailing in different areas. It is also useful in identifying the similarities and differences in the evolution of disease under different regions. Auslander et al. [19] underlined one of the most important contributions of AI: Support Vector Machines (SVM) on genome collected from different coronaviruses. The SVM were effective in identifying the high fatality rates of the affected patients [20].

Diagnosis of the COVID-19 disease can also be held by incorporating the use of Computed Tomography (CT) and X-rays. In the light of this, Rosebraock [21] provided a tutorial that explains Deep Learning's usability to diagnose COVID-19 by utilizing the X-ray images. He added that though there is a short supply of kits for undertaking COVID-19 diagnostic tests, X-ray machines present in each hospital can be favorable in this regard. Another similar contribution of AI is mentioned by Maghdid et al. [22], who proposed a unique technique through mobile phones to scan the CT images. On board smartphone sensors were used to propose the smartphone-based framework which includes algorithms, embedded sensors and smartphone. Temperature fingerprint sensor is one of its features located under the smartphone's touch screen to predict and evaluate the individual body temperature. Besides, images and videos captured through the phones are effective in detecting the individual fatigue in multiple environments through human gait analysis [23, 24].

## 2 Successful Usability of AI Features in the Global Pandemic Situation

### *2.1 AI and Deep Learning Algorithms*

Deep learning algorithms are important and are specifically good in finding small details of the visual data that are often unnoticed by the human eye [25]. One of such examples is the use of COVID-Net, which has been effectively trained on COVIDx, and serves as the public database. This database consists of 16,756 chest X-rays across 13,645 patients' cases not only for COVID-19, but also of other types of lung infections. This diversity in data enables the deep learning model to separate the characteristics which identifies each type of illness, while detecting the new X-ray. Notably, the model is effective only when it uses a sample of huge data and would contribute in improving the treatment modalities of patients [25].

Luz et al. [26] provided additional information regarding COVID-NET architecture which is entirely based on the generative synthesis technique. This technique was pre-trained on the ImageNet which was then incorporated with the COVIDx dataset. The study indicated that the model was effective in terms of effectiveness and efficiency to provide unique COVIDx database [26]. This type of data is helpful in enabling the deep learning model to separate the characteristics, which individually defines each type of illness while detecting the new X-ray images. According to Wong, through the model is not ready for production, still preliminary results are of significant value in identifying the distinction between COVID-19 and other similar infections. However, the improvisations in the model will be held by the availability of additional data [25]. Scientists further used deep learning models to examine drug candidates, specifically by making predictions regarding the drug target interactions between the present drugs and the virus's proteins. Another example of the usability of machine learning model is that it helps in assessing the probability of contamination of the novel virus. The model is effective as it helps in identifying people infected with the virus, followed by the forecasting regarding the spread of the disease. Machine learning models are further effective in predicting the viral host protein-protein interactions, which determines the reaction of human body towards pathogens. In the light of this, the machine learning models that are trained through protein have provided immense success in predicting the virus host PPIs specifically for HIV and H1N1 [27].

Wong added that inclusion of the large sample size would make a huge difference in improving COVID-Net which is used to develop new learning models in order to detect COVID-19 infections [25]. It is further suggested that chest X-rays and other CT scans are still considerable as complementary tools, and can be used in regions experiencing the short supply of the testing kits. Besides, there are certain situations which require CT scans and multiple chest X-rays for the positive diagnosis of the viral tests in order to identify the intensity of spread of disease for further treatment and care planning. Wong emphasized on the efficiency of deep learning technology in reducing the burden of radiologists by enabling different front-line health workers with minimum expertise for better diagnosis [25].

In the healthcare sector, AI contributes to save lives of the diseased patients, as held in China where radiologists are unable to quickly review a large number of CT scans in each to look for the early signs of the prevailing disease. Use of AI and deep learning augmented the work of radiologists which allowed them to diagnose any disease more easily. Another similar example is the development of DeepMind which helps in creating machines which may mimic a thought process of human brains. The application of Google's deep mind in the healthcare sector might help in reducing the time to plan treatments and to help in diagnose ailments by using machines [28].

## 2.2 *AI Through Machine Learning and COVID-19*

Mostly, computer algorithms used for AI specifically rely on Machine Learning (ML) techniques which, in a broader sense includes computer vision and natural language [29]. Wiens and Shenoy [30] conducted a recent review on the usability of ML in healthcare epidemiology and defined it as the study of methods and tools to identify patterns in data. ML techniques are developed by utilizing a set of algorithms for instance; decision trees, logistic regression, or deep learning which can be categorized into unsupervised, supervised, and reinforcement learning techniques. Contributions of each category is unique, as unsupervised learning at one end provides methods for data clustering, whereas supervised learning is centered towards disease classification. Roth et al. [31] provided a detailed analysis regarding the usefulness of ML in healthcare sector, where patients medical records have been widely stored as electronic healthcare records (EHRs) at different global healthcare institutions [31]. For instance; hospitals in most of the developed countries employs EHRs with basic functionalities including the patient's diagnosis, demographics, nursing assessments, patients' notes and medication lists, along with patients' problem lists, radiology reports, diagnostic test results, discharge summaries and order entries for medication [32]. EHRs are sometimes characterized as weak systems, since they are noisy and inconsistent, along with many missing values and unstructured text fields. However, the electronic availability of data in large volumes provides the potential for the application of ML in the field of infection management. Schaar [33] elaborated the idea further and indicated that

Machine Learning and AI help in identifying the characteristics of people who are at the maximum risks of COVID-19. This can be done by utilizing the EHRs along with a multitude of big data consisting of the human to human interactions such as; social media, airlines, road traffic etc. This type of information is critical to identify which type of infected individuals are likely to suffer more severely due to COVID-19. ML is further effective in determining the most useful treatment techniques for patients with COVID-19, depending upon their distinguished characteristics [33].

Pandey et al. [34] conducted a study to outline the methodological approaches and requirements for the optimal use of ML in future infection management. A lifelong learning application was developed in this study, to deliver the accurate information regarding the disease. Important functions are performed through the matching sources of authentic and verified information extracted from the reports such as those proposed by World Health Organization (WHO) through natural language processing and machine learning. The application is useful as it utilizes the state of art text to develop different speech engines. The approach was further useful in providing useful information regarding, sanitation, hygiene and water which is critical to the current development of the growing pandemic. Fitzpatrick et al. [35] elaborated the idea by quoting some examples from the past, where surveillance of healthcare associated infection (HAI) program functions to interpret multiple data sources to predict the future trends of the prevailing diseases. However, in the spread of Staphylococcus aureus and influenza, HAI and HER were used to track its outbreak and the potential mitigating interventions. ML was further used to predict the risks of nosocomial Clostridioides difficile infection (CDI) [36].

### 2.3 *Visual Recognition*

The development in the scientific advancements has helped in the formation of techniques that are effective in providing the visual detection of the disease among the affected patients of the COVID-19. This visual detection is held through the medical imaging such as the use of CT scans and X-ray images. Deep learning methods play a critical role here as they are being used by the radiologists to save time and efforts [37]. Besides several applications have been designed for the healthcare professionals which leverage the audio-visual information for the patients affected through the pandemic [11].

### 2.4 *CCTV and Tracking Their Movement*

Use of CCTV cameras has attracted the interest of various government officials globally, to detect and analyze the compliance of the preventive measures developed to control the spread of the virus. News18 [38] suggested the use of Heat mapping which can be favorable to assess the maintenance of social distance at different retail outlets. Following this, the spots that are generally over crowded are shown red on the camera while spots that manage to form social distancing appear green on the CCTV screen. This use of advance technology may contribute in developing the social distancing policies and other related SOPs developed to control the spread of pandemic [39].

## 2.5 *Prediction Model of AI for COVID-19 and Its Role in Curing Coronavirus*

The spread of the disease can be predicted through different models. For instance, Cao et al. [40] in their study developed the dynamic models to predict the spread of deadly disease such as the COVID-19 which will help in dealing with the prevention of the disease transmission. Following the transmission mechanism, time series model was developed by the inclusion of different mathematical formulas. The model was effective in predicting the time it takes for COVID-19 to show symptoms of the disease along with the increase in the peak size of the affected patients [41, 42].

Bragazzi et al. [43] focused on the usefulness of Big data in monitoring and predicting the spread of disease in real time. BlueDot is one of the examples that functions through the use of big data and utilized the computational techniques which contributes in visualizing the spread of the virus in different areas. In addition, big data collected through different social networks helps us to forecast the early epidemiological process of the outbreak. Qin et al. [44] incorporated the use of Big Data in order to predict the new cases of the global pandemic, which are either suspected or confirmed. According to the study, the authors analyzed the social media indexes in detailed, followed by the use of keywords. This majorly included the clinical symptoms of COVID-19 which includes chest distress, pneumonia, fever, cough etc. However, findings of the study indicated that the analysis of data was favorable in providing information regarding the confirmed and suspected cases in advance.

The development of the prediction models may serve as a greater support in making strong policies that are favorable for both, the local public and the medical community. It is further helpful in successfully differentiating between stages of the global crisis, as it helps in detection, prevention, recovery, response and to conduct successful research. Besides, deep learning models are further effective in predicting the usefulness of old and the new drugs in predicting the treatment methods for the affected patients. The usability of AI helps in initiating the knowledge share between local as well as the international experts. It also provides the access to large data sets, which is further helpful in increasing the availability of research articles regarding the disease [45, 46].

## 2.6 Contributions of AI and Evolution of BlueDot

BlueDot operates through machine learning and by processing the natural language. For instance, the natural language processing utilizes the social data of various companies' social media pages in Wuhan to track the spread of the virus, to make more accurate predictions [47]. It uses huge data to find patterns that may serve as a cautious hint against the spread of the pandemic. These results are of significant value to different data scientists, doctors, epidemiologists, and veterinarians, which are viewed to decide the point which needs further investigations. This final report, after being analyzed is sent to the BlueDot's customers including governments and businesses [25].

AI may also predict the spread of the contagious diseases by using and studying flight data along with its movement patterns. BlueDot technology was used by various nations, in order to predict the first spread of virus in different cities, right after its outbreak in Wuhan (China) [25]. Under normal conditions, BlueDot serves as the commercial application, however, in present conditions it is being used by the government of various countries in order to track the spread of COVID-19 [25]. Considering this, the usability of BlueDot and AI technology may serve as an early warning generating system to facilitate governments in preparing for the spread of deadly diseases such as; COVID-19 [25]. These predictions regarding the spread of disease are generally held by utilizing the scoured data provided through news reports, animal disease outbreaks, and airline ticketing to trigger early warnings regarding the spread of disease in various regions [48].

According to Dr. Kamran Khan (the CEO of BlueDot and an infectious disease physician) "BlueDot is grateful for the provided opportunity regarding the combination of physician's expertise in the control of infectious disease and big data analytics along with the use of digital technologies as held by the Canadian government in order to mitigate the impact of COVID-19 in Canada and globally" [25].

## 2.7 Contributions of AI Through Robotics

AI and its competent features when deployed in healthcare domain calls for highly diverse applications, along with the development of different teams with long term partnerships. Robotics is another such domain where AI technology when employed helps in fighting against COVID-19. These robots are used for cleaning, disinfecting, and logistics purposes that requires human involvement [2]. Mullane and Bischofberger [49] in their recent article in science robotics outlined that robots can be an effective resource in combating the fight against the pandemic, as their services can be utilized for medicine and food delivery, disinfecting, along with collecting information regarding the most vital signs. The practical implementation of the robotics is held in the northern Italy which was the epicenter of COVID-19 outbreak. Besides, clinicians residing in the town of Varese which is close to Italy's border with Switzerland where robotics were installed for the treatment of COVID-19 patients. These robots were equipped with a camera which allows medical staff to keep a check on their patients and to acquire readings on monitor screens [49].

The robotics have a friendly face, specifically designed to keep patients at ease, as they make communication with patients easy through their automated voice features that makes them no different from humans [49]. To protect doctors and nurses, these robotics helps in reducing the need of face masks and protective gowns specifically at the time of their short supply in these hospitals. These robots further enable patients to communicate with the medical sensors, as they are developed by utilizing over 60 sensors providing them the capacity to make meaningful voice interactions, voice localization, face recognition, and video chat, by avoiding obstacles through auto charging feature [49]. In the cases where most of the manufacturers have restricted themselves into their homes to avoid contact with virus, many manufacturers have focused in converting their production lines to the overall automation. The use of robotics is not limited to the healthcare industry only, as most of the farmers employ robotic as a helping assistant in their fields, specifically those which requires a team work [50].

Following the usability of the robots in the hospitals to treat the affected patients, China was the first to imply the use of robotics in Wuhan, where a Beijing based company known as CloudMinds sent almost robots in Wuhan in order to help China in successfully dealing with the coronavirus patients. These robots were highly functional in managing the hygiene and in providing healthcare services to the affected ones [51]. Once of such example includes, Chinese huge e-commerce JD.com deployed different self-driving robots for the proper delivery of medical goods to Wuhan [52]. Different robots were sent by the startup Shanghai TMI Rob to disinfect wards such as the intensive care units and operating rooms of the city hospitals [53]. Similarly, a robot was also used in one of the hospitals of the US in order to take vitals of patients infected with the COVID-19 virus [54]. This was specifically done to minimize the risk of exposure with the virus. Communication and other diagnostic processes were held through screen on the robot, and were further used to monitor recovering patients. These robots were also sent to Thailand to measure patients' fever, while allowing physicians to communicate with patients through them [55]. The solution was also adopted in four different hospitals of Bangkok, where robots were even called with names such as; ninja robots. In regions where people were banned to assist and visit the older ones, robots are used to break the transmission chain while keeping them in contact with their close ones remotely [49].

## 2.8 Contribution of AI Based Gadgets

According to the Medical College of Georgia, smartphone applications along with machine and artificial intelligence contributes in increasing individual access to homes based COVID-19 risk assessments, to provide medical experts with real-time information regarding target infected patients [56]. The report provided in the Infection Control and Hospital Epidemiology indicated that the application is helpful in directing health experts towards patients who are more vulnerable towards COVID-19. This prior management helps in enhancing prevention as well as treatment related initiatives [57].

Role of artificial intelligence is critical in rapidly assessing the individual's information, while providing them the risk assessment report to alert medical representatives of the nearest medical center with the testing ability [56]. According to McGrail [56], the collective information of different individuals helps in rapid aiding and the accurate identification of regions, cities, areas and villages where virus is in rapid spread along with the predicted health risks in that region. This enables healthcare providers in exhibiting better preparedness level when needed [58].

## 2.9 Digital Information and Internet of Things (IOT)

Internet of things denotes a network of the objects that are physical in nature. The internet is not only a system of computers but it has progressed into a network of devices which are of varied types and dimensions such as automobiles, communicative gadgets, home appliances, toys, cameras, medical equipment and industrial fixtures. Moreover, this system is able to ensure the deliverance of information which is based on specified protocols for the purpose of acquiring smart restructuring, locating, detecting and for controlling the safety and control [59].

The measures were purposed by the UN Department of Economic and Social Affairs that governments are required to fully operationalize all the available digital functions and technologies to square up to the unbridled spread of COVID-19. More importantly, the growth of the pandemic is invoking and dragging governments towards technological approach in order to square up the crisis-ridden situation. Gradually, owing to the widespread and indiscriminatory prevalence of COVID-19, governments seem to take a shift towards open government tactic along with making the systems effective through different form of digital tools and digital communicative channels to make available the information about any global and national trajectory related to COVID-19 [60].

Amid the lock-down approach and the measures of maintaining the social distance which has been a global practice now, people are hinged upon the digital sources in order to get themselves familiarize with every single development around the world. A precise review of the international portals of the 193 member states of United Nations, has revealed that till March 25, 2020, 57% (110 countries) had published some data on COVID-19, whereas, approximately, 43% (83 countries) had been reported not to publish data regarding the outbreak of the ominous viral disease. Conversely, by April 8, 2020, about 86% (167 countries) have been observed to publish the information and standards of procedure in the wake of novel Coronavirus on their online information portals. The information provided by governments are central to the uncontrollable prevalence of coronavirus, travel restrictions and guide, pragmatic instruction related to personal safety such as to maintain respiratory etiquettes along with the government-led efforts and responsiveness for preventing it. Through playing a critical and vital role as the first curator of the information regarding the COVID-19, governments have commenced the publication of statistics that comprise the total number of patients who have contracted with the pandemic and death toll [61].

## 2.10 *AI and Saving Lives (Review of the Wearable IOT Devices Impact Our Lives)*

The functions of the wearable medical equipment are to record and maintain the data related to health of the patients, aiming to have proper and timely knowledge about the symptoms and to get assistance in their routine chores. Growing awareness regarding fitness and health centered standard of living have also brought out the increased industrial worth. Market Research Future (MRFR) has gathered a list of drivers, challenges and prospects in the recent reinstatement of the global Wearable Medical Devices Market report that anticipated for the period of 2018–2023. The global Wearable Medical Devices Market has been sectioned in terms of type, nature of device, applicability and channels of distribution [62].

In addition to it, based on the device type, it is sectioned while distinguishing the devices for diagnostic and therapy. These therapeutic devices are prone to gather massive revenues due to the upsurge in non-invasive caring methods. Manufacturing the pain-relieving wearable devices by the industrial entities, for the sake of encountering the spread use of opioid can lead to the segment trajectory. For example, the Oska Pulse device by Oska Wellness can provide relief in pain through the use of its proprietary Pulsed Electromagnetic Field (PEMF). In terms of its applicability, it is divided on the basis of distant monitoring of the patients, sport related academics and fitness and health centered functions. Through the channels of distribution, it is separated into hypermarkets, pharmacies, and online presence. Unbridled prevalence of sedentary diseases has brought about the upsurge in personalized medicine along with distant and virtual monitoring. These devices serve in monitoring crucial metrics without having effects on the daily life patients. Additionally, patients are now more determined to have proper knowledge for their health and to maintain the record of metrics which are installed as mobile application. However, reservations related to the security of data can impede the market share and growth of wearable medical tools [63].

The West Vrigina University Rockefeller Neuroscience Institute, WVU Medicine, and smart ring maker Oura Health have proclaimed a national level research which was intended to augment detection of the COVID-19 symptoms and to timely diagnose the epidemic. In addition to initiate an artificial intelligence-led analytical model, wearable ring technology and COVD-19 monitoring app, RNI scientists and partners are progressing an innovative "digital PPE" technique that possibly recognize diseases in monitoring competences and confining the prevalence.

## 2.11 *Drone Traffic Monitoring*

The continuing COVID-19 outbreak has generated record socio-economic crises in entire earth, as the humans are compelled to spend their life confined inside their residence which has drastically changed the way of spending life. Further, it is stimulating to rapidly adopt the latest technology perceptions such as drones for assisting the governments in tackling with the situation. The manifold uses of drones have assisted law enforcement entities for floating surveillance and monitoring. Drone-based surveillance are possessed with ultra-high definition resolution cameras to capture pictures and videos to acquire information regarding a geographical area or an explicit landscape. In addition, they provide the likelihood of constructing critical and related to lock down proclamation for the over-all public by sky speakers. The trivial size and flight ability permit it to get access the areas that are not easily accessible [64].

Usually, they have computed vision, camera, face, recognition for objects and other tracing capabilities to obtain the desired outcomes. It employs the amalgamation of networking, robotics and AI to perform married function. Owing to their capabilities and characteristics, drones are now considered as significant tools to coup with the COVID-19 led crisis. The pandemic is leading governments across the world to employ the technology to execute and improved monitoring of lockdown restrictions and standards of procedures (SOPs). Apart from it, drones can dispense critical supplies for medicines, food and can make a central role for disinfecting the areas on large scale. Through the use of drones, the government machineries can diminish the level of depression on health practitioners and can assure contactless deliveries. Consequently, diminishing the risk associated to the infections [65].

In order to assure the safety guidelines, the Chinese administration is nowadays employing drones equipped with loudspeaker capabilities which can trace the defaulter who is not acting according to the guidelines. They are also used for the purpose of spreading the disinfecting spray over the affected areas i.e. bus and train stations [66].

## 2.12 *Facial Recognition*

In China, AI has been used on a large scale for the purpose of massive surveillance in which devices are used to gauge the information about the temperature as wells as performing the function of recognizing the individuals. Moreover, AI has broader functionality in assisting the law enforcement agencies in the form of "smart" helmets which have capabilities of tracing the individuals with more than normal temperature. Although, the devices of recognizing the faces have encountered some hurdles owing to the surgical masks. Nowadays, China has employed the use of technology for also the interstate service delivery in order to take necessary measures. Additionally, Hanvon has proclaimed to make a device for adding up the recognition level for the wearers of surgical masks to 95 [67].

Fourth role of AI is related to the COVID-19 is about social control. AI has been recognized as crucial to square up the pandemic through the use of thermal imaging to detect public places for people who are contained with infection and by imposing the social distance measures and lockdown restrictions. In this instance, Chun [68] described that at airports and stations in China, infrared cameras are used to detect the people with more than average body temperature. Often having the facial recognition capability, they can indicate either people have worn the surgical mask. According to the report, these cameras can detect 200 persons per minute and can evaluate about the temperature that either it is up to the normal body temperature [69].

## 2.13 AI and Gadgets for Coronavirus Outbreak

The spread of COVID-19 has indeed created a drastic impact over the individual social life as well as on economies as whole. It has further raised the bar for personal grooming standards in terms of gaining knowledge through multiple resources. The application of AI has given rise to the development of highly competent gadgets which provides maximum contribution in the management of the COVID-19. In the time where hand sanitization has been critical to save one's life, certain devices have been developed that may reduce the chances of germs transmission. One of such examples is the use of "Hygienehook" that has been developed by Steve Brooks who is a London based designer. The gadget is handy and can be carried in one's pocket easily. It is made up of the non-porous material and is easy to clean device [70].

The use of AI technology has given rise to the development of telemedicine, where National Healthcare System (NHS) England has recommended GPS to change face to face appointments between doctors and patients either through video or telephone. In certain regions, some of the telemedicine-based companies have provided their services for free, such as Doctolib in France, Kry in Sweden and Adent Health Denmark. In UK, a company named as "Push Doctor" has developed partnership with the NHS which has ultimately increased their usage up to 70%. It further initiated the development of virtual tools that are powered through AI. These tools are in the form of symptom checkers which is entirely based on the symptom checkers and was launched by the Babylon Health in UK, Mediktor in Spain etc. [71].

## 2.14 Telemedicine and Coronavirus Application

Telemedicine is term which refers to the use of ICT in improving the patient's outcomes by developing increased individual access to care and important information related to medical and healthcare. According to the World Health Organization, telemedicine refers to the delivery of healthcare services through distance [72]. The overall healthcare services are provided by the use of communication technologies which involves the exchange of medical information such as disease symptoms, diagnosis and treatment which help in the prevention of further development of disease followed by the increased contribution towards evaluation and research [73].

Four important characteristics of telemedicine have been identified, where primarily its purpose is to provide the clinical support. Its main purpose is to reduce the geographical barriers that connect users resides in distant areas. Third includes its involvement of the ICT system. Whereas, its final characteristic includes its unique purpose which is to improve the individual health outcomes. Telemedicine and telehealth are some of the critical tools that help in the preservation of medical staff and equipment, it further extends the usage of limited resources, and provides the safe, effective, and efficient healthcare delivery [74]. Following the widespread effect of the pandemic, most of the healthcare organizations are now implementing the usage of telemedicine solutions work a step ahead in providing regular healthcare services. However, the ultimate contributions of the usage of telemedicine in the existing conditions of COVID-19 is that it helps in reducing the crowds in hospitals, waiting areas, and other similar care providing institutions. It further reduces the probability of maximum admissions in the isolation wards, which ultimately reduces the chances of the spread of virus among healthcare experts. Ren et al. [75] provided the benefits of the mobile telehealth system (MTS) which was developed with an aim to facilitate the provision of information to patients. The study findings indicated that most of the people encouraged the usage MTS, since most of the people used the system [76].

## 2.15 Smartphone Apps for Fighting Coronavirus COVID-19

These technologies help in the identification of different COVID-19. Digital technologies such as mobile phone applications which have functionalities can be of high-value in identifying the affected patients by the pandemic. This is of greater benefit when used in the area where majority patients are suspected. This further helps in straightening the rising curve of the affected patients. The functionality of these applications is of ultimate benefit for different Member states to detect the contact of cases, collection of information, while providing information regarding the need to test and follow up of patient when required [77].

The applications can be further useful specifically for the COVID-19 cases, regarding the ways on transmission of disease along with an advice to control the developing symptoms. Warning and Contract tracing provided by the mobile applications plays a critical role in controlling the spread of the pandemic. However, the impact of the pandemic can be controlled by maximizing the resources for individual testing, specifically of those showing the minor symptoms of the disease. This type of technology is further beneficial to trigger warnings for the public health authorities to identify people with the confirmed cases of COVID-19 resulting in self-quarantine and isolations [78]. Some of the common examples of the smartphone applications designed to combat COVID-19 cases are described below in brief [79].

An application, developed by the Austrian Red Cross along with the partnership of Uniqa Stiftung launched an application, with an aim to track the patients that are in contact with COVID-19. On the identification of any such issue, the app notifies the user if they have recently contacted a COVID-19 positive patient. The application therefore served as one of the most amazing online platforms which offers real time monitoring of the evolution of COVID-19. Another such application includes WeChat apps and Alipay, which allows citizen to assess their interaction with any of the person affected by the pandemic. A QR code is provided to the people, which alerts them either to be in quarantine or not. It further raises important concerns regarding the individual privacy [80].

Globally, government organizations are now providing significant interest regarding the usability of the technology in tackling the coronavirus. One of the major goals in this regard is to track individuals who have been affected or are at a greater risk of being affected. In most of the member states of the European Union, location tracking measures have been adopted in response to the spread of the disease. Besides, health organizations and researchers of the present day are also making important measures to develop such mobile applications which help individuals in sharing their health-related data to different health authorities. In addition, different companies are now incorporating the usage of the AI empowered data to map and predict the occurrence of disease through social media content, and search queries [81].

## 2.16 *Successful Stories of AI in COVID-19 and Lesson Learned*

AI systems are efficient as they cover most of the routine-based tasks which includes daily diagnosis and other related treatment processes, while leaving decision making related tasks to humans. The intention behind the intervention of AI technology in human clinicians is not to reduce human clinicians, rather it helps in delivering an extremely high-quality healthcare delivery process. Among all the AI based domains, its contributions in the field of healthcare are significant due to the development of multiple AI based technology. AI through its deep learning and machine learning features help in predicting the spread of any such disease while opening ways for appropriate measures through timely diagnosis and development of treatment methods. Besides, mobile applications that are developed on the basis of AI features are effective in guiding and transmitting information important for the safety of people.

The implementation of the AI technology has provided some of the most prominent and successful results. For instance, in South Korea, the installation of the location-based messaging has served as the most important tool in battling against the transmission of the disease. Another such example is the use of AI algorithm, as announced by the Alibaba that the use of AI algorithms will help in diagnosing the suspected cases of the coronavirus within 20 s. Besides, the use of robots provided greater assistance in performing tasks such as; providing medicines, food and goods to patients that were in quarantine [82].

Before the spread of this pandemic, the usability and the usefulness of the AI technology was often underestimated. One of the reasons behind major unpreparedness of global countries towards the pandemic is the limited use of AI technology, which failed to trigger alerts for the disease's spread. However, on analyzing the present conditions it is important to realize the usefulness of collecting and analyzing big data obtained from different resources to forecast the spread of any disease which might resultantly collapse different economies [83].

## 3 Conclusion

The delivery of quality healthcare services has been significantly challenging and complex. Major part of these complexities is due to the voluminous data generated for the timely process of quality healthcare delivery. For quick processing, this data needs to be interpreted through intelligent means. The role of AI is critical here, as it can address this need through its problem-solving approach. Besides, its intelligent architecture can incorporate reasoning and the smart ability to act anonymously without being directed through human attention.

Though AI application in healthcare delivery shows promising results, there is still a greater probability of both technical and ethical limitations. Computer scientists primarily drive research based on AI technology without including significant medical training, which has led to a very problem-focused yet technology-based approach in applying AI technology in the healthcare sector. Moreover, contemporary healthcare models are largely dependent on human reasoning, communication between clinicians and patients while establishing professional relationships with patients to ensure maximum compliance. These aspects are sometimes difficult to replace through AI technology. The use of robotic assistants in the healthcare sector is further problematic due to the development of issues related to the healthcare mechanism specifically in vulnerable situations where interaction and intervention are more required. Also, many clinicians are reluctant to adopt AI technology as this will eventually replace the human role.

## References


1. Liu, Y.C., Kuo, R.L., Shih, S.R.: COVID-19: the first documented coronavirus pandemic in history. Biomed. J. (2020)
2. Bullock, J., Pham, K.H., Lam, C.S.N., Luengo-Oroz, M.: Mapping the landscape of artificial intelligence applications against COVID-19. arXiv:2003.11336 (2020)
3. Nong, N.B., Ha, V.H.T.: Impact of Covid-19 on Airbnb: evidence from Vietnam. J. Sustain. Finance Investment **11**(1), 1–22 (2021). https://doi.org/10.1080/20430795.2021.1894544
4. Obeidat, S.: How artificial intelligence is helping fight the COVID-19 pandemic (2020). Retrieved from https://www.entrepreneur.com/article/348368. Accessed on 30 June 2020
5. Metsky, H.C., Freije, C.A., Kosoko-Thoroddsen, T.S.F., Sabeti, P.C., Myhrvold, C.: CRISPR-based surveillance for COVID-19 using genomically-comprehensive machine learning design. bioRxiv (2020)
6. Farooq, M., Hafeez, A.: Covid-resnet: a deep learning framework for screening of COVID-19 from radiographs (2020). arXiv:2003.14395
7. Alansari, H., Gerwe, O., Razzaque, A.: Role of artificial intelligence during the Covid-19 era. In: The Big Data-Driven Digital Economy: Artificial and Computational Intelligence, 974, pp. 157–173 (2021)
8. Elali, W.: The importance of strategic agility to business survival during corona crisis and beyond. Int. J. Bus. Ethics Gov. **4**(2), 1–8 (2021). https://doi.org/10.51325/ijbeg.v4i2.64
9. Alrabba, H., Almahameed, T.: The impact of board's characteristics on risk disclosure in Jordanian industrial corporations. Jordan J. Bus. Adm. **16**(4), 790–811 (2020)
10. Youssef, J., Diab, S.: Does quality of governance contribute to the heterogeneity in happiness levels across MENA countries? J. Bus. Socio-economic Dev. **1**(1), 87–101 (2021). https://doi.org/10.1108/JBSED-03-2021-0027
11. Dr. Jordan, S.: Artificial Intelligence and the COVID-19 Pandemic (2020). Retrieved from https://fpf.org/2020/05/07/artificial-intelligence-and-the-covid-19-pandemic/. Accessed on 30 June 2020
12. Hao, K.: Doctors are using AI to triage COVID-19 patients. The tools may be here to stay (2020). Retrieved from https://www.technologyreview.com/2020/04/23/1000410/ai-triage-covid-19-patients-health-care/. Accessed 20 June 2020
13. Naudé, W.: Artificial Intelligence against COVID-19: an early review (2020). Retrieved from http://ftp.iza.org/dp13110.pdf. Accessed on 28 June 2020
14. Bogoch, I.I., Watts, A., Thomas-Bachli, A., Huber, C., Kraemer, M.U., Khan, K.: Pneumonia of unknown etiology in Wuhan, China: potential for international spread via commercial air travel. J. Travel Med. (2020)
15. Albinali, E.A., Hamdan, A.: The implementation of artificial intelligence in social media marketing and its impact on consumer behavior: evidence from Bahrain. Lecture Notes in Networks and Systems, 194 LNNS, pp. 767–774 (2021)
16. Aminova, M., Marchi, E.: The role of innovation on start-up failure vs. its success. Int. J. Bus. Ethics Gov. **4**(1), 41–72 (2021). https://doi.org/10.51325/ijbeg.v4i1.60
17. Hao, K.: This is how the CDC is trying to forecast coronavirus's spread. Technol. Rev. (2020). Available at: https://www.technologyreview.com/2020/03/13/905313/cdc-cmu-forecasts-coronavirus-spread/. Accessed on 20 June 2020
18. Zhang, J., Xie, Y., Li, Y., Shen, C., Xia, Y.: Covid-19 screening on chest x-ray images using deep learning-based anomaly detection. arXiv preprint (2020)

19. Auslander, N., Gussow, A.B., Wolf, Y.I., Koonin, E.V.: Genomic determinants of pathogenicity in SARS-CoV-2 and other human coronaviruses. bioRxiv (2020)
20. Mseer, I.N.: Ethics of artificial intelligence and the spirit of humanity. Stud. Comput. Intell. **935**, pp. 327–340 (2021)
21. Rosebraock, A.: Detecting COVID-19 in X-ray images with Keras, TensorFlow, and Deep Learning. Image Search (2020). https://www.pyimagesearch.com/2020/03/16/detecting-covid-19-in-x-ray-images-with-keras-tensorflow-and-deep-learning/. Accessed on 28 June 2020
22. Maghdid, H.S., Ghafoor, K.Z., Sadiq, A.S., Curran, K., Rabie, K.: A novel ai-enabled framework to diagnose coronavirus COVID-19 using smartphone embedded sensors: design study (2020). arXiv:2003.07434
23. Awad, I.M., Al-Jerashi, G.K. Alabaddi, Z.A.: Determinants of private domestic investment in Palestine: time series analysis. J. Bus. Socio-economic Dev. **1**(1), 71–86 (2021). https://doi.org/10.1108/JBSED-04-2021-0038
24. Ahmed, N., Hamdan, A., Alareeni, B.: The contribution of healthcare middle managers as change agents in the era of covid-19: critical review. Lecture Notes in Networks and Systems, 194 LNNS, pp. 670–678 (2021)
25. Deckson, B.: How AI is helping in the fight against COVID-19. Pc Mag (2020). https://www.pcmag.com/news/how-ai-is-helping-in-the-fight-against-covid-19. Accessed 21 June 2020
26. Luz, C.F., Vollmer, M., Decruyenaere, J., Nijsten, M.W., Glasner, C., Sinha, B.: Machine learning in infection management using routine electronic health records: tools, techniques, and reporting of future technologies. Clin. Microbiol. Inf. (2020)
27. Schmitt, M.: How to fight COVID-19 with machine learning (2020). Retrieved from https://towardsdatascience.com/fight-covid-19-with-machine-learning-1d1106192d84. Accessed on 30 June 2020
28. Marr, B.: 27 Incredible Examples of AI and Machine Learning in Practice (2018). Retrieved from https://www.forbes.com/sites/bernardmarr/2018/04/30/27-incredible-examples-of-ai-and-machine-learning-in-practice/. Accessed on 20 June 2020
29. Kaplan, A., Haenlein, M.: Siri, Siri, in my hand: Who's the fairest in the land? On the interpretations, illustrations, and implications of artificial intelligence. Bus. Horiz. **62**(1), 15–25 (2019)
30. Wiens, J., Shenoy, E.S.: Machine learning for healthcare: on the verge of a major shift in healthcare epidemiology. Clin. Infect. Dis. **66**(1), 149–153 (2018)
31. Roth, J.A., Battegay, M., Juchler, F., Vogt, J.E., Widmer, A.F.: Introduction to machine learning in digital healthcare epidemiology. Infect. Control Hosp. Epidemiol. **39**(12), 1457–1462 (2018)
32. Adler-Milstein, J., Holmgren, A.J., Kralovec, P., Worzala, C., Searcy, T., Patel, V.: Electronic health record adoption in US hospitals: the emergence of a digital "advanced use" divide. J. Am. Med. Inf. Assoc. **24**(6), 1142–1148 (2017)
33. Schaar, M.V.: Responding to COVID-19 with AI and machine learning (2020). Retrieved from https://www.id-hub.com/2020/04/08/responding-to-covid-19-with-ai-and-machine-learning/. Accessed on 30 June 2020
34. Pandey, R., Gautam, V., Bhagat, K., Sethi, T.: A machine learning application for raising wash awareness in the times of covid-19 pandemic (2020). arXiv:2003.07074
35. Fitzpatrick, F., Doherty, A., Lacey, G.: Using artificial intelligence in infection prevention. Curr. Treat. Options Infect. Dis. 1–10 (2020)
36. Li, B. Y., Oh, J., Young, V.B., Rao, K., Wiens, J.: Using machine learning and the electronic health record to predict complicated Clostridium difficile infection. In: Open Forum Infectious Diseases, vol. 6, No. 5, p. 186. Oxford University Press, US (2019)
37. Nguyen, T.T.: Artificial intelligence in the battle against coronavirus (COVID-19): a survey and future research directions. Preprint 10 (2020)
38. News18.: AI-Powered CCTV cameras to help shops maintain social distancing in times of COVID-19 (2020). Retrieved from https://www.news18.com/news/buzz/ai-powered-cctv-

cameras-to-help-humans-maintain-social-distancing-in-times-of-covid-19-2603473.html. Accessed 28 June 2020
39. Alves, G.: The impact of culture and relational quality in the cooperation between export companies and local distributors. Int. J. Bus. Ethics Gov. **1**(2), 1–19 (2018). https://doi.org/10.51325/ijbeg.v1i2.13
40. Cao, J., Jiang, X., Zhao, B.: Mathematical modeling and epidemic prediction of COVID-19 and its significance to epidemic prevention and control measures. J. Biomed. Res. Innovation **1**(1), 1–19 (2020)
41. Nassar, S.: The impact of intellectual capital on corporate performance of IT companies: evidence from Bursa Istanbul. Int. J. Bus. Ethics Gov. **1**(3), 1–10 (2018). https://doi.org/10.51325/ijbeg.v1i3.17
42. Ali, M.H., Hamdan, A., Alareeni, B.: The implementation of artificial intelligence in organizations' systems: opportunities and challenges. Lecture Notes in Networks and Systems, 194 LNNS, pp. 153–163 (2021)
43. Bragazzi, N.L., Dai, H., Damiani, G., Behzadifar, M., Martini, M., Wu, J.: How big data and artificial intelligence can help better manage the COVID-19 pandemic. Int. J. Environ. Res. Public Health **17**(9), 3176 (2020)
44. Qin, L., Sun, Q., Wang, Y., Wu, K.F., Chen, M., Shia, B.C., Wu, S.Y.: Prediction of number of cases of 2019 novel coronavirus (COVID-19) using social media search index. In. J. Environ. Res. Public Health **17**(7), 2365 (2020)
45. Ramaano, A.I.: Potential of ecotourism as a mechanism to buoy community livelihoods: the case of Musina Municipality, Limpopo, South Africa. J. Bus. Socio-economic Dev. **1**(1), 47–70 (2021). https://doi.org/10.1108/JBSED-02-2021-0020
46. Awwad, B., Zidan, J.: The role of the clearance crisis on public expenditure and budget deficit in Palestine. Int. J. Bus. Ethics Gov. **4**(1), 1–40 (2021). https://doi.org/10.51325/ijbeg.v4i1.59
47. Allam, Z., Dey, G., Jones, D.S.: Artificial intelligence (AI) provided early detection of the coronavirus (COVID-19) in China and will influence future urban health policy internationally. AI **1**(2), 156–165 (2020)
48. Allam, Z., Jones, D.S.: On the coronavirus (COVID-19) outbreak and the smart city network: universal data sharing standards coupled with artificial intelligence (AI) to benefit urban health monitoring and management. Multi. Digit. Publishing Inst. Healthc. **8**(1), 46 (2020)
49. Mullane, M.A., Bischofberger, C.: Robots in the frontline of the fight against COVID-19. Technology Focus, 1. E-tech News (2020). https://iecetech.org/issue/2020-02/Robots-in-the-frontline-of-the-fight-against-COVID-19. Accessed 28 June 2020
50. Al Kurdi, O.F.: A critical comparative review of emergency and disaster management in the Arab world. J. Bus. Socio-economic Dev. **1**(1), 24–46 (2021). https://doi.org/10.1108/JBSED-02-2021-0021
51. Emergency Live: Coronavirus, treating COVID-19 patients with robots? (2020). Retrieved from https://www.emergency-live.com/news/coronavirus-treating-covid-19-patients-with-robots/. Accessed on 25 June 2020
52. Yang, T., Liu, R.: JD'S Robot Delivers First Order in Wuhan In Coronavirus Aid Support (2020). Retrieved from https://jdcorporateblog.com/jds-robot-delivers-first-order-in-wuhan-in-coronavirus-aid-support/. Accessed 3 May 2020
53. Xinhua.: Disinfection robots deployed on frontlines to combat coronavirus (2020). Retrieved from https://global.chinadaily.com.cn/a/202002/07/WS5e3cfe0fa310128217275cd2.html. Accessed on 30 June 2020
54. Magill, J.: This Time, They're on Our Side: Meet the Robots Confronting COVID-19 (2020). Retrieved from https://www.usnews.com/news/healthiest-communities/articles/2020-05-12/this-time-theyre-on-our-side-meet-the-robots-confronting-coronavirus. Accessed on 30 June 2020
55. Monocski, S., Muresan, D.: Digital Health and the Fight Against the COVID-19 Pandemic (n. d.). Retrieved from https://www.matrc.org/wp-content/uploads/2020/04/Digital-Health-and-COVID19.pdf?9b3fb7&9b3fb7. Accessed on 28 June 2020

56. McGrail, S.: AI-Powered Smartphone App Offers Coronavirus Risk Assessment. mHealth Intelligence. (n.d.)
57. Al-Mohaisen, F., Al-Kasasbeh, M.: Impact of succession planning on talent retention at Orange – Jordan. Jordan J. Bus. Adm. **17**(1), 126–146 (2021)
58. Sisaye, S.: The influence of non-governmental organizations (NGOs) on the development of voluntary sustainability accounting reporting rules. J. Bus. Socio-economic Dev. **1**(1), 5–23 (2021). https://doi.org/10.1108/JBSED-02-2021-0017
59. Patel, K.K., Patel, S.M.: Internet of things-IOT: definition, characteristics, architecture, enabling technologies, application and future challenges. Int. J. Eng. Sci. Comput. **6**(5) (2016)
60. Areiqat, A.Y., Hamdan, A., Alheet, A.F., Alareeni, B.: Impact of artificial intelligence on e-commerce development. Lecture Notes in Networks and Systems, 194 LNNS, pp. 571–578 (2021)
61. United Nations: Digital technologies critical in facing COVID-19 pandemic (2020). Retrieved from https://www.un.org/development/desa/en/news/policy/digital-technologies-critical-in-facing-covid-19-pandemic.html. Accessed on 28 June 2020
62. Razzaque, A.: Artificial intelligence and IT governance: A literature review. Stud. Comput. Intell. **974**, 85–97 (2021)
63. Market Research Future.: COVID-19 Impact on Wearable Medical Devices Market Growth, Size Estimation, Future Trends, Share Analysis, Sales Statistics and Key Insights By 2023 (2020). Retrieved from https://www.medgadget.com/2020/05/covid-19-impact-on-wearable-medical-devices-market-growth-size-estimation-future-trends-share-analysis-sales-statistics-and-key-insights-by-2023.html. Accessed on 20 June 2020
64. Awwad, B.: Market power and performance: An Islamic banking perspective. Corporate Ownership & Control, vol 15, no 3, pp. 163–171 (2018)
65. Shivaramaiah, P.: Drones are enabling authorities to implement an effective COVID-19 lockdown. Retrieved from https://www.cyient.com/blog/drones-are-enabling-authorities-to-implement-an-effective-covid-19-lockdown. Accessed 28 June 2020
66. Kapoor, A., Guha, S., Das, M.K., Goswami, K.C., Yadav, R.: Digital healthcare: the only solution for better healthcare during COVID-19 pandemic? Indian Heart J. (2020)
67. Council of Europe.: AI and control of Covid-19 coronavirus (2020). Retrieved from https://www.coe.int/en/web/artificial-intelligence/ai-and-control-of-covid-19-coronavirus. Accessed on 21 May 2020
68. Chun, A.: In a time of coronavirus, Chinas investment in AI is paying off in a big way. South China Morning Post 18 (2020)
69. Dickson, B.: Why AI might be the most effective weapon we have to fight COVID-19. The Next Web 21 (2020). Accessed on 21 June 2020
70. Beech, P.: These new gadgets were designed to fight COVID-19. Retrieved from https://www.weforum.org/agenda/2020/04/coronavirus-covid19-pandemic-gadgets-innovation-technology/. Accessed on 25 June 2020/
71. Dr. Crismin, M.: Digital health and AI in the time of COVID-19 (2020). Retrieved from https://www.ie.edu/building-resilience/knowledge/digital-health-ai-time-covid-19/. Accessed on 21 June 2020
72. Al Azemi, M., Al Omari, A.M., Al Omrani, T.: The reality of financial corruption in Kuwait: a procedure research according to corruption perception index & related rules. Int. J. Bus. Ethics Gov. **2**(2), 64–86 (2019). https://doi.org/10.51325/ijbeg.v2i2.52
73. WHO, A.: Health telematics policy in support of WHO's Health-For-All strategy for global health development. Report of the WHO group consultation on health telematics, pp. 11–16 (1997)
74. Scott, B.K., Armaignac, D.L., Hravnak, M., Afifi, S., Palmer, C., Everhart, S., Danner, O., et al.: Tip Sheet: Application of Telemedicine and Telecritical Care to Emergency Management of COVID-19 (2020). Retrieved from https://www.sccm.org/getattachment/Disaster/Telemedicine-and-COVID-19/SCCM-COVID-19-Telemedicine-Tip-Sheet.pdf?lang=en-US. Accessed on 25 June 2020

75. Ren, X., Zhai, Y., Song, X., Wang, Z., Dou, D., Li, Y.: The application of mobile Telehealth system to facilitate patient information presentation and case discussion. Telemedicine and e-Health (2020)
76. Adnan, S.M., Hamdan, A., Alareeni, B.: Artificial intelligence for public sector: chatbots as a customer service representative. Lecture Notes in Networks and Systems, 194 LNNS, pp. 164–173 (2021)
77. Al-Taee, H., Al-Khawaldeh, K.: Impact of health marketing mix on competitive advantage: the case of King Hussein cancer center. Jordan J. Bus. Adm. **16**(1), 125–152 (2020)
78. E-Health network: Mobile applications to support contact tracing in the EU's fight against COVID-19 (2020). Retrieved from https://ec.europa.eu/health/sites/health/files/ehealth/docs/covid-19_apps_en.pdf. Accessed on 25 June 2020
79. AL-Hashimi, M., Hamdan, A.: Artificial intelligence and coronavirus covid-19: applications, impact and future implications. Lecture Notes in Networks and Systems, 194 LNNS, pp. 830–843 (2021)
80. Davidson, H.: China's coronavirus health code apps raise concerns over privacy (2020). Retrieved from https://www.theguardian.com/world/2020/apr/01/chinas-coronavirus-health-code-apps-raise-concerns-over-privacy Accessed on 31 May 2020
81. Dumbrava, C.: Tracking mobile devices to fight coronavirus. European Parliamentary Research Service (2020). Retrieved from https://www.europarl.europa.eu/RegData/etudes/BRIE/2020/649384/EPRS_BRI(2020)649384_EN.pdf. Accessed on 30 May 2020
82. Lee, K.: Covid-19 will accelerate the AI health care revolution (2020). Retrieved from https://www.wired.com/story/covid-19-will-accelerate-ai-health-care-revolution/
83. Talbot, D., Ordonez-Ponce, E.: Canadian banks' responses to COVID-19: a strategic positioning analysis. J. Sustain. Finance Investment **11**(1), 58–89 (2020). https://doi.org/10.1080/20430795.2020.1771982

# System of Systems as a Solution to Mitigate the Spread of Covid-19

**Hana Yousuf, Asma Y. Zainal, and Said A. Salloum**

**Abstract** System of systems is a compilation of systems that focus on a specific task or a system by combining multiple resources to create complicated task-oriented or dedicated systems that combine their resources and skills to develop a new, more complicated system with higher capability and performance than the individual systems. The paper will define the System of Systems in terms of its architecture and the importance of having a System of System; then, it will define Covid-19's spread and effect on economics. After that, it will describe System of Systems in the medical sector as a case study in terms of mitigating Covid-19 and the challenges. Finally, the solutions are provided, where different applications can be combined to operate in real-time and assist the government agencies and medical staff in operating effectively.

**Keywords** System of systems · Mitigation · Solution · Corona · Covid-19

## 1 Introduction

Around the turn of the new millennium, most of the world did not have any official policies with respect to best practices and recommendations for implementing, in case of a new pandemic or a new strain of a virus that could damage the economy of the entire world. There has been an instance of such deadly pandemics in the past,

H. Yousuf
Faculty of Engineering & IT, The British University in Dubai, Dubai, UAE

A. Y. Zainal
Faculty of Business Management, The British University in Dubai, Dubai, UAE

S. A. Salloum (✉)
School of Science, Engineering, and Environment, University of Salford, Salford, UK
e-mail: ssalloum@sharjah.ac.ae

S. A. Salloum
Machine Learning and NLP Research Group, Department of Computer Science, University of Sharjah, Sharjah, UAE



A.-E. Hassanien et al. (eds.), *Advances in Data Science and Intelligent Data Communication Technologies for COVID-19*, Studies in Systems, Decision and Control 378, https://doi.org/10.1007/978-3-030-77302-1_5

right from the well documented Black Death during the middle ages and multiple plagues that followed it [1, 2]. The most recent large scale pandemic was the 1918 Spanish flu, which killed more people than those dead in the Great War [1]. Influenza that affects many people worldwide during the flu season is less dangerous since many people have the immunity to cope with it, and the virus has been present for a long time [2]. However, newer strains are sudden and leave everyone vulnerable to the disease since no one had immunity to the disease.

The emergence of H5N1 influenza between 2004 and 2009 is an example of newer strains. Even though this strain affected poultry and swine, there have been many human deaths related to the virus [3]. H5N1 spread quickly across domesticated and wild birds' species from Southeast Asia sources: individuals in close contact with the infected birds may become infected; more than 50% of the infected individuals died. Fortunately, at that period, H5N1 hadn't gained the ability to transmit between humans [4]. Forensic epidemiologists had ascertained, as was suspected, that the Spanish influenza virus of 1918 needed only a limited collection of essential mutations from its original avian type to leap species and enable human transmission [5]. It has been estimated to kill around 50 million people globally. Several of such mutations have already been reached with H5N1. We had a worldwide question about life or death. It attained global spread since the novel strain was different on the genetic level and was easily spread globally. It was a rapidly mutating, quick-spreading influenza that mainly targeted birds and was considered panzootic and epizootic [6].

## 2 System of Systems

For several years the idea of a "system of systems" has been applied in different scientific applications. The definition is made use of for both the psychological, biological, and physical sciences. Nevertheless, there is no generally agreed concept or consensus about whether an SOS varies from more traditional methods, as applied to information systems [7]. More mass acceptance of the term "system," which further developed the process of architectural structures, has provided a characterization that other domains will familiarize with:

> A system is defined as a set of different elements so connected or related as to perform a unique function not performable by the elements alone.

This explanation conveys that a structure has some basic ingredients—components and partnerships, and essentially a limit, which distinguishes it from the remainder of the world. The communications network's products and elements refer to hardware, applications, and even individuals. Associations are protocols, connections between software applications and hardware devices, or associations between individuals impacted [8]. The external world can also be interfaced. A system produces outcomes that the elements themselves are unachievable.

What exactly is an SOS? There is no generally agreed definition, as large; complex structures are categorized differently as SOS. The more traditional concept of SoS is

> A system of systems is a set of different systems so connected or related as to produce results unachievable by the individual systems alone [9].

## 3 The Architecture of the SOS

An SOS deployment case may be the landmark event comically if it happens without anticipating the appropriate network architecture and engineering. The architectural framework and engineering efforts are relevant. As relevant material, history on their architectures is given in the summary of the two case studies presented in the next part [10]—the design of a device in the form of an organization. It can be represented using several and specific perspectives, referred to as perspectives, each offering different facts. Medical applications like curbing the spread of pandemic are typically used to describe the defense enterprises. As a general characterization, the architecture of the structures has been established using the requirements described in the network infrastructure to meet the operational architecture requirements. Figure 1 shows the SoS architecture in each implemented system.

### 3.1 *Operational Architecture*

Operational architecture typically describes what the customer wants to accomplish and what knowledge the operational structures may require to share. From the medical perspective, it is the team of medical staff like doctors, nurses, and other government agencies [11]. Operational architecture characterizations communicate the disease's factors like the number of days suffered, hotspot contacted from, and other people interacted with. It determines the type of knowledge, the level of communication, and which tasks these exchanges help.

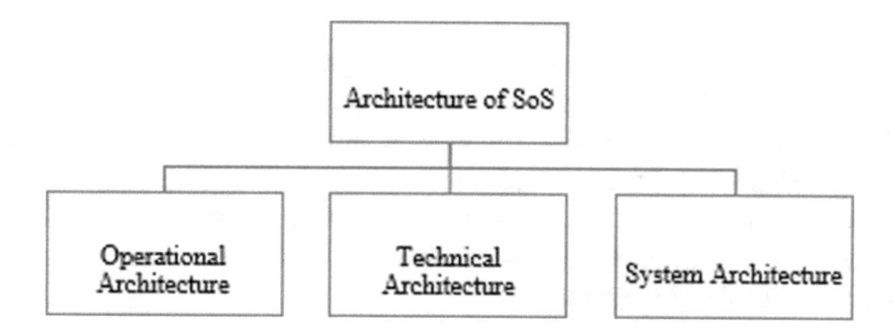


**Fig. 1** The architecture of SoS

### 3.2 Technical Architecture

The technical architecture defines the requirements and rules that the device will conform to, such as protocols for content, retrieval, and transportation: tools, interfaces, specifications, and their connections. It offers basic guidance to execute structures on which architecture requirements are cantered, constructing traditional building blocks, and designing product lines. It used particular criteria for its architecture as defined later in the two case studies in this text, whereas the Army established a comprehensive technical architecture.

### 3.3 System Architecture

The system architecture offers a graphical perspective that defines the individual elements of the device as designed, which is applied, i.e., a summary of the spatial relation, position, and recognition of main nodes, loops, networks, combat platforms, etc., and determines device and function output parameters. The system architecture demonstrations how lots of systems connect and interoperate within a subject area but may describe the institutional structures or operational activities within the architecture of particular systems. Any of the pandemic SOS component structures would have in the same period introduced innovations.

## 4 The Requirement for a System of Systems

Applications were historically designed to operate on specific hardware implemented in a particular environment to perform specific actions. This strategy still holds true for many modern installations, particularly those categorized as critical systems with common nominator legislative regulations. Third-party entities usually impose the rules on these systems. A third party agent may be an authority overseeing the consideration of extra-system intervention to an appropriate degree, for example, the maintenance interval of the platform screen doors of the mass transport network or eliminating a pandemic [12].

These governed systems are generally simulated to completion, even though the framework may depend on constituent subsystems. They are the kind of algorithmic-like implementations for which appropriate techniques were created [13]. Nevertheless, apart from key components and hegemonic system architecture, technical developments have created an abundance of computer devices' implementation opportunities. Applications that rely on these systems offer a sense of security and make our lives simpler, making the support systems an essential component of our daily interactions.

The autonomy and its distributed design are the notable properties of these systems. The autonomy properties are attributed to such a system conducting a function of its own irrespective of its environment; for example, a disease prediction system predicts the number of affected patients irrespective of how this data would be utilized, if at all. Systems management is driven by load balancing, optimization, and physical position, among others. It together allows for what is commonly known as the technology revolution. The overwhelming amount of knowledge available from these systems comes with a trade-off regarding controllability and reliability, especially when a fault can spread hierarchically [14]. Thus, a system based on such separate constituent systems that conceal information can merely incorporate and not define them.

The aim of this work is to find solutions to challenges within the immensely diverse and increasingly relevant environmentally friendly-socio-economic-technical structures, all of which are inherently Structures of Systems (SoS). Some examples are interdependent infrastructures, the public sector, political systems, educational institutions, healthcare facilities, banking institutions, monetary institutions and their distribution network, and the current energy system, a framework within the global climatic system. Technology covers a large functional area within the SoS concept. It is nuanced, frequently complicated, broad, and irreducible; its complexities have a broad range of time scales, rendering it challenging to understand, change, and measure alteration impacts. SoS is flexible and repetitive, and it is difficult to construct understanding by research since replicable initial conditions are usually not possible, and simultaneous research is sometimes not separate. SoS is made up of structures that cannot be substituted by a single individual, which can be incredibly complicated. Many critically relevant complexities related to financial and political aspects involve a wide variety of technology to tackle scientific issues, economic problems, political motivations, and interfaces among them. Since SoS embeds individuals almost exclusively, experimenting among them is dangerous and costly, sometimes leaving simulation to be the only realistic choice to find possible alternatives to adverse circumstances. All of these aspects promote greatly varying perspectives about what the challenges of SoS are, how large the concerns are, and how to fix them.

Our motivation as engineers is to impact and influence the SoS to resolve issues, make use of the prospects, and to attain the objectives. An emphasis on the expectations tries to make us deviate from conventional scientific practices of learning ever more rigorously on the details of each SoS. These must be understood for the effective design of the system. The expectations come under a series of readily identifiable engineering domains, which are predicting, preventing, preparing, and monitoring the control. Comparable between the categories, the three major elements of intention should be described, which are decision, the resiliency of making decisions, and adaptability. The decision tells us the choices made for the implications, and adaptability discusses the conditions for producing. In the context of SoS, these concepts clearly distinguish possibilities for informing policy for the actions taken to impact the SoS.

## 5 Covid-19 and Its Spread

In early December 2019, an outbreak of 'pneumonia of uncertain origin in Wuhan, China, has developed into a pandemic that ravaged China and is now threatening other countries, especially the Americas [15]. A modern beta-coronavirus similar to the Middle East Respiratory-Syndrome Virus (MERS-CoV) and Extreme Acute Respiratory Syndrome Virus (SARS-CoV) soon proved to be the potential cause. The novel coronavirus SARS-CoV-2 disease was branded "COVID-19" by the World Health Organization (WHO), and in January, the epidemic of COVID-19 was announced by the Director-General of the WHO as a public health emergency of international concern [16].

Amid Wuhan's shutdown and the shutdown of all public transit, planes, and railways on 23 January, the National Health Commission registered a total of 40,000 recorded cases in China and then stabilizing to around 82,000 cases after which there were very few new cases. However, the cases started to extrapolate in European countries like France, Italy, Spain, and Germany before the USA became a hotspot. The nature of COVID-19 has still not been established, although initial studies indicate a zoonotic nature, probably from a bat [17, 18].

Compared to SARS-CoV and MERS-CoV, according to the US-CDC, the novel virus is spread between people mainly through droplets in the air, inducing flu-like symptoms, coughing, and difficulty breathing during a duration expected to extend from 2 to 14 days after infection [19]. Preliminary findings indicate that older males with COVID-19 comorbidity could be at greater risk for serious disease [20]. And the exact virological and epidemiological features of this latest zoonotic coronavirus, including infectivity and survival, are still unclear. At the time of such a public crisis, it is necessary to use advanced engineering techniques to mitigate the long term effects of the pandemic.

Around 80% of the confirmed patients recover completely from the infection without any grave consequences. The rest of the patients develop some amount of complications and get seriously sick, with few of them even dying. Untreated symptomatic cases develop difficulty in breathing and pneumonia, which may require higher facilities like ventilators. It is estimated that around 15% of those over the age of 80, who were infected with the virus, has passed away from the disease, while only 1% of them died among those below 50 [21]. However, these numbers are not final since each country shows a different amount of infection and mortality. Hence, the numbers would keep changing based on the stage of the disease.

With the number of fatalities mostly happening in elderly adults, it is apparent that even young people with the virus are likely to experience a significant illness that needs hospitalization. The research to date suggests that the virus is transmitted from person to person by tiny respiratory droplets. Such droplets may also fall on surrounding surfaces when a human coughs or sneezes. There is also data that the COVID-19 virus will survive for up to 3 days on some surfaces-especially plastics or metals. That is why education to stop getting COVID-19 has concentrated on

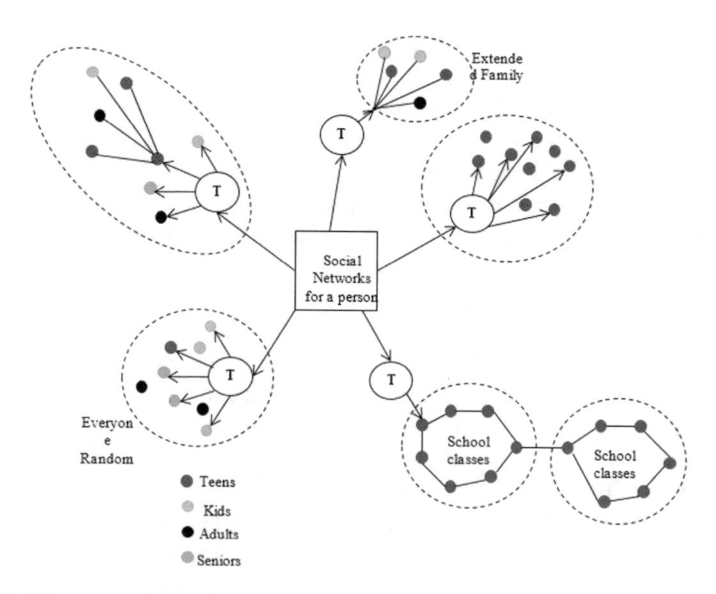


**Fig. 2** Social contact of a sample person

washing with soap, utilizing alcohol-based hand sanitizing gels, and maintaining a buffer from symptomatic individuals [22].

There are currently no antivirals or vaccinations to cure or avoid COVID-19, though at least 44 new coronavirus vaccinations are under production [23]. Several antiviral drugs are being tested, such as those against flu and HIV, to see if they can be used against the new coronavirus, a prevalent antimalarial, chloroquine. Vaccines will take a long time to evolve except in an emergency—no matter how hard experts sprint through the initial process of finding potential vaccines and bringing their vaccines into clinical trials. This is because it would usually take many years to send the vaccine into the lengthy phases of health and effectiveness research. A sample epidemic network is shown in Fig. 2 for a sample person. A person getting infected would spread it to other people, those in contact with the person. It can be spread to the extended family when they meet them at home. In schools, colleges, or offices, everyone in their class is vulnerable if they don't take the necessary precautions. Also, the elderly are most commonly susceptible to the disease.

## 6 Economic Effects of Covid-19

The trending yet troublesome coronavirus pandemic (COVID-19) has a huge impact on countries' economic and financial status. The level of impact varies across the developmental strata of the nation. Without a doubt, developing and underdeveloped countries are getting affected the most. We are not new to get financial damages from a virus [24]. But then, it deems to be a pandemic when it comes to a novel disease like coronavirus. And hence, the demand for resources, food, medicines, insurance, etc., is much higher than usual. Mandatory quarantine and closing the nation's borders are some of the impactful strategies followed by the world's poorest countries, such as Latin America, Asia, and Africa. Some of the minor objectives that government sectors from various countries were managing include reducing the immediate and needful resources to poor and needy and preventive measures from macro-level specialists. Preventive strategies such as self-quarantine, social distancing, and avoiding physical contact are indirect means to protect oneself from this disease. Yet, these steps are more of a primary level focus—to avoid getting affected by the coronavirus. With the immense count of efforts and dedication from several leaders and workers, almost the majority of the world nations have been financially and economically hit hard. About 1,853,464 cases [25] of COVID-19 has been estimated, as of now. Out of which, many have lost lives counted in the trajectory. Now, think about the underdeveloped countries underdeveloped; from basic medicines to serious cases of getting ventilators for survival is no good for people who fall below poverty. Moreover, even the highly developed and technologically advanced nations like The United States of America has serious issues when there was not enough count with their ventilation devices!

According to the data of the 'World Health Organization' (WHO), it is said that spending money on resilient health system protocols is both cost-effective and also useful to fight back in the long-run through sustainability, instead of pinning cash over emergency resources.

When a pandemic gets released, regardless of its developmental milestone, the effect on a country is adverse in results. The following pointers are to give you a general idea about how a country or a specific community falls economically [26]:

- In the stance of quantifying the cost of inaction, government agencies spend a huge amount on experts and domain specialists to get accurate results.
- There will be a reduction in the total trade revenue value.
- High crises in the form of liquidity, along with financial market shocks, are possible.
- Many wages and loads get wasted, thus not benefitting both the seller or vendor and consumer or customer.

As the severity of coronavirus is getting higher, the less resourceful countries will suffer the majority due to a lack of governance and resource allocation processes.

According to the DESA statistics (Department of Economic and Social Affairs), with international trade getting severely affected, the supply chain is greatly reduced due to the impact of the coronavirus pandemic [27].

The experts also gave a warning that without the support of enough fiscal responses, economic activities should not be extended. If that is the case, then from the reversal of their earlier prediction and forecasting, countries might experience about 1% shrinkage in the total global economy. Due to the rise of coronavirus cases, world lockdown has come to a screeching halt, thus reducing the overall economy. Furthermore, the Organization for Economic Cooperation and Development (OECD) has estimates that:

- There will be a less severe impact of coronavirus over countries in agriculture and mining areas.
- Economic costs are adding to regular households of Americans.
- The Bank of Canada has lower rates on their key interest value.
- Low-income families in Singapore will get relative vouchers for buying their groceries.

## 7 Creating SOS for Medical Sectors

Modeling and research are at the center of our commitment. Still, the realization of the information we obtained from that initiative took engagement, motivation, spontaneity, dedication, and support for utilizing the frameworks to choose stable policies in the face of considerable uncertainty. Multiple previous researchers created the framework that will be developed, constructed, and implemented to this issue and the string of the subsequent studies. More to come; descriptions of the model and the findings can be found. These studies became predominant after the turn of the millennium after the diseases like H1N1 and H5N1 started to be prominent. Nevertheless, the period for public policy impact was far shorter than the publication cycle and demanded direct engagement in the course of its development.

Throughout the literature, basic models of dynamic structures, vital management dimensions of electricity grid outages, financial problems, protests, species destruction, forest fires, and several other cascading disruptions have been researched and clarified. A pandemic's transmission has certain parallels to a forest fire: you are spreading disease from your contacts. Such simulation had queried two completely different strategies to checking a forest fire: constructing fire checkpoints to halt the fire and thinning the woods so that no matter where a fire is started, it would not spread quickly.

The second option is much superior because cigarettes are often tossed in unlikely locations, and sparks from a blazing fire will jump the checkpoint and further spread the fire. Since a pandemic is based on people's communications, the contact of an individual during this period is reduced. Within each group, as the pandemic moves around the globe, the spread can be checked. If successful, the mitigation of contacts would prevent the spread widely and might even stop the global pandemic's continuing ride. That strategy just wouldn't hinder the free flow of people and products and would thus encourage industry to continue to function within the globalized market until the production of a vaccine is feasible. It will also eradicate the introduction of quarantines at some size. The "thinning" method takes advantage of our perception of hierarchical structures and uses this knowledge to change society's social structure. The main research problem of this work is identifying whether the mitigation of social contacts reduces the social and economic impacts during pandemics. A study recognized as Applied Methods and Technologies Investigations (AMTI) has been carried out over the past decade. The AMTI initiative has been to be a long-term commitment to recognizing essential infrastructure and its interdependencies. The goal was to define and establish hypotheses, approaches, and theoretical techniques that would help understand the layout, role, and evolution of large, interdependent critical infrastructures. A large number of services communicate within complicated systems with essential infrastructures. In reaction to their multifaceted physical, environmental, cultural, and political contexts, individual infrastructures affect others and develop, change, and thereby evolve. Simply stated, dynamic adaptive networks are vital infrastructures. Complexity makes it incredibly challenging to grasp and model vital infrastructures using classical methods. Fortunately, most of the existing research has focused on comprehending the complicated adaptive systems and creating theories to explicate how they act under stress. AMTI uses this viewpoint to uncover approaches that would render vital facilities more durable and/or resilient, and require long-term policies to be implemented that would encourage reliability and durability to develop over time.

AMTI built the Loki toolkit to devise and adapt Network Models to complex structures quickly. Loki collects computational artifacts that can be chosen, adapted, and combined to construct models of different networks, including power grids, reservoirs, social networks, and financial networks, and interactions between them. With important and fast performance, this can be used. Within Loki, a disease-transmitting model can be developed that can be modified to research influenza spread. Looking at people's societies as a network of nodes and links, The Loki model is used within such a social media network to spread influenza. People were nodes; links were their interactions. Influenza will branch widely from individual to individual through these ties. A simplified representation of the natural history of influenza inside a model human may be built to minimize the economic and medical consequences of Covid-19. It can also reduce probabilistic transmission across linkages depending on a person's condition of the disease.

## 8 Challenges of Implementing SOS for Covid-19

A domain of major challenges in SoS, there seems to be an unresolved issue almost anywhere one impacts, and a lot of researchers and medical staff require immense attention. It is necessary to tackle the problems in the medical field with great urgency [28]. The "SoS engineering" is at the forefront of engineering problems in SoS, leading to new challenges and solutions. Its principles, such as analysis, regulation, estimate, architecture, modeling, controllability, observability, stabilization, filtration, simulation, etc. can be applied to SoS.

It can be noted that SoS continues to exist within a spectrum that includes quick-lived and comparatively simplistic SoS towards one end and complicated, continuously evolving SoS on the other side. However, biotic SoS tends to adopt less complex innovation and growth approaches, generally speaking, helping them to constantly learn and adjust, mature and evolve, overcome evolving disputes, and have more stable behavior. Depending on the biotic SoS, it is evident that these structures adopt robust modular configurations that enable them to focus adequately on approaches and practices for the adoption of open systems such as modular structure, streamlined interfaces, emergence, coordination, conservation, synergism, symbiosis, homeostasis, and self-organization. The infrastructure of the SoS provides services such as energy, transportation of medical supplies, communications, and clean and safe water is vital for the modem organization to exist. The main challenges regarding current and future infrastructure systems include security and reliability, accessibility, and sustainability transitions. The complexity of the infrastructure system precludes simple responses to those challenges. While every one of the subsystems may be considered in itself as a complicated system of systems, increasing interdependence between these systems adds a complexity layer.

One strategy to improving comprehension of complicated subsystems that has so far gained little consideration in the engineering community is concentrating on the similarities of the various sectors and establishing generalized hypotheses and solutions such that concepts from one field can be effectively extended to other domains. In that same regard, the framework of systems offers interesting perspectives. A relatively basic three-level model that separates the physical/technological processes, the organizational and management structures, and the processes and organizations that provide infrastructure-related goods and services can be implemented. As a conceptual structure, the model can be used to identify a number of significant similarities and differences between transportation, medical resources, food distribution, safety equipment, and medicines. The significant challenges are:

- Communication and coordination between the different departments involved.
- Drafting laws and regulations in a timely manner in order to ensure that there further contraction.
- Effective transportation of medical equipment between the demand and supply areas.
- Ensuring that the people maintain social distances to avoid viral contraction from one another.
- Instituting and delivering social programs and welfare to socially and an economically weak section of the population, and to certain essential professions to avoid the work.
- Providing economic assistance to small and medium scale industries to avoid economic collapse
- Ensure that financial institutions accept special laws and work closely with the concerned government.

## 9 Suggested Solutions

The major solution to identify the backbone of the transfers. By analyzing various cases, we can find that school children are the most vulnerable, transmitting the disease between the people. Even though young children are not affected badly, they mostly remain asymptomatic and transmit the infection to others. The most obvious solution is to close schools and colleges during the transmitting period. The other main source is large gatherings, such as communal and religious gatherings, which must be avoided. However, this cannot go on forever, and such places will start to reopen. Once schools and gatherings are opened, there must be a system in place that can identify and mitigate the problem.

A series of applications must be created for mobiles, which would be carried around by everyone at all times. This application or device would continuously keep track of the location so that when a person gets tested positive, all the others who were in close proximity to the person can be tested and mitigated. This would help to mitigate the spread of the virus and flatten the curve. A special government agency must constantly monitor these applications. Similarly, the hospitals and volunteers must have a common application or network that can identify the availability of ventilators, medicine, hospital vacancy, and other medical resources. Since many patients would be discharged and new patients are admitted on a daily basis, it is necessary to maintain a systematic real-time network of steaming data with a dedicated agency to oversee the process. When these SoS resources are shared between different inter-governmental agencies, it would be easy to provide assistance to nearby regions that are badly affected. E.g., If region A has many vacant beds and ventilators, while region B Is highly stressed with lots of new patients, region A can provide the required assistance to region B easily and also plan for the future implications. These applications can be combined into a single System of systems applications such that there is more interdependency on the government and helps enforce the rules. A software framework in the form of a software application that combines all the concerned departments' functionalities may be created with a singular database. Maintaining a single database is the primary solution to avoid disorientation between the departments.

When a person is tested positive, it takes time for the confirmed test results to the municipal body, identifying the contacts of the affected. An automated real-time updating system would immediately update the news to the local body. This would enable faster contact tracing. The police department may immediately seal the entire apartment or the affected street to create a containment zone. The sick and the elderly population may be identified directly among the contacts and tested for faster updates. This may effectively reduce the time taken for contact tracing and hinder the rapid spread of the virus. It would also be easier to provide financial assistance to the concerned families since the database may be shared with government financial institutions to provide a relief package to the affected families. The availability of the medicines and other resources in a particular medical facility may be updated in real-time so that the government agencies may continuously

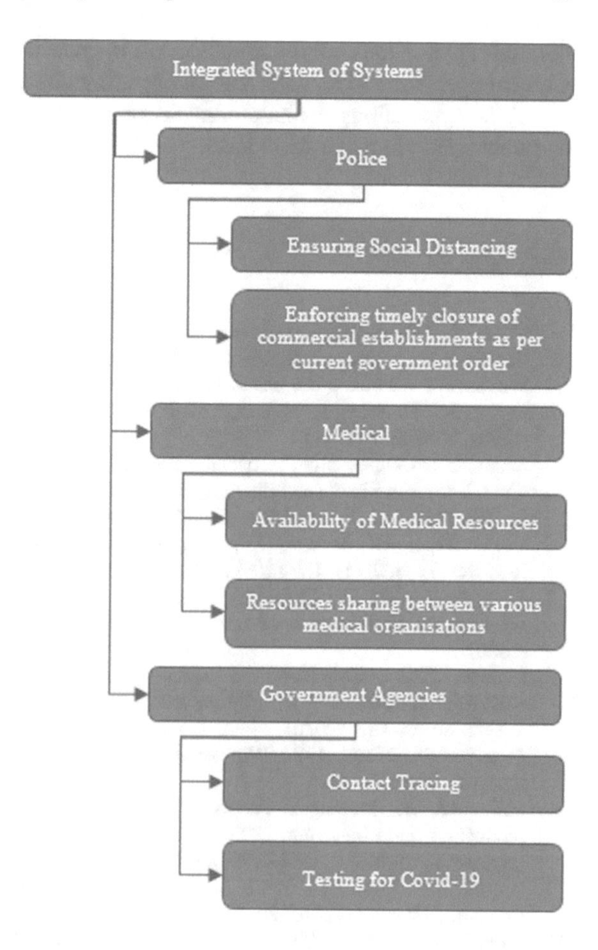


**Fig. 3** SoS solution to mitigate the spread of covid-19

monitor all the hospitals' stocks. This would avoid medicines getting over. The number of patients admitted to a facility can also be considered a factor in calculating the number of drugs required to predict the cases for the next two weeks. A predictive analysis tool would help the medical sector make preparations and provisions for the future to stock up the medicines and make room for more patients. Figure 3 summarizes the SoS solution in the case to mitigate the spread of Covid-19.

## 10 Conclusion

The paper presented a brief description of SoS concept and Covid-19 and how SoS in the medical sector mitigated the spread of Covid-19 besides the challenges that may face the health care sector while implanting. The solutions are provided, where different applications can be combined to operate in real-time and assist the government agencies and medical staff in operating effectively. Continuous and careful use of the applications can help mitigate the effect of the virus and flatten the curve, and may even eliminate the spread.

## References


1. CDC: 1918 Pandemic (H1N1 virus) (2020)
2. Lemon, S.M., Mahmoud, A., Mack, A., Knobler, S.L.: The Threat of Pandemic Influenza: Are We Ready? Workshop Summary. National Academies Press (2005)
3. Lycett, S.J., Duchatel, F., Digard, P.: A brief history of bird flu. Philos. Trans. R. Soc. B **374** (1775), 20180257 (2019)
4. Read, M.C.: CDC-Highly Pathogenic Asian Avian Influenza A (H5N1) in People, 3 Feb 2015
5. Song, L.: It is unlikely that influenza viruses will cause a pandemic again like what happened in 1918 and 1919. Front. Public Heal. **2**, 39 (2014)
6. Taubenberger, J.K., Morens, D.M.: Pandemic influenza–including a risk assessment of H5N1. Rev. Sci. Tech. **28**(1), 187 (2009)
7. Silva, E., Batista, T., Oquendo, F.: On the verification of mission-related properties in software-intensive systems-of-systems architectural design. Sci. Comput. Program. **192**, (2020)
8. Hachem, J.E.L., Chiprianov, V., Babar, M.A., Khalil, T.A.L., Aniorte, P.: Modeling, analyzing and predicting security cascading attacks in smart buildings systems-of-systems. J. Syst. Softw. **162**, (2020)
9. Luzeaux, D., Ruault, J.-R., Wippler, J.-L.: Large-Scale Complex System and Systems of Systems. Wiley (2013)
10. Mahmood, A., Montagna, F.: System of systems architecture framework (SoSAF) for production industries. In: 2012 7th International Conference on System of Systems Engineering (SoSE), pp. 543–548 (2012)
11. Zeigler, B.P., Mittal, S., Traore, M.K.: MBSE with/out simulation: state of the art and way forward. Systems **6**(4), 40 (2018)
12. Turgut, Z., Aydin, G.Z.G., Sertbas, A.: Indoor localization techniques for smart building environment. Proc. Comput. Sci. **83**, 1176–1181 (2016)
13. Chaabane, M., Rodriguez, I.B., Colomo-Palacios, R., Gaaloul, W., Jmaiel, M.: A modeling approach for Systems-of-Systems by adapting ISO/IEC/IEEE 42010 standard evaluated by goal-question-metric. Sci. Comput. Program. **184**, (2019)
14. Wang, H., Fei, H., Yu, Q., Zhao, W., Yan, J., Hong, T.: A motifs-based maximum Entropy Markov model for realtime reliability prediction in system of systems. J. Syst. Softw. **151**, 180–193 (2019)
15. Siettos, C.I., Kevrekidis, I.G., Kevrekidis, P.G.: Focusing revisited: a renormalization/bifurcation approach. Nonlinearity **16**(2), 497 (2003)
16. WHO, WHO Statement Regarding Cluster of Pneumonia Cases in Wuhan, China (2020)
17. Zhou, P., Lou, Y.X., Wang, X.G., Hu, B., Zhang, L., Zhang, W.: A pneumonia outbreak associated with a new coronavirus of probable bat origin. Nature **579**(7798), 270–273 (2020)
18. Lu, R.: XZ (30 de Enero de 2020) Caracterización genómica y epidemiología del nuevo coronavirus 2019: implicaciones para los orígenes del virus y la unión al receptor. Lancet **395** (10224), 566–568
19. Chen, N., et al.: Epidemiological and clinical characteristics of 99 cases of 2019 novel coronavirus pneumonia in Wuhan, China: a descriptive study. Lancet **395**(10223), 507–513 (2020)

20. Yang, X., et al.: Clinical course and outcomes of critically ill patients with SARS-CoV-2 pneumonia in Wuhan, China: a single-centered, retrospective, observational study. Lancet Respir. Med. (2020)
21. Cheatley, J., et al.: The effectiveness of non-pharmaceutical interventions in containing epidemics: a rapid review of the literature and quantitative assessment. medRxiv (2020)
22. Berardi, A., et al.: Hand sanitisers amid CoViD-19: a critical review of alcohol-based products on the market and formulation approaches to respond to increasing demand. Int. J. Pharm. 119431 (2020)
23. Liu, C., et al.: Research and Development on Therapeutic Agents and Vaccines for COVID-19 and Related Human Coronavirus Diseases. ACS Publications (2020)
24. Kieny, M.P., et al.: Strengthening health systems for universal health coverage and sustainable development. Bull. World Health Organ. **95**(7), 537 (2017)
25. ASN: Making Health and Nutrition a Priority During the Coronavirus (COVID-19) Pandemic (2020)
26. Schar, D.L., Yamey, G.M., Machalaba, C.C., Karesh, W.B.: A framework for stimulating economic investments to prevent emerging diseases. Bull. World Health Organ. **96**(2), 138 (2018)
27. The Economic time: Global Economy Could Shrink by Almost 1% in 2020 Due to COVID-19 Pandemic: United Nations (2020)
28. Thatcher, A., Guibourdenche, J., Cahour, B.: Sustainable system-of-systems and francophone activity-centered approaches in ergonomics: converging and diverging lines of dialogue. Psychol. Française **64**(2), 159–177 (2019)

# Data Classification Model for COVID-19 Pandemic

**Sakinat Oluwabukonla Folorunso, Joseph Bamidele Awotunde, Nureni Olawale Adeboye, and Opeyemi Emmanuel Matiluko**

**Abstract** A significant worldwide pandemic disease that has shut the whole world's economy and put the health care services personnel into anxiety is COronaVIrus Disease 2019 (COVID-19). It is difficult to model as it shared closely related characteristics/symptoms with other pneumonia diseases like SARS, MERS, ARDS, and Pulmonary Tuberculosis (PTB). Health practitioners use images (CT scan, Chest X-Ray (CXR)), timely occurrences (daily), audio (Cough), text (clinical and laboratory data) to detect, predict and treat patients with this disease. But machine learning has been proven by researchers when it can effectively and precisely detect, predict, classify, recommend treatment. This chapter discusses and implements a data classification task for early diagnosis and prognosis of the COVID-19 pandemic using CXR image. Classification is a supervised learning task that uses labeled data to assign items to different classes. The indicators that define a good classification task and assess classification models' performance are Receiver Operating Characteristic (ROC), Precision-Recall Curve (PRC), Recall, F1-Score Precision.

**Keywords** Ensemble · COVID-19 · Classification · Pulmonary tuberculosis

S. O. Folorunso (✉)
Department of Mathematical Science, Olabisi Onabanjo University, Ago-Iwoye, Nigeria
e-mail: sakinat.folorunso@oouagoiwoye.edu.ng

J. B. Awotunde
Department of Computer Science, University of Ilorin, Ilorin, Nigeria
e-mail: awotunde.jb@unilorin.edu.ng

N. O. Adeboye
Department of Mathematics & Statistics, Federal Polytechnic Ilaro, Ilaro, Ogun State, Nigeria
e-mail: nureni.adeboye@federalpolyilaro.edu.ng

O. E. Matiluko
Department of Computer Science, Landmark University, Omu Aran, Nigeria
e-mail: matiluko.opeyemi@lmu.edu.ng



A.-E. Hassanien et al. (eds.), *Advances in Data Science and Intelligent Data Communication Technologies for COVID-19*, Studies in Systems, Decision and Control 378, https://doi.org/10.1007/978-3-030-77302-1_6

## 1 Introduction

The new coronavirus (SARS-CoV-2) spread widely in China, and a large count of people got infested from the epidemic in December 2019. The World Health Organization (WHO) declared the epidemic is a public health emergency of international concern on 30 January 2020 [1, 2]. In February, WHO announced a name for the new coronavirus disease as the COVID-19 [3], and on 11 March, the COVID 19 outbreak as pandemic [4]. The new COVID-19 outbreak has a great threat to the health and safety of people worldwide due to its alarming spreading power and potential harms [5, 6]. As of 24 September 2020, more 31,664,104 cases have been reported across 188 countries and territories, resulting in more than 972,221 deaths [1, 7].

COVID-19 epidemic is the highest global health crisis we have faced for a long time. The pandemic has also created an unprecedented socio-economic crisis besides a health crisis. It can create disastrous political social, political, and economic effects that will leave a lifelong scar on every country it spreads to [8]. Faced with this unprecedented situation, governments worldwide have been focusing on taking the pandemic under control and reviving their economies again after the lockdown period.

The crucial question which the researchers (actually all people) have been seeking the answer since the beginning of the pandemic is how many people will be possibly infected and die of COVID-19 until the end of the pandemic. We still have limited knowledge about the epidemic for example whether the virus has a seasonal effect and a second wave will come. A way to answer these questions is to interpret the pandemic behaviors through modeling of the COVID-19 data. There are numerous mathematical models and methods in the literature. We should keep in mind that these models aren't a magic wand to find out everything's about the pandemic. Another important point to consider the lack of correct data. This is caused by the misreporting of the cases for many reasons. The studies are so important for decision-makers and government leaders to allocate resources to health care facilities and to take precautions against the spreading of the virus.

COVID-19 is extremely infectious and can spread complications before and after the onset. Monitoring and lockdown have to encompass anyone with symptoms and properly isolate persons who have been infected from those who are not, to allow good containment. Patients carrying the virus could either be minor symptomless (like fever, sore throat, and sneezing) or have serious clinical signs (such as pneumonia, respiratory failure, and eventually death) [9]. The transmittable SARS-CoV-2 condition is called "coronavirus disease" (COVID-19) [10]. Gratitude to the recent developments in analytical methods and Information and Communication Technologies (ICTs), Machine Learning (ML) and big data will aid manage the immense, unparalleled volume of data generated from patient monitoring, real-time tracking of disease outbreaks, now-casting/predicting patterns, daily situation briefings and public updates [11].

The use of machine learning and statistical methods to forecast uncertain or potential effects is referred to as predictive analysis [12, 13]. It responds to the question as what is the next step? It also uses past and present data to predict actions, patterns, and activity in the future. The prediction is rendered using quantitative questions, automated machine learning, and statistical analysis [13]. Experts need to construct predictive models that are used for forecasting in predictive analytics [13]. ML is an umbrella word that depicts various tools for knowledge and control. Without an ML instrument, the data from a wearable would be without any worth to the merchant as well as the consumer. For this purpose, wearable app inventors are progressively adding ML engine inside wearable medical apps. Furthermore, ML aided data mining is also fundamental to the attainment of an intellectual medical care platform that connects many smartphones, website, IoT gadgets, and wearables together to congregate data and benefit fascinating medical comprehensions of an individual [14].

Health professionals are in desperate need of technology for decision-making to tackle this epidemic and allow them to get timely feedback in real-time to prevent its transmission. ML works to simulate the human intellect competently. This may also play a crucial role in interpreting and recommending the creation of a COVID-19 vaccine. This result-driven engineering is used to better scan, evaluate, forecast, and monitor current clinicians and patients expected to be future. The relevant technologies relate to the monitoring of verified, recovered, and deaths cases. The application of ML in COVID-19 can be expediting the diagnoses and monitoring of COVID-19 and minimizes the burden of these processes. Therefore, this chapter discusses the areas of applicability of AI during the COVID-19 pandemic.

Discusses several extraordinary opportunities brought by ML in the COVID-19 outbreak and the research challenges of ML during the outbreak. In medical detection, the use of ML has increased tremendously. This has been commonly used to achieve relatively precise recognition accuracy and to reduce the burden on health systems by reducing the time of evaluation associated with conventional approach detection procedure. The ML techniques are seen as a major aspect in identifying the risk of infectious diseases in enhancing the forecasting and identification of potential world health threats. The continued expansion of ML for the Covid-19 disease outbreak has dramatically improved monitoring, diagnosis, monitoring, analysis, forecasting, touch trailing, and medications/vaccine production process and minimized human involvement in nursing treatment.

The data science analysis using ML is newly evolving, intending to empower health care systems and organizations to connect to harness information and convert it to usable knowledge and preferably personalized clinical decision making. Utilizing deep learning, the implementation of ML in the field of infectious diseases has implemented a range of improvements in the modeling of knowledge generation. Big data can be interpreted, stored, and collected in healthcare through the constantly emerging field of ML models, thereby allowing the understanding, rationalization, and use of data for various reasons.

Therefore, this chapter discusses the concept of ML models, its applicability and challenges in fighting the COVID-19 epidemic. The chapter also defines indicators that defines a good classification task and assesses the performance of classification models as Receiver Operating Characteristic (ROC), Precision Recall Curve (PRC), Recall, Precision and F1-Score. The main contribution of the chapter is modelling COVID-19 CXR image as a data classification task and assessing the performance of the model. The hope of using ML in COVID-19 will have a great impact on the quality of outbreak diagnosis, prediction, and treatment and can be delivered quality care to patients across socioeconomic and geographic boundaries. Therefore, this chapter presents ML applicability and the challenges in fighting the COVID-19 epidemic.

The chapter is organized as follows. Section 2 explains ML to combat the COVID-19 epidemic. Section 3 discusses the applicability and challenges of using ML for fighting COVID-19. Section 4 presents prediction and classification for the fighting COVID-19 outbreak, Sect. 5 discusses the taxonomy of ML models for data classification while Sect. 6 presents the practical case of COVID-19 pandemic classification with ML technique. Finally, Sect. 6 concludes the chapter and discusses future works for the realization of efficient uses of ML methods in fighting the COVID-19 pandemic.

## 2 Machine Learning in Fighting COVID-19 Pandemic

Computational Intelligence (CI) centers on complications that hypothetically only individuals and animals can unravel, difficulties needing intelligence. It is a section of computer science considering complications for which there are no operational computational algorithms. The term acts as patronage under which further approaches have been added over time. Computational Intelligence methods have pulled an increased level of concentration in the investigation community. As described in several studies, machine learning approaches can provide excellent precision in classification as likened to alternative algorithms for data classification [15–18]. Accomplishing conspicuous correctness in forecasting is significant because it could lead to an appropriate precaution system. Forecasting correctness may differ varying on diverse studying systems methodologies. Hence, it is fundamental to recognize devices proficient in offering extreme accuracy of projection in diabetes outbursts.

Information technology is constantly evolving and is now being applied to other disciplines that give way to multidisciplinary research, including medicine, agriculture, etc. Using Fuzzy Logic and Artificial Intelligence approaches [19], the advent of electronic health records has also made available information that can be analyzed and inferences in disease diagnosis and treatment. The introduction of machine learning for prediction and classification of COVID-19 data to detect COVID-19 outbreak is new and few authors have work in this direction, only the clinical approaches have been used to established related symptoms for COVID-19 using genomics data. ML is the capacity of devices that are normally in the computer hardware and software format to imitate or surpass individual intelligence in daily engineering and systematic undertakings connected with identifying, intellectual, and performing. ML is multifaceted since individual intelligence is multifaceted, hence including objectives that vary from knowledge representation and intellectual to acquiring, to ophthalmic observation and language understanding [20, 21].

ML-based techniques remain a hot science and technology space research subject, and it increases its dissemination into other jurisdictions such as industry, healthcare, and gaming. ML-based approaches have emerged as a capable tool for designing and implementing smart systems in the healthcare system. Use ML in healthcare can enhance clinical disease management by incorporating smart approaches for prevention, diagnosis, treatment, and follow-up as well as administrative process review. By taking into account the complexity that characterizes health data and procedures, ML-based systems can model according to changes in the environment.

The use of data mining and ML approaches was used to produce metrics that help the decision-making process through the use and identification of relevant information from collected hospital data using computer devices. ML-based paradigm, through the numerical representation of data, incorporated many methods and approaches to model information. Applications can be built using an ML-based approach that can classify high-risk patients with chronic diseases such as COVID-19, thus becoming useful tools for implementing effective treatments and scheduling. Therefore, the number of hospitalizations of patients will be reduced. This can also result in a better diagnosis and rehabilitation system for appropriate evaluation of readmissions, thereby enhancing the use of resources in the provision of healthcare. Therefore, it can be used to determine the best treatment options by using the available evidence from ML applications that endorse medical decisions. The ML-based approach has been used to solve various health sector problems, particularly the most urgent problems, without reducing the quality of care it has reduced the patient's cost diagnosis and treatment. It is used to reduce the length of stay in hospitals for patients, thus helping to prevent medical problems arising from hospital infections.

Using ML-based applications can also provide valuable information to medical scientists and enhance the outcome of patients, which improves service quality and also reduces costs [22]. It is also helpful in recognizing and knowing, among others, patients with chronic signs and symptoms resulting in high health expenses [23, 24]. ML techniques can add value to financial and health-related costs to healthcare management [25]. Evolutionary ML approaches are effective in promoting and diagnosing chronic diseases such as Yang's heart [26–28] and Metabolic diseases [29–31], diagnosis of cancer [32–34].

ML can scrutinize the abnormal symptoms as well as other 'obvious signs' swiftly and alerts patients and medical officials [35, 36]. It provides great quicker policy-making, which is cost-effective. It tends to help, through valuable techniques, establish a new treatment and monitoring skills for the COVID 19 cases.

Using diagnostic imaging inventions like Computed Tomography (CT), Positron Emission Tomography (PET) and Magnetic Resonance Imaging (MRI) scanning of human organs, AI helps diagnose highly contagious cases.

ML will provide relevant information with the aid of real-time analysis, which is useful in preventing this infection. It could be used to forecast possible sites of an outbreak throughout this epidemic, the proliferation of the epidemic, the need for spaces, and health care providers. ML is useful for the potential detection of viruses and illnesses, with the aid of prior coached data over data that are present at various times. It describes the features, sources, and factors why infection has dispersed. This will be a promising technique in the future for fighting the other outbreaks and infectious diseases. It can provide precautionary action and tackle various illnesses. In the future, ML can play a crucial role in making healthcare more informative and reactive.

ML will create a smart framework to automatically track and forecast the spreading of this outbreak [37–39]. A genetic algorithm may also be built to remove the visual characteristics of this infection. It has the potential to provide patients with daily alerts and also to offer better options for COVID-19 epidemic follow-up. ML can easily determine this virus' level of transmission by recognizing the fragments and 'hot spots' and can effectively track the individuals' contacts and even monitor them. It can foresee this spread of the disease future path, and possibly reoccurrence. These technologies can also monitor and predict the existence of the epidemic from the data, social media, and broadcasting channels present, about the threats of the outbreak and its probable spreading. It can also forecast the count of use cases and deaths in any area. ML will help recognize the areas, citizens, and communities most affected and take effective measures.

Analysis of available data on COVID-19 allows the use of ML for drug testing. It's suitable for the production and creation of therapeutics. This innovation is used to expedite drug screening in real-time, where normal screening takes a lot of time and thus enables to greatly achieve a better result, which a person would not be able to accomplish [38, 39]. Identifying useful drugs for treating COVID-19 patients can help. This has become a strong instrument for designing diagnostic tests and improving vaccinations [40, 41]. AI helps to produce vaccines and therapies much quicker than normal and is also effective in clinical trials during vaccine production.

Healthcare workers have such a heavy workload because of a rapid and unprecedented rise in the number of patients throughout the period COVID-19 outbreak. Here ML is being used to minimize health professionals' duties [42, 43]. It assists with initial screening and effective medication using digital methodologies and decision-making skills, provides the best exercise for students and health professionals on this new virus [44, 45]. ML can greatly impact clinical outcomes and address more future problems that reduce the physicians' burden.

## 3 The Applicability and Challenges of Machine Learning for Fighting COVID-19

ML developments in recent years have continued to attract a lot of healthcare workers. The digitization of health-related data and the use of computer hardware and software applications have brought about the growth and use of ML in medicine, thereby creating new possibilities and challenges and offering guidance for ML's future in healthcare. In the COVID-19 pandemic, ML-based approaches have been used for various medical purposes such as diagnosis, identification, forecasting, and control of various diseases and diseases [46, 47]. ML data, including clinical, behavioral, environmental, drug, and biomedical data, were used to establish a decision support system, diagnosis, prediction, and patient classification from various diseases. Due to its ability to systematize several activities that currently require human intervention, ML has drawn significant interest from various fields [48, 49]. Recently, ML techniques and methodologies are used to support, among others, in the processing of natural language, speech recognition, computer vision, and imagery.

ML provides new opportunities for human intelligence and mathematical models to evolve to extract information from health data. ML systems can, therefore, be used to simulate human intelligence in the COVID-19 outbreak at various levels [1]. ML has been providing rapid development of computer system hardware and software applications in recent years to promote the digitization of health data to be used to build automated medicine systems for various uses. ML-based systems in medicine have always been considered both to reflect medical knowledge and to derive new knowledge from stored clinical data and to help clinical decision making. The healthcare system has increasingly leveraged ML-based technologies to improve the delivery of healthcare at a reduced cost. ML's application to diagnosis, perdition, and identification is well established in recent years, and ML is increasingly being used to inform decisions on health care management.

The advantages of ML-based healthcare have been extensively discussed in recent years to the point that it demonstrates that it is possible in the nearest future to replace human doctors with AI, thereby showing that AI can be used in healthcare in many ways: the ability of ML to learn features from a large volume of medical information and then to use the knowledge gained to assist clinical practice. The ML-based system is shown to be effective in collecting useful information from large patient populations and used to make real-time health risk warning inferences, diagnosis, and prediction of health outcomes. AI-based systems can also easily perform repetitive tasks such as X-Rays, CT scans, or data entry for the COVID-19 outbreak. ML can be used to reduce unavoidable human medical practice clinical and diagnostic errors [49, 50].

ML-based can support doctors by presenting up-to-date medical information from clinical procedures, textbooks, and reviews to provide proper patient care. ML has shown significant improvements in the fields of record-keeping and assessing the quality of an individual organization as well as the entire health care system [51,

52]. AI leads to the development of new medicines and precision medicine based on the faster processing of mutations and disease linkages. Ultimately, health monitoring services and online consultations are provided by ML-based to the degree that they are medical bots or virtual nurses [53, 54].

To accept ML fully for clinical care, it will need to meet specific high standards in order to meet the needs and desires of physicians and patients. Though the ML approach has shown superiority over other approaches, it has shown some degree of error and it is not and never will be flawless, no matter how infrequent, it will drive major, negative perceptions [55]. ML-enabled system errors regardless of how small it would have a major worrying impact on any medical matter [55], so a suitable level of regulation and supervision is very important when incorporating ML into the clinical field. It is also important to evaluate the cost-effectiveness of ML-based clinical efficacy [55, 56]. Huge investments in ML were made, similar to robotic surgery, with expected efficiencies and anticipated cost reductions in return. It is unclear, however, whether ML techniques can significantly reduce costs with their associated data storage requirements, data curation, design maintenance, upgrading, and data visualization. Such resources and associated needs can simply replace current costs with costs that are unique and theoretically lower [55, 56].

The following problems in the COVID-19 outbreak system were still highlighted by ML-based techniques. Security is one of the healthcare issues that have to apply aggressively when it comes to medical information as it is well established that security is very relevant in every sector [56–58]. For example, patient data is not allowed to leave Europe in European countries, most infirmary and research institutes are cautious about cloud platforms and would rather use their databases, making it difficult to access patient data to build AI-based systems [57].

In some cases, the use of standard application procedures to facilitate research-based on patient clinical data is much simpler for medical researchers. For example, ML algorithms intended to be used in healthcare (in Europe) must be licensed and controlled for proper use, CE labeling must be applied for. More precisely, according to the Medical Device Directive, they need to be marked. The Global Data Protection Regulation (GDPR) guidelines adopted in May 2018 will also produce several new regulations that need to be adhere to and that are not clear-cut in some situations [59].

Deep Neural Network (DNN) has performed poorly throughout healthcare, given their emerging capabilities. The technologies are still in the infant stage and resources needed to support them are still young with few experts to handle the volume of information and software engineering issues. ML solutions are often plagued with data scarcity and quality, particularly in medicine. As new data are procured, predictive models will need to be re-trained, keeping a close eye on changes in data creation processes and other practical issues that can cause the data distributions to drift over time. If multiple data sources are used to train models, additional "data dependencies" types are added, which are rarely reported or specifically treated [56, 57, 59].

For the ML-based medical system, the consistency of decision aid is very important. A doctor needs to understand the medical ML-based system and be able to explain why an algorithm has suggested a certain treatment. This includes the creation of methods for predictive analysis and more intuition. There is often a trade-off between predictive accuracy and accessibility of the model, particularly with the latest generation of ML techniques using neural networks, which makes this problem even more urgent. Also, the data contained in electronic health records are not appropriate for AI-based techniques in many cases [56, 60].

ML works best with high-quality data sources, while with large, restricted categories, electronic health records, and medical billing claims appear to be ill-defined. Sociocultural is another healthcare-based ML challenge because it can be challenging to get doctors to accept advice from an automated system, as doctors make decisions based on previous experience and instinct, expertise gained, and problem-solving skills [56, 60]. It is therefore important to incorporate certain elements of ML literacy into medical curricula so that ML is not perceived as a threat to physicians, but as a medical knowledge aid and amplifier. Nevertheless, if ML is implemented in a way that empowers rather than displaces human workers, it could free up their time to perform more important tasks or provide more capital to hire more workers [56, 57, 60].

Owing to a lack of evidence, crowded social platforms and outlier data, big data hubris, and computational complexities, COVID-19 spread ML predictions are still not very reliable and valid [61]. The majority of frameworks used for monitoring and forecasting so far do not use AI approaches. Rather, most analysts consider epidemiologically proven models, the so-called SIR models. The Robert Koch Institute in Berlin, for example, uses an immunological SIR method that captures into account government control steps, such as lockdowns, quarantines, and social distancing recommendations. This method was established in China to demonstrate that containment would succeed in limiting the incidence to slower than the [62] exponential value. Table 1 presents overviews of ML problems in the COVID-19 epidemic.

The latest outbreak of COVID-19 has brought the difficulty of securing personal data to a head in a transnational sense [63], this is because COVID-19 spreads fast with the international travel of people [64]. Many countries require international travelers to disclose their personal information such as the name, gender, date of birth, travel history, the purpose of travel and residence, among other and impose quarantine requirements accordingly [65].

Using a genuine case where Chinese media secretly published the sensitive information of a foreign traveler, the study describes that multiple patterns for LEX cause emerged at each point of dispute-of-law analysis: (1) the EU, the US, and China vary in characterizing the right to personal data; (2) The expanding centralized approach to relevant legislation lies in the fact that all three territories either find the law on personal information privacy to be a contractual law or follow linking factors leading to the law of the forum, and (3) actively support the de-Americanization of meaningful data privacy legislation. The patterns and their mechanisms have important consequences in the application of regulations for transnational information [63].

**Table 1** Limitations of machine learning

| S/No | Challenges of ML | Explanations |
|---|---|---|
| 1 | Requires proper supervision | It doesn't have any other information like a person to represent the exact result<br>It needs serious monitoring and data collection; else, practitioners and physicians are not producing accurate findings |
| 2 | Not provide creative thinking | Only precise information will yield enough performance<br>It cannot satisfy the definition of innovative thinking in COVID-19 like a human being, because a human being can feel, learn to make a valuable decision about the computer system<br>This new technology does not allow for new decision without any data |
| 3 | Only learns from the given data | ML always operates from the patient's provided data<br>Precise diagnosis and forecasting of COVID-19 relies on the data collected |
| 4 | Crucial for treatment by AI algorithm used | One of the major drawbacks is the proportion of accuracy of the estimation used by the ML algorithm, which is quite important for the care and even more so<br>The potential scenario of the predictions would be whether an ML's function is an absolute way or whether it requires more data and train over time |
| 5 | Does not understand the emotion | ML approaches do not grasp the feelings, human reasoning, and motives for making the right decision<br>ML computers do just what they are configured to do |

# 4 Classification Task for Combating COVID-19 Pandemic

Prediction is the process of modelling historic data to make estimates about the forthcoming or unknown event. For health care, predictive analytics will enable the best decisions to be made, allowing for care to be personalized to each individual. When the prediction is done right in healthcare will help the medical scientist to see an increase in patient access, lower cost, increased revenue, increased asset utilization, and also improved patient experience. Predictive analytics tried to apply what doctors have been doing on a larger scale. The most changes brought to healthcare by predictive analytics are the ability to make sense of previously non-existent behavioral, to better measure, quantitative, psychosocial, and biometric data, so the wheel is not reinvented [66–68]. It allows individual treatment to be personalized to each person and allows the best health care decisions to be made. The medium goal of healthcare predictive analytics is to reliably predict the unpredictable. Confidence in predictions affects the type of question posed with a high certainty of response. For example, it can be answered with a high degree of certainty to ask a historical question like "what did I eat today."

Applying predictive analytics in healthcare through data analytics helps improve the operational efficiency, planning, and execution of key care delivery systems, use of resources, staff schedules, and admission and discharge of patients. All of this leads to increased patient access to good health care delivery and income, lower costs, increased use of assets, and improved patient experience. New data sources are used for predictive analytics, including those directly from patients. The simultaneous use of two or more forms of statistical analysis is called a predictive model and is a subset of concurrent analytics. Predictive modeling's main objective is to forecast action, outcome, or occurrence using a multivariate array of predictors [69]. The models were further used in healthcare by analyzing patient lifestyles, demographics, psychographics, and preferences that can then be used by medical scientists to produce results that are useful to both patient and physician, and theses can be delivered through specific channels.

Classification refers to the recognition of a group description of a novel inspection. Fundamentally, an established data is utilized as teaching data. The group of participation data and the equivalent outputs are specified to the algorithm. Hence, the teaching dataset comprises the input data and their related class descriptions. Employing the teaching dataset, the system obtains a replica or the classifier then this obtained replica can be a decision tree, arithmetical method, or a neural network. In the classification process, once an undescribed data is specified to the system, it would then discover the class which it fits into. The latest data delivered to the prototypical is the test dataset [70, 71].

Classification can also be defined as the procedure of categorizing a report. An instance of classification is to scrutinize whether it is raining or not and the response can either be yes or no. Hence, there is a precise number of selections. Although occasionally there could be more than two groups to categorize and this is called multiclass classification [70, 71]. An instance in real life is a bank requiring to consider whether granting a mortgage to a specific customer is uncertain or not. In this instance, a replica is created to discover the uncompromising label.

The objective of the data classification task is to develop a machine learning model that will accurately and effectively classify or group a set of inputs (features) into their respective classes. The classification problem could be binary i.e. with two different classes as "0" or "1" [72], Yes or No or True or False. It could be a multiple class problem where there are three (3) or more classes as "COVID-19+", "COVID-19−" and "Healthy lungs" [73]. A classification problem could also be multiple label problem where a class could have two or more labels [74].

## 5 Taxonomy of Data Classification Models for COVID-19

Data classification algorithms vary from the base classifier, ensemble of classifiers to complex deep learning models. The following section presents some selected data classification models and related work for COVID-19. ML deals with the design and development of biologically and linguistically motivated computational models. ML covers methods that are primarily based on Artificial Immune Systems (AIS), Artificial Neural Networks (ANN), Fuzzy logic, Swarm intelligence, Evolutionary algorithms, Support Vector Machines (SVM), Rough sets, chaotic systems as well as hybrid methods that combine two or more techniques. Also, ML has been extended to other computational methods including ambient intelligence, cultural learning, artificial endocrine networks, and social reasoning. ML involves computational models and tools that deal with learning, adaptation, and heuristic optimization to provide solutions to problems that are difficult to solve using traditional computational algorithms. ML techniques have played a major role in developing intelligent healthcare systems such as systems for COVID-19 pandemic detection and diagnosis for personalized health monitoring. Figure 1 presents the taxonomy of ML techniques used for COVID-19 detection and diagnosis.

## 5.1 Classification Techniques

For the data classification model, the target class is known apriori from the data samples collected. Health practitioners use data samples as images [CT scan, Chest X-Ray (CXR)], timely occurrences (daily), audio (Cough), text (clinical and laboratory data) to detect, predict and treat patients with this disease. The goal is to predict the health status (i.e. the target class) of patients taken into consideration

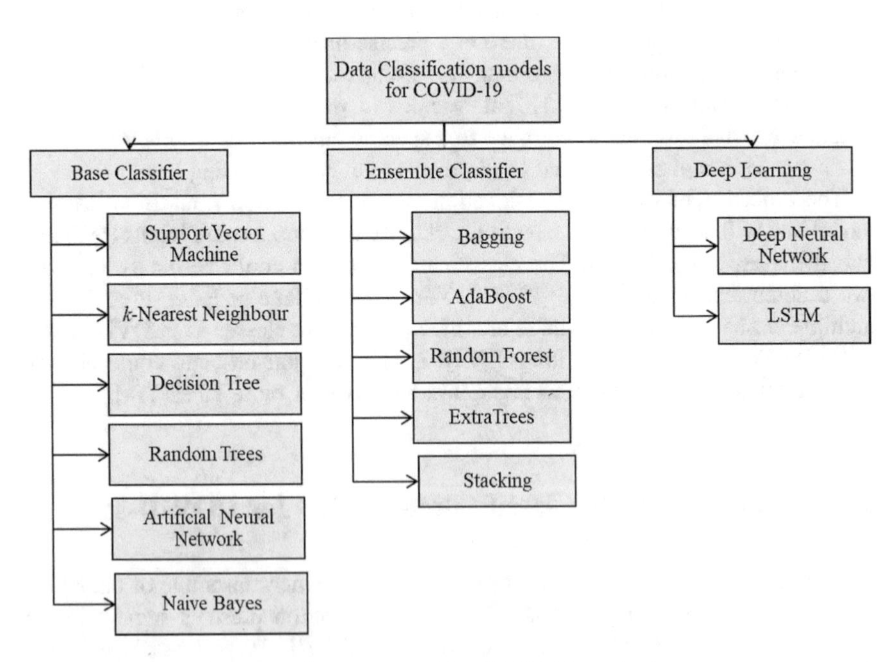


**Fig. 1** Taxonomy of data classification model for COVID-19 pandemic

important features that describe the type of COVID-19 under analysis. In some cases, patient health condition is diagnosed to provide alert for personalized healthcare services.

## 5.2 Data Classification Workflow

The generic workflow for data classification usually begins with problem identification as shown in Fig. 2. The type of task precedes the type of data to collect. Then, depending on the type of data obtained, data preprocessing as feature extraction and/or selection might be needed. Some of the common feature extraction that has been applied to COVID-19 analysis are: Local binary pattern (LBP), Histogram of Oriented Gradients (HOG), Elongated quinary patterns (EQP), first-order statistical features (FOSF), Local directional number (LDN), Mel-Frequency Cepstral Coefficients (MFCC), Locally encoded transform feature histogram (LETRIST), VGG 16 and 19, Binarized statistical image features (BSIF), Local phase quantization (LPQ), inception-V3, Oriented basic image features (oBIFs) and grey level co-occurrence matrix (GLCM). Other features such as morphological, statistical and transform features have also been employed for COVID-19 prediction and diagnosis [75–77].

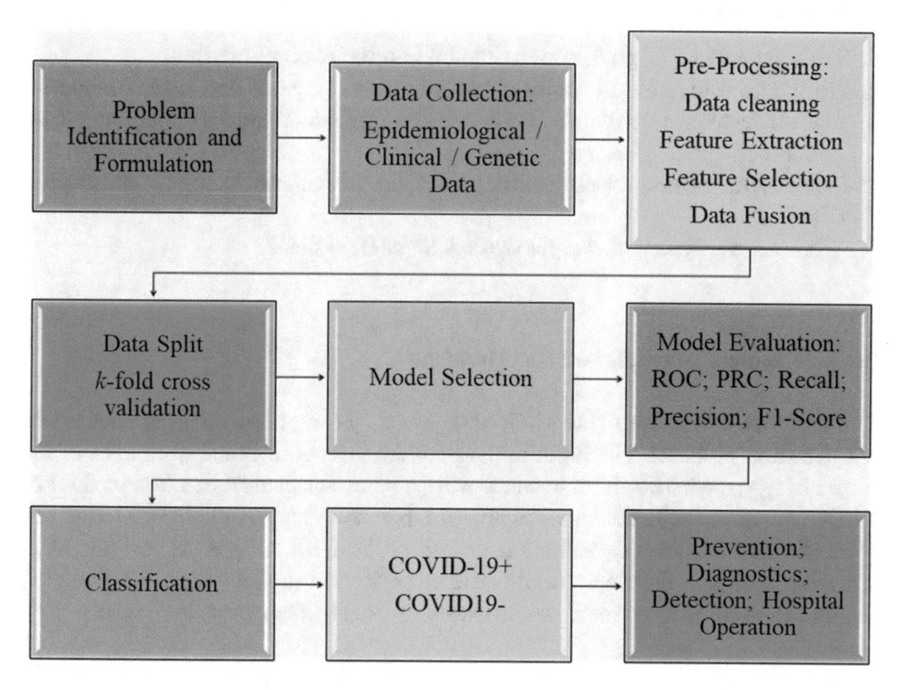


**Fig. 2** Data classification workflow

For instance, Discrete Cosine Transform (DCT), Discrete Wavelet Transform (DWT), and Empirical Mode Decomposition (EMD) techniques have been widely used for signal decomposition of the collected data into different frequency bands [78, 79]. These techniques can be combined with projection techniques like Principal Component Analysis (PCA), Multiscale Principal Component Analysis (MSPCA), and Independent Component Analysis (ICA) among others [80, 81]. Filter techniques are applied to remove noise and to prepare the samples for further processing.

To select the most discriminative features for COVID-19 prediction and diagnosis, several feature selection approaches, which include Analysis of Variance (ANOVA) [82], information gain [83] as well as swarm intelligence optimization algorithm like Whale Optimization Algorithm (WOA) [84] and Grey Wolf Optimization (GWO) [85] have been used. WOA has been combined with the Naïve Bayes classification algorithm to produce a robust classification model for infectious disease prediction and diagnosis.

## 5.3 *Problem Identification and Formulation*

Problem identification is a vital initial phase of building a classification model. It describes the problem and aid in answering some basic questions when choosing an appropriate model.

1. What type of input goes into the model and the expected output?
2. Do we need to extract features depending on the type of data on the model?
3. Are all features important? Or do we need to select some features based on schemes?
4. What types of data classification model do we choose based on the dataset obtained?
5. What are the type and sample size of the curated data?

### 5.3.1 Problem Formulation for Modeling

The features mined from the CXR images can be represented as a multi-class classification problem. The input to the problem is a set of training samples of the form of $(x_i, y_i)$, where $x_i$ is an instance with $y_i$ label, selected from a bounded set of labels $\{y_1, y_2, \ldots, y_n\}$. For this chapter, $y_i$ labels are denoted by COVID-19/PTB/NORMAL and $x_i$ are the features mined from the CXR images. Hence, the classification task is to develop a robust data classification model with a function $f(x)$ whose objective is to connect the instances of $(x's)$ to their respective labels $(y's)$.

## 5.4 Data Classification Models

Data classification models for COVID-19 varies from base classifier, ensembles to deep learning. This section presents some popular models that have implemented a variety of pandemic diseases.

### 5.4.1 Support Vector Machine (SVM)

Support vector machines are risk-based models that learn hidden patterns in a dataset by creating a borderline with maximum margin between data samples of each of the classes [86]. The goal is to find an objective function $f(x)$ with a minimum margin error mapping the input/output features together. The optimal model is determined by the parameters $w$ and $b$ through solving the convex optimization problem as shown by Eqs. (1) and (2):

$$minimize\ \frac{1}{2}\|w\|^2 + C\sum_{i=1}^{n}\xi_i \tag{1}$$

$$\begin{aligned} subject\ to\ y_i\left(w^T x_i + b\right) \geq \xi_i, \quad i = 1, \ldots, n \\ \xi_i \geq 0, \qquad i = 1, \ldots, n \end{aligned} \tag{2}$$

with corresponding linear prediction function $f(x) = wx + b$.

The magnitude of penalization for misclassification is controlled by the parameter C and slack variables $\xi_i$ [87].

### 5.4.2 Artificial Neural Network (ANN)

ANN models were built to imitate the functioning logic of the human brain and how it handles convoluted, non-linear pattern recognition tasks efficiently. ANNs are developed from processing units called neurons, which permits the network to model groups of input-output mappings to solve classification problems. Individual neuron entails three (3) components: a collection of synapses representing the input signals, an adder that sums up the input signals and an activation function which bounds the level of a neuron's output. Furthermore, weight is assigned to each input, which is tuned through training stage and denotes the comparative contribution of that input (positive or negative) to the entire neuron output. The output function of neuron $k$ is shown in Eq. (3)

$$y_k = f\left(\sum_{j=1}^{n} w_{kj}x_j + b_k\right) \tag{3}$$

where $y_k$ is the output of neuron $k$, $b_k$, $i$th bias, $x_j$ is the inputs into the neural networks ($i = 1, \ldots, n$), $w_{ij}$ represents the synaptic weight for input $j$ on neuron $k$ and $f(x)$ is the neuron activation function [88].

### 5.4.3 Extremely Randomized Trees (Extra Trees)

Extra-Trees is an ensemble of classifiers produces a set of un-pruned decision trees using the top-down approach. A random subset of $k$ features is selected at each node for splitting. Each of the trees are constructed from the entire dataset without bootstrapping and for each of the randomly selected features at a node, a discretization threshold (cut-point) is selected at random to define a split, instead of choosing the best cut-point based on the local sample. Consequently, when $k = 1$, the resulting tree structure is actually selected independently of the output labels of the training set [89, 90].

### 5.4.4 Random Forest

This model is a variant of Bagging ensemble proposed by [91].

For this model, a tree is built using bootstrap replication of the training set without pruning. The classification performance of a base classifier is improved by combining the bootstrap aggregating method and randomization in the selection of data nodes during the construction of a decision tree

$$f(x) = \arg\max_{y \in y} \sum_{j=1}^{J} I\big(y = h_j(x)\big) \tag{4}$$

For $j = 1, \cdots, J$.

## 5.5 *Indicators to Model Performance*

Assessing the performance of a classifier, a binary or multi-class is selected class, for example, the class of COVID-19 infected patients. A good classifier categorizes all COVID-19 infested patients as COVID-19 infested patients and all non-COVID-19 infested patients as Normal. A misclassification can either be that a Normal patient is classified as COVID-19 infested patients, or a PTB patient is classified as COVID-19 infected patients. These two types of scenarios are tagged classification errors and are known as Type I and II respectively [92].

**Table 2** Confusion matrix

| Actual class | Predicted class | |
|---|---|---|
| | Class A | Class B |
| Class A | TP | FP |
| Class B | FN | TN |

The classification report contains four cases with two classification errors. This is presented in Table 2 named Confusion Matrix (CM). CM sum up correctly and incorrectly samples belonging to each class [93]. The four groupings in CM are explained as follows:

True Positive (TP) $y = i, f(x) = i$ (COVID-19 patient correctly classified as COVID-19)
True Negative (TN) $y \neq i, f(x) \neq i$ (NORMAL patient correctly classified as NORMAL)
False Positive (FP) $y \neq i, f(x) = i$ (NORMAL patient is classified as COVID-19)
False Negative (FN) $y = i, f(x) \neq i$ (COVID-19 patient is classified as NORMAL)

This groping could also be extended to multi-class problems. The derived indicators from CM are shown in Eqs. (5)–(11)

$$Accuracy = \frac{TP + TN}{TP + TN + FP + FN} \tag{5}$$

$$Precision = \frac{TP}{TP + FP} \tag{6}$$

$$Recall(TPR) = \frac{TP}{TP + FN} \tag{7}$$

$$FPR = \frac{FP}{FP + TN} = 1 \tag{8}$$

$$FNR = \frac{FN}{TP + FN} = 1 - Recall \tag{9}$$

$$F1 - Score = 2 \times \frac{Precision \times Recall}{Precision + Recall} \tag{10}$$

$$ROC = \frac{1 + (TPR - FPR)}{2} \tag{11}$$

## 6 Modeling and Analysis of COVID-19 with Data Classification Model

The content of the proposed framework to detect and classify COVID-19 entails the data collection and synthesis by image processing technique, transformation of RGB images into grayscale images, applying Histogram of Oriented Gradient (HOG) [94] textual feature descriptor stage, Principal Component Analysis (PCA) [95] feature selector stage, train_test split of the data, and performance evaluation stage of the proposed model. The aim here is to build a robust data classification model for COVID-19 and Pulmonary Tuberculosis (PTB) diseases using their frontal CXR images. A 4-phased methodology is proposed and shown in Fig. 3.

For clear data classification, a sample frontal CXR image dataset has been obtained [96, 97] consisting of three classes and their sample size: COVID-19 (527), NORMAL (406) and PTB (394). The dataset is an image, so a feature extraction method (HOG) is applied to remove textual features from it. Figure 4 shows the original image and the corresponding transformed HOG image. Table 3 presents the parameters for the model, HOG and PCA used in the study (Table 3).

## 7 Results and Discussions

The data classification experiment was performed on COVID-19, PTB, and NORMAL lungs. The dataset currently contains 1, 327 of frontal CXR images, and classification of 3 classes. Since the dataset was split, the test result obtained was based on the test set which is 25% of the dataset. This experiment has been carried out on Intel Core i5-7200 CPU @ 2.50 GHz to 2.70 GHz Pentium Windows computer with 8 GB RAM and the application software for implementation of programming code is Anaconda (Jupyter 3.7). An accuracy result of 79.22% and

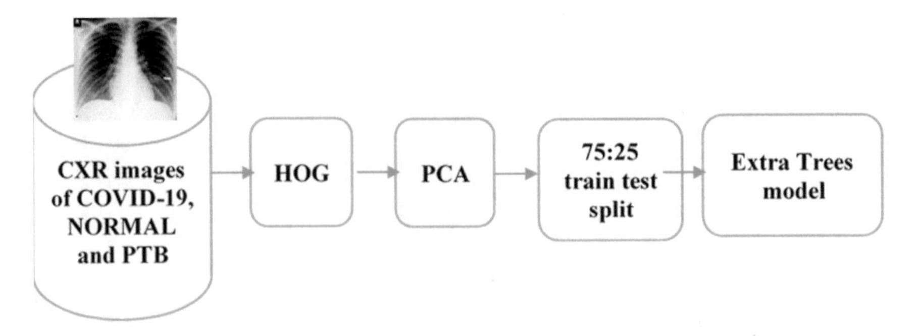


**Fig. 3** The workflow of the framework

**Fig. 4** CXR images and their corresponding HOG image

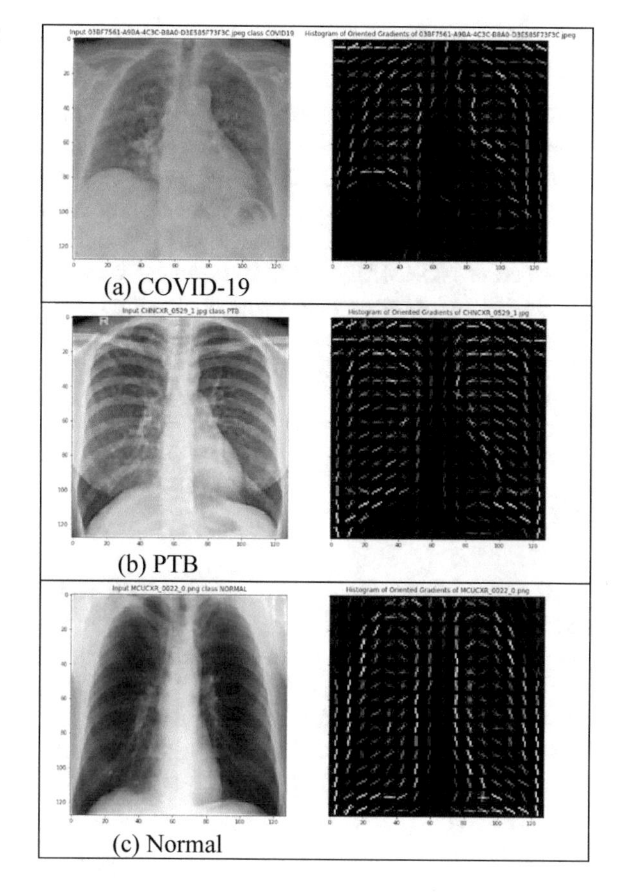


**Table 3** Model parameters used in this study

| Models | Parameters |
|---|---|
| HOG | Block_norm = 'L2-Hys'<br>Cells_per_block = (8,8)<br>Transform_sqrt = True<br>Pixels_per_cell = (8, 8)<br>Rescaled for better display in_range = (0, 10) |
| PCA | 95% |
| Extra Trees | n_estimators=10, max_depth=None, min_samples_split=2, random_state= (0) |

precision of 0.81, recall, and F1-score of 0.79 was achieved based on the ExtraTrees algorithm. This improved result was achieved based on the PCA method selection the best features out of the generated HOG features.

## 7.1 Confusion Matrix

The confusion matrix as shown in Fig. 5 shows the performances of each class. It is observed that out of 134 instances of COVID-19 disease, 129 (96%) instances were correctly classified as COVID-19 disease, 4 (3%) instances were incorrectly classified as NORMAL while 1 (1%) instance were incorrectly classified as PTB disease. For NORMAL lungs cases, 65 (64%) out of 102 instances were correctly classified as NORMAL, 23 (23%) instances were wrongly classified as COVID-19 and 14 (14%) instances were wrongly classified as PTB. Also, for the class PTB, 22 (23%) instances were misclassified as COVID-19, 5 (5%) was wrongly classified as NORMAL lungs while 69 (72%) out of 96 instances were correctly classified as PTB.

## 7.2 ROC Curves

ROC curves show the trade-off between TPR and FPR as shown by Fig. 6. It is used to compare the training and recall performance of each of the classes. An ideal data classification model returns 100% TPR and 0% FPR which corresponds to the top leftmost corner of the ROC layout. So, an able data classification should give a point closer to the top left corner of the ROC layout [98]. The ROC values for all classes are close to 1 showing a very good classification performance. The ROC value for class for COVID-19 is 0.96, NORMAL is 1.00 while for PTB is 0.97.

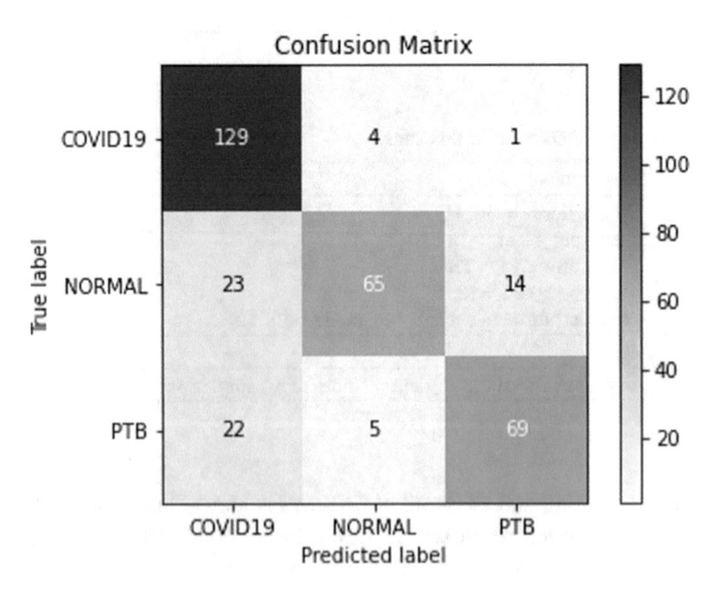


**Fig. 5** Confusion matrix for ET model

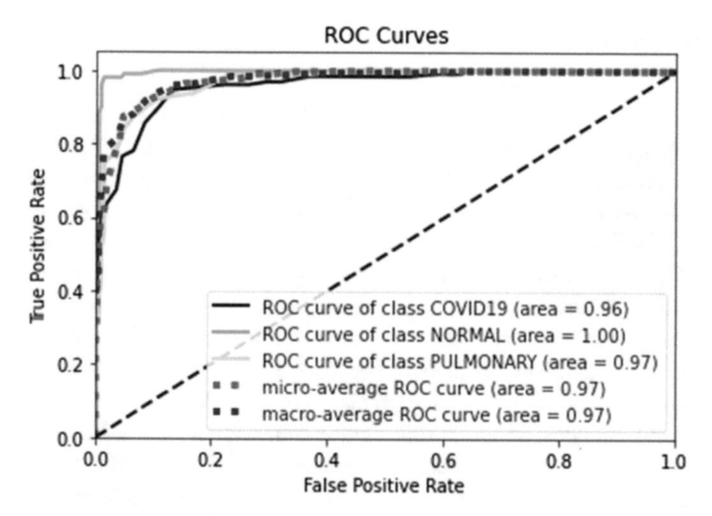


**Fig. 6** ROC curve for ET model

## *7.3 Precision-Recall Curve (PRC)*

This indicator shows the trade-off between precision and recalls for each class as shown in Fig. 7. Typically, high precision can be obtained for a low recall or vice versa. An ideal data classification model should sustain a high precision with high recall [98]. The PRC value for class for COVID-19 is 0.938, NORMAL is 0.991 while for PTB is 0.929.

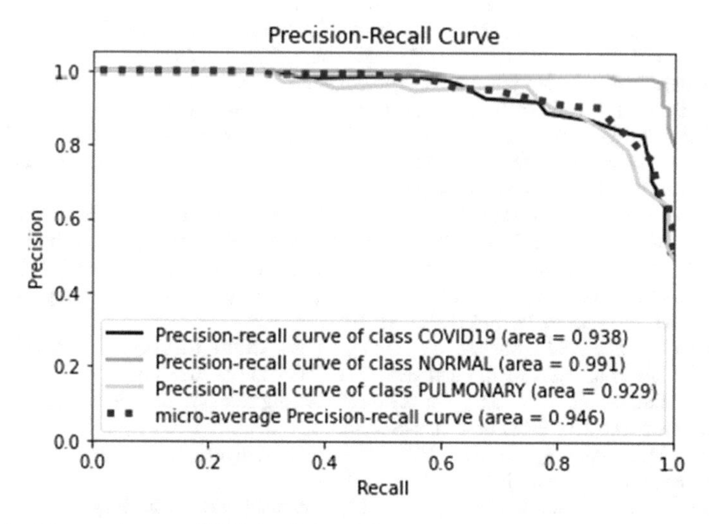


**Fig. 7** Precision-recall curve (PRC) for ET model

# 8 Conclusion

The novel COVID-19 is newly declared and belongs to the pneumonia family. A good method in detecting COVID-19 in infested patients is significant and necessary and CXR image has been proved useful in this direction. A CXR data classification based on feature extraction, selection, and ensemble modeling is presented in this chapter. The use of HOG, a feature descriptor, extracts useful features from the CXR image. At the same time, PCA is applied to select the most salient features to aid the model in producing an optimal classification. The CXR image dataset used for this chapter entails three classes namely COVID-19, NORMAL, and PTB. The result obtained on the ensemble classifier, Extra Trees, achieved an accuracy of 79.22% and precision of 0.81, recall, and F1-score of 0.79. Modeling and analysis of the coronavirus pandemic can be achieved with Data classification models. Machine learning techniques are considered a key factor in recognizing the risk of epidemic diseases in improving the prediction, monitoring, and detection of future global health risks.

# References

1. World Health Organization: COVID 19 Public Health Emergency of International Concern (PHEIC). Global research and innovation forum: towards a research roadmap (2020)
2. Ogundokun, R.O., Lukman, A.F., Kibria, G.B., Awotunde, J.B., Aladeitan, B.B.: Predictive modelling of COVID-19 confirmed cases in Nigeria. Infect. Dis. Model. **5**, 543–548 (2020)
3. Rodrigues-Pinto, R., Sousa, R., Oliveira, A.: Preparing to perform trauma and orthopaedic surgery on patients with COVID-19. J. Bone Joint Surg. American Volume (2020)
4. Ebrahim, S.H., Ahmed, Q.A., Gozzer, E., Schlagenhauf, P., Memish, Z.A.: Covid-19 and Community Mitigation Strategies in a Pandemic (2020)
5. Li, L., Yang, Z., Dang, Z., Meng, C., Huang, J., Meng, H., et al.: Propagation analysis and prediction of the COVID-19. Infect. Dis. Model. **5**, 282–292 (2020)
6. Awotunde, J.B., Adeniyi, A.E., Ogundokun, R.O., Ajamu, G.J., Adebayo, P.O.: MIoT-Based big data analytics architecture, Opportunities and Challenges for Enhanced Telemedicine Systems. Stud. in Fuzziness and Soft Computing. **410**, 199–220 (2021)
7. WHO: Coronavirus Disease (COVID-19) Dashboard. 2020. Accessed 24.09.2020. 291 Available: https://covid19.who.int/292
8. Arthi, V., Parman, J.: Disease, downturns, and wellbeing: economic history and the long-run impacts of COVID-19 (No. w27805). National Bureau of Economic Research (2020)
9. Perrella, A., Carannante, N., Berretta, M., Rinaldi, M., Maturo, N., Rinaldi, L.: Editorial–novel coronavirus 2019 (Sars-CoV2): a global emergency that needs new approaches. Eur. Rev. Med. Pharmacol. **24**, 2162–2164 (2020)
10. Kannan, S., Ali, P.S.S., Sheeza, A., Hemalatha, K.: COVID-19 (novel coronavirus 2019)-recent trends. Eur. Rev. Med. Pharmacol. Sci. **24**(4), 2006–2011 (2020)
11. Wong, Z.S., Zhou, J., Zhang, Q.: Artificial intelligence for infectious disease big data analytics. Infect. Dis. Health **24**(1), 44–48 (2019)
12. Brown, D.E., Abbasi, A., Lau, R.Y.: Predictive analytics: predictive modeling at the micro-level. IEEE Intell. Syst. **30**(3), 6–8 (2015)
13. Jayanthi, N., Valluvan, K.R.: A review of performance metrics in designing protocols for wireless sensor networks. Asian J. Res. Soc. Sci. Humanit. **7**(1), 716–730 (2017)

14. Ahmed, M.B., Boudhir, A.A., Santos, D., El Aroussi, M., Karas, İ.R. (eds.): Innovations in Smart Cities Applications Edition 3: The Proceedings of the 4th International Conference on Smart City Applications. Springer Nature (2020)
15. Oladipo, I.D., Babatunde, A.O., Awotunde, J.B., Abdulraheem, M.: An improved hybridization in the diagnosis of diabetes mellitus using selected computational intelligence. Commun. Comput. Inf. Sci, **1350**, pp. 272–285 (2021)
16. Awotunde, J. B., Jimoh, R. G., Oladipo, I. D., Abdulraheem, M.: Prediction of malaria fever using long-short-term memory and big data. Commun. Comput. Inf. Sci. **1350**, 41–53 (2021). Springer
17. Ameen, A.O., Olagunju, M., Awotunde, J.B., Adebakin, T.O., Alabi, I.O.: Performance evaluation of breast cancer diagnosis using radial basis function, C4. 5 and adaboost. Univ. Pitesti Sci. Bull. Series Electron. Comput. Sci. **17**(2), 1–12 (2017)
18. Liberti, L., Lavor, C., Maculan, N., Mucherino, A.: Euclidean distance geometry and applications. Siam Rev. **56**(1), 3–69 (2014)
19. Ayo, F.E., Awotunde, J.B., Ogundokun, R.O., Folorunso, S.O., Adekunle, A.O.: A decision support system for multi-target disease diagnosis: a bioinformatics approach. Heliyon **6**(3), (2020)
20. Bone, D., Lee, C.C., Chaspari, T., Gibson, J., Narayanan, S.: Signal processing and machine learning for mental health research and clinical applications [perspectives]. IEEE Signal Process. Mag. **34**(5), 196–195 (2017)
21. Kalaiselvi, K., Karthika, D.: Identifying diseases and diagnosis using machine learning. In: Machine Learning with Health Care Perspective, pp. 391–415. Springer, Cham (2020)
22. Ayo, F.E., Ogundokun, R.O., Awotunde, J.B., Adebiyi, M.O., Adeniyi, A.E.: Severe acne skin disease: A fuzzy-based method for diagnosis. Lect. Notes Comput. Sci. (including subseries Lect. Notes in Artif. Intell. Lect. Notes in Bioinformatics), 12254 LNCS, 320–334 (2020)
23. Yadav, S.S., Jadhav, S.M.: Detection of common risk factors for the diagnosis of cardiac arrhythmia using a machine learning algorithm. Expert Syst. Appl. **163**, (2020)
24. Jamshidi, A., Pelletier, J.P., Martel-Pelletier, J.: Machine-learning-based patient-specific prediction models for knee osteoarthritis. Nat. Rev. Rheumatol. **15**(1), 49–60 (2019)
25. Dos Santos, B.S., Steiner, M.T.A., Fenerich, A.T., Lima, R.H.P.: Data mining and machine learning techniques applied to public health problems: a bibliometric analysis from 2009 to 2018. Comput. Ind. Eng. **138**, (2019)
26. Jia, F., Lei, Y., Lin, J., Zhou, X., Lu, N.: Deep neural networks: a promising tool for fault characteristic mining and intelligent diagnosis of rotating machinery with massive data. Mech. Syst. Signal Process. **72**, 303–315 (2016)
27. Johnson, K.W., Soto, J.T., Glicksberg, B.S., Shameer, K., Miotto, R., Ali, M., Dudley, J.T.: Artificial intelligence in cardiology. J. Am. Coll. Cardiol. **71**(23), 2668–2679 (2018)
28. Krittanawong, C., Zhang, H., Wang, Z., Aydar, M., Kitai, T.: Artificial intelligence in precision cardiovascular medicine. J. Am. Coll. Cardiol. **69**(21), 2657–2664 (2017)
29. Hasan, M.K., Alam, M.A., Das, D., Hossain, E., Hasan, M.: Diabetes prediction using ensembling of different machine learning classifiers. IEEE Access **8**, 76516–76531 (2020)
30. Namkung, J.: Machine learning methods for microbiome studies. J. Microbiol. **58**(3), 206–216 (2020)
31. Guan, W.J., Liang, W.H., Zhao, Y., Liang, H.R., Chen, Z.S., Li, Y.M., et al.: Comorbidity and its impact on 1590 patients with Covid-19 in China: a nationwide analysis. Eur. Respir. J. **55**(5) (2020)
32. Ganggayah, M.D., Taib, N.A., Har, Y.C., Lio, P., Dhillon, S.K.: Predicting factors for survival of breast cancer patients using machine learning techniques. BMC Med. Inform. Decis. Mak. **19**(1), 48 (2019)
33. Nair, J.K.R., Saeed, U.A., McDougall, C.C., Sabri, A., Kovacina, B., Raidu, B.V.S., et al.: Radiogenomic models using machine learning techniques to predict EGFR mutations in non-small cell lung cancer. Can. Assoc. Radiol. J. 0846537119899526 (2020)

34. Islam, M.M., Haque, M.R., Iqbal, H., Hasan, M.M., Hasan, M., Kabir, M.N.: Breast cancer prediction: a comparative study using machine learning techniques. SN Comput. Sci. **1**(5), 1–14 (2020)
35. Ai, T., Yang, Z., Hou, H., Zhan, C., Chen, C., Lv, W., et al.: Correlation of chest CT and RT-PCR testing in coronavirus disease 2019 (COVID-19) in China: a report of 1014 cases. Radiology **200642** (2020)
36. Luo, H., Tang, Q.L., Shang, Y.X., Liang, S.B., Yang, M., Robinson, N., Liu, J.P.: Can Chinese medicine be used for prevention of corona virus disease 2019 (COVID-19)? A review of historical classics, research evidence, and current prevention programs. Chin. J. Integr. Med. 1–8 (2020)
37. Haleem, A., Vaishya, R., Javaid, M., Khan, I.H.: Artificial intelligence (AI) applications in orthopaedics: an innovative technology to embrace. J. Clin. Orthop. Trauma **11**, S80–S81 (2020)
38. Biswas K, Sen, P.: Space-Time Dependence of Coronavirus (COVID-19) Outbreak (2020). arXiv preprint arXiv:2003.03149
39. Stebbing, J., Phelan, A., Griffin, I., Tucker, C., Oechsle, O., Smith, D., Richardson, P.: COVID-19: combining antiviral and anti-inflammatory treatments. Lancet. Infect. Dis **20**(4), 400–402 (2020)
40. Sohrabi, C., Alsafi, Z., O'Neill, N., Khan, M., Kerwan, A., Al-Jabir, A., et al.: World Health Organization declares global emergency: a review of the 2019 novel coronavirus (COVID-19). Int. J. Surg. (2020)
41. Chen, S., Yang, J., Yang, W., Wang, C., Bärnighausen, T.: COVID-19 control in China during mass population movements at New Year. Lancet **395**(10226), 764–766 (2020)
42. Fix, O.K., Hameed, B., Fontana, R.J., Kwok, R.M., McGuire, B.M., Mulligan, D.C., et al.: Clinical best practice advice for hepatology and liver transplant providers during the COVID-19 pandemic: AASLD expert panel consensus statement. Hepatology (2020)
43. Debnath, S., Barnaby, D.P., Coppa, K., Makhnevich, A., Kim, E.J., Chatterjee, S., et al.: Machine learning to assist clinical decision-making during the COVID-19 pandemic. Bioelectron. Med. **6**(1), 1–8 (2020)
44. Gupta, R., Ghosh, A., Singh, A.K., Misra, A.: Clinical considerations for patients with diabetes in times of COVID-19 epidemic. Diab. Metab. Syndr. **14**(3), 211 (2020)
45. Hussain, A., Bhowmik, B., do Vale Moreira, N.C.: COVID-19 and diabetes: knowledge in progress. Diab. Res. Clin. Pract. **108142** (2020)
46. Rajula, H.S.R., Verlato, G., Manchia, M., Antonucci, N., Fanos, V.: Comparison of conventional statistical methods with machine learning in medicine: diagnosis, drug development, and treatment. Medicina **56**(9), 455 (2020)
47. Ali, F., El-Sappagh, S., Islam, S.R., Kwak, D., Ali, A., Imran, M., Kwak, K.S.: A smart healthcare monitoring system for heart disease prediction based on ensemble deep learning and feature fusion. Inform. Fus. **63**, 208–222 (2020)
48. Ahmed, Z., Mohamed, K., Zeeshan, S., Dong, X.: Artificial intelligence with multi-functional machine learning platform development for better healthcare and precision medicine. Database (2020)
49. Silverston, P.: SAFER: a mnemonic to improve safety-netting advice. Pract. Nurs. **31**(1), 26–28 (2020)
50. Newman-Toker, D.E., Wang, Z., Zhu, Y., Nassery, N., Tehrani, A.S.S., Schaffer, A.C., Siegal, D.: Rate of diagnostic errors and serious misdiagnosis-related harms for major vascular events, infections, and cancers: toward a national incidence estimate using the "Big Three". Diagnosis **1**(ahead-of-print) (2020)
51. Graber, M.L., Franklin, N., Gordon, R.: Diagnostic error in internal medicine. Arch. Intern. Med. **165**(13), 1493–1499 (2005)
52. Winters, B., Custer, J., Galvagno, S.M., Colantuoni, E., Kapoor, S.G., Lee, H., Pronovost, P.: Diagnostic errors in the intensive care unit: a systematic review of autopsy studies. BMJ Qual. Saf. **21**(11), 894–902 (2012)

53. Jiang, F., Jiang, Y., Zhi, H., Dong, Y., Li, H., Ma, S., Wang, Y.: Artificial intelligence in healthcare: past, present, and future. Stroke Vasc. Neurol. **2**(4), 230–243 (2017)
54. Lee, C.S., Nagy, P.G., Weaver, S.J., Newman-Toker, D.E.: Cognitive and system factors contributing to diagnostic errors in radiology. Am. J. Roentgenol. **201**(3), 611–617 (2013)
55. Maddox, T.M., Rumsfeld, J.S., Payne, P.R.: Questions for artificial intelligence in health care. JAMA **321**(1), 31–32 (2019)
56. Panch, T., Mattie, H., Celi, L.A.: The "inconvenient truth" about AI in healthcare. NPJ Digital Med. **2**(1), 1–3 (2019)
57. Panch, T., Szolovits, P., Atun, R.: Artificial intelligence, machine learning, and health systems. J. Global Health **8**(2) (2018)
58. Collins, F.S., Varmus, H.: A new initiative on precision medicine. N. Engl. J. Med. **372**(9), 793–795 (2015)
59. Hoofnagle, C.J., van der Sloot, B., Borgesius, F.Z.: The European Union general data protection regulation: what it is and what it means. Inf. Commun. Technol. Law **28**(1), 65–98 (2019)
60. Shaban-Nejad, A., Michalowski, M., Buckeridge, D.L.: Health Intelligence: How Artificial Intelligence Transforms Population and Personalized Health. Nature Publishing Group (2018)
61. Naudé, W.: Artificial Intelligence Against COVID-19: An Early Review (2020)
62. Maier, B.F., Brockmann, D.: Effective containment explains subexponential growth in recent confirmed COVID-19 cases in China. Science **368**(6492), 742–746 (2020)
63. Huang, J.J.: COVID-19 and Applicable Law to Transnational Personal Data: Trends and Dynamics. Sydney Law School Research Paper (20/23) (2020)
64. Chen, B.: Historical foundations of choice of law in fiduciary obligations. J. Private Int. Law **10**(2), 171–203 (2014)
65. Douglas, M.: Characterization of breach of confidence as a privacy tort in private international law. UNSWLJ **41**, 490 (2018)
66. Matta, D.M., Saraf, M.K.: Prediction of COVID-19 Using Machine Learning Techniques (2020)
67. Ayyoubzadeh, S.M., Ayyoubzadeh, S.M., Zahedi, H., Ahmadi, M., Kalhori, S.R.N.: Predicting COVID-19 incidence through analysis of google trends data in Iran: data mining and deep learning pilot study. JMIR Public Health Surveill. **6**(2), (2020)
68. Narin, A., Kaya, C., Pamuk, Z.: Automatic detection of coronavirus disease (COVID-19) using x-ray images and deep convolutional neural networks (2020). arXiv preprint arXiv:2003.10849
69. Newman, S.C.: Prediction and Privacy in Healthcare Analytics (2016)
70. Radha, P., Srinivasan, B.: Predicting diabetes by cosequencing various data mining classification techniques. Int. J. Innovative Sci. Eng. Technol. **1**(6), 334–339 (2014)
71. Oladipo, I.D., Babatunde, A.O.: Framework for genetic-neuro-fuzzy inferential system for diagnosis of diabetes mellitus. Annals Comput. Sci. Series **16**(1) (2018)
72. Fitkov-Norris, E., Folorunso, S.O.: Impact of sampling on neural network classification performance in the context of repeat movie viewing. In: International Conference on Engineering Applications of Neural Networks, pp. 213–222. Springer, Berlin (2013)
73. Pereira, R.M., Bertolini, D., Teixeira, L.O., Silla Jr., C.N., Costa, Y.M.: COVID-19 identification in chest X-ray images on flat and hierarchical classification scenarios. Comput. Methods Programs Biomed. **194**, (2020). https://doi.org/10.1016/j.cmpb.2020.105532
74. Folorunso, S.O., Fashoto, S.G., Olaomi, J., Fashoto, O.Y.: A multi-label learning model for psychotic diseases in Nigeria. Inf. Med. Unlocked **19**(100326), 11 (2020). https://doi.org/10.1016/j.imu.2020.100326
75. Liu, H., Ren, H., Wu, Z., Xu, H., Zhang, S., Li, J., et al.: CT radiomics facilitates a more accurate diagnosis of COVID-19 pneumonia: compared with CO-RADS (2020)
76. Zebin, T., Rezvy, S.: COVID-19 detection and disease progression visualization: deep learning on chest X-rays for classification and coarse localization. Appl. Intell. 1–12 (2020)
77. Raj, V., Renjini, A., Swapna, M.S., Sreejyothi, S., Sankararaman, S.: Nonlinear time series and principal component analysis: a potential diagnostic tool for COVID-19 auscultation. Chaos Solitons Fractals **110246** (2020)

78. Huang, B., Yang, F., Yin, M., Mo, X., Zhong, C.: A Review of multimodal medical image fusion techniques. Comput. Math. Methods Med. (2020)
79. Sharma, R., Vignolo, L., Schlotthauer, G., Colominas, M.A., Rufiner, H.L., Prasanna, S.R.M.: Empirical mode decomposition for adaptive AM-FM analysis of speech: a review. Speech Commun. **88**, 39–64 (2017)
80. Alickovic, E., Subasi, A.: Medical decision support system for diagnosis of heart arrhythmia using DWT and random forests classifier. J. Med. Syst. **40**(4), 108 (2016)
81. Nilashi, M., bin Ibrahim, O., Ahmadi, H., Shahmoradi, L.: An analytical method for diseases prediction using machine learning techniques. Comput. Chem. Eng. **106**, 212–223 (2017)
82. Gil, D., Díaz-Chito, K., Sánchez, C., Hernández-Sabaté, A.: Early Screening of SARS-CoV-2 by Intelligent Analysis of X-Ray Images (2020). arXiv preprint arXiv:2005.13928
83. Motwani, M., Dey, D., Berman, D. S., Germano, G., Achenbach, S., Al-Mallah, M.H., et al.: Machine learning for prediction of all-cause mortality in patients with suspected coronary artery disease: a 5-year multicentre prospective registry analysis. Eur. Heart J. **38**(7), 500–507 (2017)
84. Agrawal, R.K., Kaur, B., Sharma, S.: Quantum based whale optimization algorithm for wrapper feature selection. Appl. Soft Comput. **89**, (2020)
85. Wiharto, W., Suryani, E., Cahyawati, V.: The methods of duo output neural network ensemble for the prediction of coronary heart disease. Indonesian J. Electr. Eng. Inform. (IJEEI) **7**(1), 51–58 (2019)
86. Vapnik, V.: The Nature of Statistical Learning Theory. Springer Science & Business Media, New York, NY (2013)
87. Bhavsar, P., Safro, I., Bouaynaya, N., Polikar, P., Dera, D.: Machine Learning in Transportation Data Analytics. Data Analytics for Intelligent Transportation Systems, pp. 283–309 (2017). http://doi.org/10.1016/B978-0-12-809715-1.00012-2
88. Fitkov-Norris, E., Folorunso, S.O.: Impact of sampling on neural network classification performance in the context of repeat movie viewing. In the Proceedings of Engineering Applications of Neural Networks (EANN 2013). In: Iliadis L., Papadopoulos H., Jayne C. (eds.) Communications in Computer and Information Science (CICS), vol. 383, pp. 213–222. Springer, Berlin (2013)
89. Geurts, P., Ernst, D., Wehenkel, L.: Extremely randomized trees. Mach. Learn. **63**(1), 3–42 (2006)
90. Geurts, P., Louppe, G.: Learning to rank with extremely randomized trees. JMLR Workshop Conf. Proc. **14**, 49–61 (2011)
91. Breiman, L.: Random forests. Mach. Learn. **45**, 5–32 (2001)
92. Runkler, T.A.: Data Analytics: Models and Algorithms for Intelligent Data Analysis, 2nd edn. Springer Vieweg, Munich (2016). https://doi.org/10.1007/978-3-658-14075-5
93. Folorunso, S.O., Adeyemo, A.B.: Alleviating classification problem of imbalanced dataset. African J. Comput. ICT **6**(1), 137–144 (2013)
94. Dalal, N., Triggs, B.: Histograms of oriented gradients for human detection. In: Proceedings of the IEEE Computer Society Conference on Computer Vision and Pattern Recognition (2005)
95. Hotelling, H.: Analysis of a complex of statistical variables into principal components. J. Educ. Psychol. **24**(6), 417 (1933)
96. Cohen, J.P., Morrison, P., Dao, L., Roth, K., Duong, T.Q., Ghassemi, M.: COVID-19 Image Data Collection: Prospective Predictions Are the Future (2020). arXiv:2006.11988v1 [q-bio. QM], 25. Retrieved from https://github.com/ieee8023/covid-chestxray-dataset
97. Candemir, S., Jaeger, S., Musco, J., Xue, Z., Karargyris, A., Antani, S.K., et al.: Lung segmentation in chest radiographs using anatomical atlases with nonrigid registration. IEEE Trans. Med. Imaging **33**(2), 577–590 (2014). https://doi.org/10.1109/TMI.2013.2290491. PMID: 24239990
98. Davis, J., Goadrich, M.: The relationship between precision-recall and ROC curves. In: International Conference on Machine Learning, pp 233–240 (2006)

# A Hybrid Automated Intelligent COVID-19 Classification System Based on Neutrosophic Logic and Machine Learning Techniques Using Chest X-Ray Images

**Ibrahim Yasser, Aya A. Abd El-Khalek, Abeer Twakol, Mohy-Eldin Abo-Elsoud, Ahmed A. Salama, and Fahmi Khalifa**

**Abstract** To facilitate timely treatment and management of COVID-ap patients, efficient and quick identification of COVID-19 patients is of immense importance during the COVID-19 crisis. Technological developments in machine learning (ML) methods, edge computing, computer-aided medical diagnostic been utilized for COVID-19 Classification. This is mainly because of their ability to deal with *Big data* and their inherent robustness and ability to provide distinct output characteristics attributed to the underlying application. The contrary transcription-polymerase chain reaction is currently the clinical typical for COVID-19 diagnosis. Besides being expensive, it has low sensitivity and requires expert medical personnel. Compared with RT-PCR, chest X-rays are easily accessible with highly available annotated datasets and can be utilized as an ascendant alternative in COVID-19 diagnosis. Using X-rays, ML methods can be employed to identify COVID-19 patients by quantitively examining chest X-rays effectively. Therefore, we introduce an alternative, robust, and intelligent diagnostic tool for automatically

I. Yasser (✉)
Engineering Faculty, University of Mansoura, Mansoura 35516, Egypt
e-mail: Ibrahim_yasser@mans.edu.eg

A. A. Abd El-Khalek
Communications and Electronics Engineering Department, Nile Higher Institute for Engineering and Technology, Mansoura 35524, Egypt

A. Twakol
Computer Engineering Department, Engineering Faculty, Benha University, Banha 13518, Egypt

M.-E. Abo-Elsoud · F. Khalifa
Electronics and Communications Engineering Department, Engineering Faculty, University of Mansoura, Mansoura 35516, Egypt
e-mail: fahmikhalifa@mans.edu.eg

A. A. Salama
Mathematics and Computer Science Department, Faculty of Sciences, University of Port Said, Port Said 42522, Egypt

 
A.-E. Hassanien et al. (eds.), *Advances in Data Science and Intelligent Data Communication Technologies for COVID-19*, Studies in Systems, Decision and Control 378, https://doi.org/10.1007/978-3-030-77302-1_7

detecting COVID-19 utilizing available resources from digital chest X-rays. Our technique is a hybrid framework that is based on the fusion of two techniques, Neutrosophic techniques (NTs) and ML. Classification features are extracted from X-ray images using morphological features (MFs) and principal component analysis (PCA). The ML networks were trained to classify the chest X-rays into two classes: positive (+ve) COVID-19 patients or normal subjects (or −ve). The experimental results are performed based on a sample from a collected comprehensive image dataset from several hospitals worldwide. The classification accuracy, precision, sensitivity, specificity and F1-score for the proposed scheme was 98.46%, 98.19%, 98.18%, 98.67%, and 98.17%. The experimental results also documented the high accuracy of the proposed pipeline compared to other literature techniques.

**Keywords** COVID-19 · Artificial intelligence · Machine learning · Neutrosophic techniques · Computer-aided diagnostic tool

## 1 Introduction

Coronavirus Disease-2019 (COVID-19) caused by the so-called severe acute respiratory syndrome coronavirus2 (or SARS-CoV-2) is a devastating worldwide epidemic. In March 2020, the world health organization (WHO) announced that SARS-CoV-2 (or COVID-19) had become an unprecedented public health crisis. SARS-CoV-2 is known for its high transmutability and pathogenicity. Many research laboratories worldwide are working on developing a clinical antiviral drug and/or vaccine to help control the spread of the virus. Governments in various countries have implemented flight restrictions and limits on the border crossing. Also, public awareness of hygiene and social distancing has been imposed. However, the virus continues to unfold at a rapid pace. While the general population encountered mild to severe respiratory disease with COVID-19, others have had fatal pneumonia [1]. Several analysis within medicine, clinical, and computing are failing to organize proactive plans of action for COVID-19 with established research goals. As this illness is very infectious, the foremost fascinating interference technique is to spot infected patients to contain the spread; therefore, the transmission chain is interrupted [2]. Medical screening tool that is currently in use for COVID-19 diagnosis is RT-PCR, which stands for reverse transcription enzyme chain reaction. RT-PCR is complex manually employed, and it is also time consuming. Furthermore, it only has a rate of positivity 63%. Additionally, the swab take a look at involves separation for the test protocol, whereas chest X-ray detection is simply managed [3]. In addition, there's a severe lack of supplies, that results in an interruption in tries to eradicate disease [4]. Delays in take a look at results will result in infected patients communication with and infecting healthy patients within the process. The other diagnostic approaches of COVID-19 embrace the study of medical specialty signs, epidemiologic histories and positive pradiographic images [computed tomography (CT)/chest radiography (CXR)] also as positive morbific checks. The majority of COVID-19 cases have similar decisions for X-ray imaging, together with longitudinal, multifocal, ground-glass natural process with peripheral or point distribution, chiefly within the lower lobes of the earlying process and late stage respiratory organ consolidation [5]. Although CT chest may be a lot of economical imaging procedure for lung-related dis-ease diagnosis, CXR is favored as a result of its pronto quicker, accessible and cheaper than with CT. Since infection with COVID-19 attacks epithelial tissue our metastasis tract, X-ray images is accustomed diagnose respiratory organ pneumonia, lung inflammation, abscesses, Associate in nursing/or swollen bodily fluid nodes. The studies advocate the necessity for a computerised classification for speedy diagnosis. Not only manual interpretations of X-rays is time overwhelming, it also unable to satisfy the perform because of the restricted offer of X-ray radiologists or machine operators [6].

Classification of biomedical images is an emerged area of study to form the care sector additional promising. Amid the aftermath of the epidemic of COVID-19, many recent international studies round the world have adopted subtle ML algorithms and AI based mostly technologies to higher perceive the pattern of infective agent transmission, more improve diagnostic speed and accuracy. Not only that, but it would also support health care staff by reducing their interaction with COVID-19 patients, also. Furthermore, ML/AI-based system help to develop new, successful treatment interventions and theoretically classify the most vulnerable people. ML/AI-based techniques are expected and to improve the accuracy forecast for both infectious and non-infectious disease screening and identification COVID-19 on chest X-ray scans. In the medical sector, AI is not used to eliminate human experiences, but to provide decision-making guidance for physicians in what they model. Figure 1 depicts how AI and ML techniques can be used in the fight against Covid-19.

The rest of this chapter is structured as follows. Section 2 shall precede the work involved and shall discuss the nature of the work given. Section 3 introduces the suggested solution to the characterization of COVID-19 and its stages using chest x-rays. The experimental findings shall be addressed in Sect. 4. Finally, the conclusion are laid out in Sect. 5.

## 2 Related Work

At present, in progress tries are created to develop novel diagnostic strategies victimization ML algorithms. During this regard, studies demonstrate the potential of AI and ML tools by proposing new models for speedy and valid SARS-CoV-2 diagnostic methods, such as: Apostolopoulos and Mpesiana [7] in their research, advanced convolutionary neural network structures were used to identify CXR photos. Transfer learning has been introduced to cope with a number of gift anomalies within the dataset. 2 datasets from slightly separate repositories are used

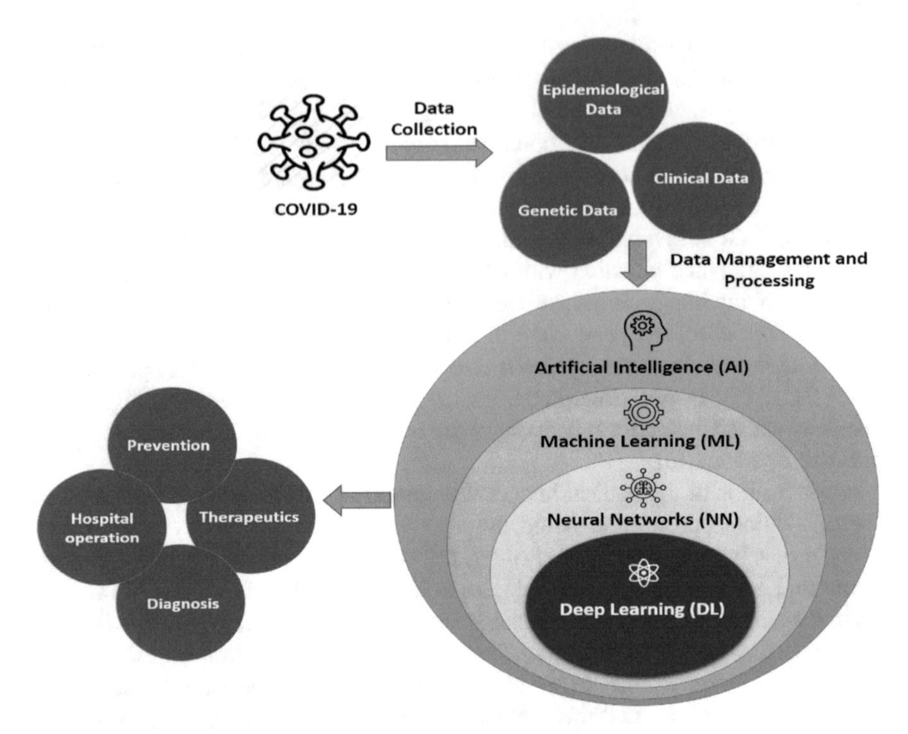


**Fig. 1** Schematic illustration of artificial intelligence (AI) and machine learning (ML) techniques applied in the fight against covid-19

to research photographs in two classes: COVID-19 and the traditional state with a ~97% accuracy (sensitivity of 98.66% and specificity of 96.46%). Sethy and Behera [8] proposed a CXR-b intelligent system for COVID-19 identification. Deep aspects of the CXRs have been identified and vector support machine (SVM) classifier was used. The study concluded that ResNet50 combined with SVM achieved best performances (accuracy, F1 score, Matthew's correlation coefficient (MCC) and Kappa were 95.38%, 95.52%, 91.41% and 90.76%, respectively). Another research by Khalifa et al. used generative adversarial networks (GAN) [9] to detect the respiratory disease from CXRs. They addressed the topic of over-processing and proclaimed their power by generating additional GAN images. The data collection contained 5863 CXR images for both natural and nuclear purposes. Well-known deep learning structures (e.g., GoogLeNet, AlexNet, Squeznet and ResNet) were used to diagnose pneumonia. Their work highlighted the fact that ResNet18 along with GAN outperformed other deep transport models. Afshar et al. [10] suggested a capsule network called COVID-CAPS instead of a conventional CNN network to communicate with a data set. COVID-CAPS are optimized to obtain an accuracy of 95.7%, sensitivity of 90%, and specificity of 95.8%.

Wang et al. [11] provided a COVID-Net for COVID-19 cases detection from ~14,000 CXRs, with a precision of 83.5%. Abbas et al. [12] rendered a pipeline using the pre-trained modified CNN architecture (DeTraC Decompose, Transfer, and Compose) for COVID-19 detection. They tested their approach on a database of 105 COVID-19, 80 normal subjects and 11 SARS X-rays. The method attained an accuracy of 95.12%, sensitivity of 97.91%, while the specificity was 91.87%.

A systematic study concluded that CXR images throughout the chest can be used for the early diagnosis of COVID-19-infected patients. The literature review also documented the role of modern AI/ML technology to solve the challenges during the outburst. Therefore, statistical models are used in this chapter to distinguish COVID-19 patients from CXR images in the chest; Therefor, the present work proposes a new automated intelligent ML algorithm combined with Neutrosophic techniques (NTs) to detect and classify COVID-19 pneumonia cases using X-ray images. The main objective is to reduce the error in detection while obtaining a more accurate diagnosis. The contributions are summarized as follows:

(i) Building a novel ML model in automatically and efficiently assist in early diagnosis of patients with COVID-19;
(ii) Achieving an empirical analysis of the proposed ML image classifier with NTs in the task of classifying COVID-19 disease using conventional chest X-rays with lower cost than other imaging modalities like CT;
(iii) Comparing our technique with different models to document the robustness and the most accurate results of our method using the available X-rays;
(iv) Our analysis pipeline supports multidisciplinary researchers in developing advanced AI and ML techniques that can help fight the outbreak of COVID-19.

## 3 The Proposed COVID-19 Diagnostic System

The proposed architecture to accomplish the diagnostic procedure and for detecting the existence of COVID-19 using X-ray images as seen in Fig. 2, consisting of four phases, (i) preprocessing step; (ii) Feature extraction phase; (iii) operation of the NT process; and (iv) classification phase of the ML. These steps are listed in detail in the following paragraphs.

### *3.1 Pre-processing*

The CXR images within the collective knowledge set accommodates uncalled-for objects resembling bright text, icons, differing resolutions, varied shapes, textures and morphological characteristics, usually of various colours, proportions, positioning, component level noise so on, that necessitate its pre-processing. Various pre-processing techniques are employed in this work to supply COVID-19 coaching files. Every image is regenerate as follows: (i) noise extracted to eliminate

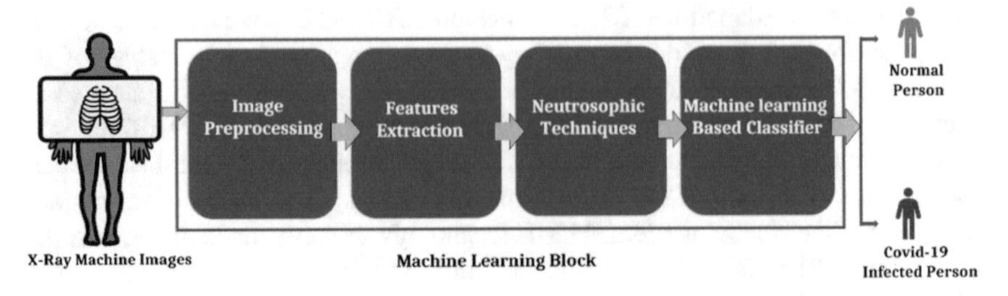


**Fig. 2** The proposed block diagram for the COVID-19 classification pipeline

constituent level noise for many ac-curate images, so illuminated textual and semantic noise are curbed by a photomask generated using binary thresholds, as indicated by Eq. 1 [13], (ii) turned bound images that aren't utterly set in a very horizontal position. The rotation method was meted out by rotating the pictures in a horizontal direction with the most rotation angle of ±5 degrees, al-so, (iii) resized the CXR images to a hard and fast resized to a standard dimension, (iv) cropped to isolate the essential a part of the image and to delete certain orthogonal background objects, even to attenuate the commonly-found matter detail within the CXR images, (v) histogram adjustment technique was accustomed improve the distinction of every image, additionally to enhance the lighting conditions and therefore improving the training model. As seen in Fig. 3 as seen within the section on experimental outcomes. The pre-processed images are then divided into the training and testing assortment to evaluate the classification models.

$$M(x,y) = \begin{cases} max_{th}, & i(x,y) \geq min_{th} \\ 0, & otherwise. \end{cases} \quad (1)$$

where $i(x, y)$ represents the input image, $max_{th}$ and $min_{th}$ are maximum and minimum thresholds mask design.

## 3.2 Feature Extraction

Extraction of options performs any transformation of the initial features by permitting variations and transformations of the original features set to develop new, additional important features to enhance the exactness of learning models by extracting features from input data. This method decreases the spatial property of the info by eliminating the redundant data and, thus, by protective the foremost wonderful elements, improves the speed and training. Extraction of feature is commonly used for describing the creation of the covariance matrix of incessant features with respectable discriminatory power between classes [14]. The salient and efficient features were extracted using nine simple morphic features (MFs) in

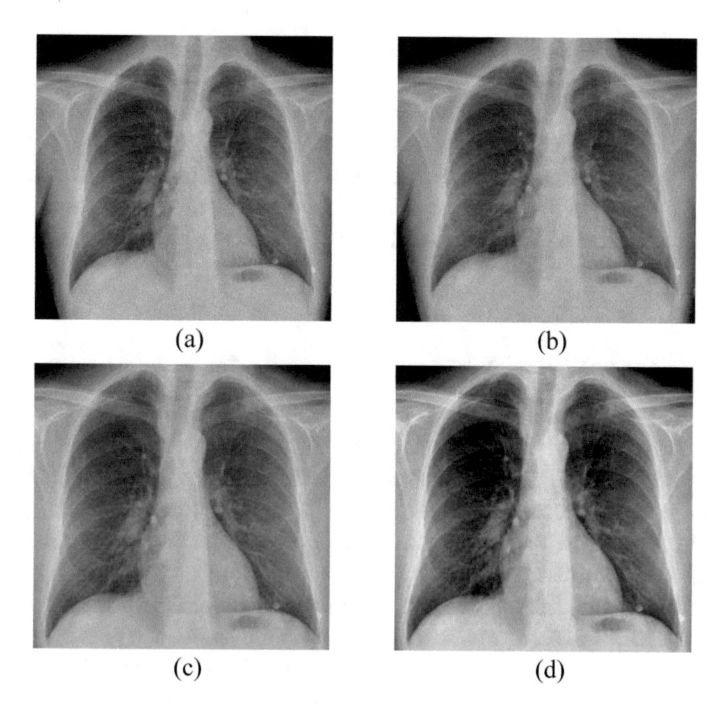


**Fig. 3** Operation in data preprocessing: **a** original, **b** resized, **c** cropped, and **d** contrast-enhanced CXR image

addition to a collection of main components (PCs) were collected for the chosen X-ray image database, 4, 8, 16 and 24 PCs. PCs were derived from the intensive training collection. Namely, a given each image was interpreted as a row vector and its coefficients were then projected into a set of features; a new re-statement of the COVID-19 detail.

1. **Principal Component Analysis (PCA)**
   Principal component analysis (PCA) may be a second-order applied mathematics approach that turns a group of related factors to lower range of non-correlated factors known as PCs. PCA is primarily accustomed cut back the size of the information set whereas maintaining the maximum amount details as possible rather than all the variance matrix PCs, the data are often expressed in mere a couple of straightforward vectors [15]. Computers were extracted from the complete training kit, where each image was presented as a carrier of a chain. Several characteristics have reduced the coefficients of each image; a new definition of the COVID-19 detail.
2. **Morphological Features (MFs)**
   Morphological characteristics are derived from the apparent visual characteristics of a given image: (i) area: real number of pixels in the region; (ii) bounding box: represents the smallest measure box that contain all points; (iii) centriod:

reflects the intersection of all straight lines separating a given image into two sections of the same moment of the line informally; (iv) convex area: represents a line segment that is completely enclosed within the form of any two points; (v) orientation: represents regions that can be defined as locally 1D (e.g. using lines or edges); and (vi) extreme: represents the largest value (maximum) or the smallest value (minimum) that takes place at a point, either within a neighbourhood or in a on the function domain in its entirety [16].

### 3.3 Neutrosophic Techniques (NTS)

After extracting the features using MFs and PCA, we employed the Neutrosophic techniques to extract three components: (i) membership degree (or T), (ii) indeterminacy degree (or I) and (iii) non-membership degree (or F). Those three components are correlated with the variables and are of immense importance for our classification algorithm in order to get more accurate and efficient diagnostic result.

Neutrosophic technique (NT) is a technique that would use the Neutrosophic logic rules and the Neutrosophic sets for grouping. Particularly, NT integrates a Neutrosophic rule-based method (e.g., IF X and Y THEN Z) for problem-solving instead of mathematically modeling the underlying system, i.e., NT is equivalent to a fuzzy technique [17]. The architecture of a Neutrosophic classification inference scheme using fuzzy terminology utilizes the Mamdani fuzzy inference method [18]. The block diagram of the Neutrosophic Classification Scheme is shown in Fig. 4. The membership functions (T, I, and F) are mutually independent. Thus, using the Matlab fuzzy toolbox, three modules have been constructed: one for each of the Neutrosophic reality variable, Indeterminacy component, and the falsity component. Although independently for each other's components, a connection is drawn among the T, I, and F components to capture the truthfulness, in-determinacy and falseness of the input and the output [19].

In Neutrosophic-based classification (i.e., NRCS), rule-based systems using Neutrosophic logic are used to describe various types of information about the underlying topic and for modeling the connections and relationships that occur between its variables [20]. The generic structure of the NRCS is seen in Fig. 5.

Suppose U represents a discourse world that contains a subset W compromising bright pixels. Neutrosophical pictures are distinguished by three subsets T, I, and F. In the image, the pixel P in the image is defined as P (T, I, F) that relates to W by its t% is true in the bright pixel, i% is undetermined, and f% is false. Here, t, i, and f vary in T, I, and F, respectively. Mathematically, a given pixel p (x, y) in the image domain, is converted to:

$$NDP_{Ns}(x, y) = \{T(x, y), I(x, y), F(x, y)\} \quad (2)$$

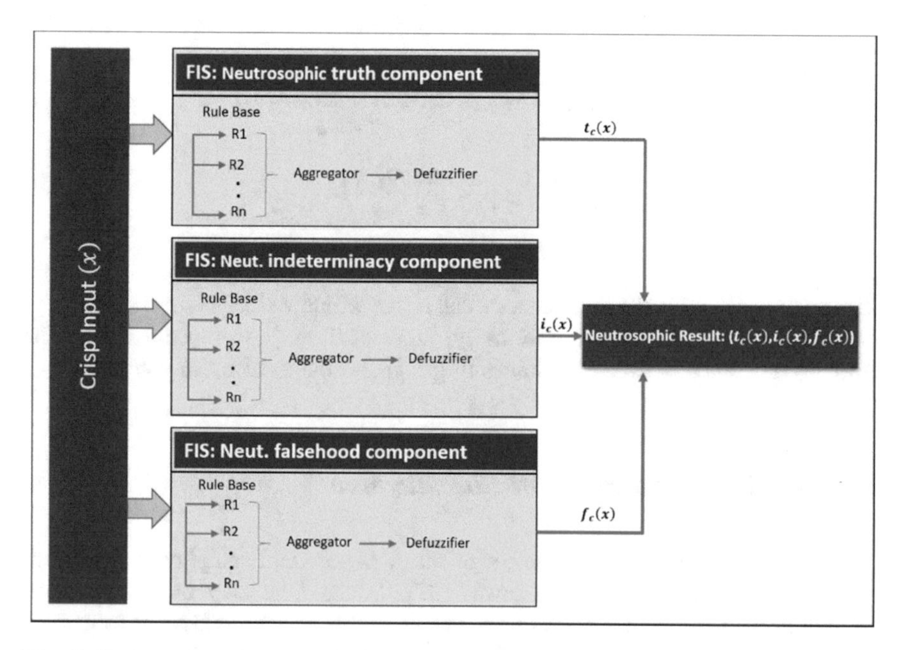


**Fig. 4** Block diagram for Neutrosophic components

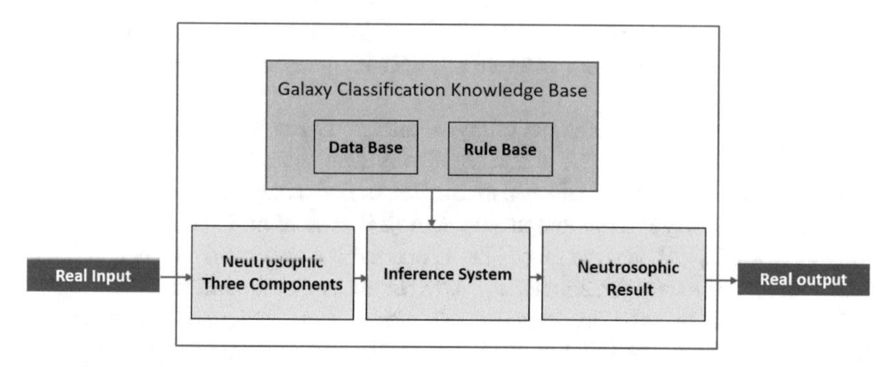


**Fig. 5** The basic structure of a Neutrosophic rule-based classifier system

It refers to a white set, it applies to an indeterminate class and refers to a non-white set. Which could be described as [21]:

$$P_{NS}(x, y) = \{T(x, y), I(x, y), F(x, y)\} \quad (3)$$

$$T(x, y) = \frac{\overline{g(x, y)} - \overline{g}_{min}}{\overline{g}_{max} - \overline{g}_{min}} \quad (4)$$

$$I(x,y) = 1 - \frac{H_o(x,y) - H_o}{H_{O_{max}} - H_{O_{min}}} \tag{5}$$

$$F(x,y) = 1 - T(x,y) \tag{6}$$

$$H_o(x,y) = abs(g(x,y)) - \overline{g(x,y)} \tag{7}$$

where g$(x, y)$ signifies the local mean value of the window size pixels, and Ho $(x, y)$ that could be distinct as the uniformity value of T at $(x, y)$ represented by the absolute difference between the intensity g(i, j) and the local mean value g$(x, y)$.

### 3.4 Machine Learning and Classification

Classification is often a part of the Diagnosis (CAD) system and plays a significant role in the medical imaging diagnosis [22]. In the final step of the proposed algorithm; we have used a multilayer perceptron (MLP) based classifier to make a learning process from the extracted features. After extracting the three components (membership, indeterminacy and non-membership) for each MFs and PCs features using NTs, the most robust component of the three components of every MFs and PCs namely membership component were fed to MLP classifier to categorize all the image into one of two cases: positive COVID-19 (+ve) or normal case negative COVID-19 (−ve).

MLP is front-feeding networks of layers that are typically trained on fixed back spread. These networks have made their way into a multitude of applications involving the Classification of set patterns. Its key advantage is that it's simple to use and round every internal/output diagram [23]. The multi-layer neural network consists of the input layer, the output layer, and more than a hidden layer capable of generating more teamwork effects. The key benefit is that it can reflect any logical and continuous function as long as the hidden units are appropriate and an acceptable activation function is used. The method of updating neural weights to accomplish goals is called positional propagation. It attempts in such a manner that it generates the least number of errors. Error is only apparent on the output layer, and this error is returned to the previous layers. The mistake is proportional to the combination of layers on the neural network. Finally, the new weights are revised and replicated again. Since the error size is high in the output layer, the same amount (proportion) of error is propagated back to the previous layer [24].

# 4 Experimental Results and Discussion

## *4.1 Data Set Collection*

In this chapter, images of chest X-rays were utilized since this reading of radiography is often utilized by radiologists in clinical diagnosis. We tend to use a mix of two separate databases to construct our evaluation set. COVID-19 databases have been collected from in public access and picked up databases, whereas usual, and pneumonia databases have been compiled from publicly out there Kaggle databases:

- ***COVID-19 Dataset***: A public database in GitHub have been built [25] by aggregating 319 X-rays of COVID-19, Middle East Respiratory Syndrome (MERS), Extreme Acute metabolism Syndrome (SARS) and respiratory disease from online resources and revealed articles. This database includes 250 positive x-rays of COVID-19 and 25 COVID-19 positive X-rays of lung CT images with a different image resolution. The COVID-19 database was extracted from this database to construct a new database.
- ***Normal (or Healthy) Subjects Database***: The Kaggle Chest X-ray info may be an extremely popular database containing 5247 normal, infective agent and microorganism pneumonic x-rays of the chest with preciseness starting from 400p to 2000p [26]. Of the 5247 chest X-rays, 3906 are from separate subjects with pneumonia (2561 footage of bacterial pneumonia and 1345 pictures of virus infection) and 1341 pictures of standard people. In this database, chest x-rays of traditional and viral infection is employed to create an entire new database.

Table 1 shows the number of training and testing images for each type of chest X-ray. Figure 6 shows a sample of images taken from the database of X-ray images of the pulmonary chest COVID-19.

As shown in Fig. 7, X-rays are chest images taken from various patients with positive COVID-19 virus, and experienced radiologists examined the imaging results over a span of 20 years. Radiological findings are classified concerning [27] (i) peripheral predominance; a 72-year-old man complains of fever and cough. Initial chest x-rays revealed a two-terminal opacity in the central and lower oceanic airspace regions (yellow arrows). The degree of severity was 2 per lung, therefore TSS was four (Fig. 7a), (ii) perihilar predominance, and 35-year-old males complained of coughing and fever. Primary chest x-ray revealed opacity to consolidate real oceanic air (yellow arrow). The darkness of the substitute lower-left airspace

**Table 1** Summary of the training and testing data used for the evaluation of the proposed framework

| Type | Data set | Training set | Testing set |
|---|---|---|---|
| COVID-19 subjects | 250 | 140 | 110 |
| Normal subjects | 320 | 170 | 150 |

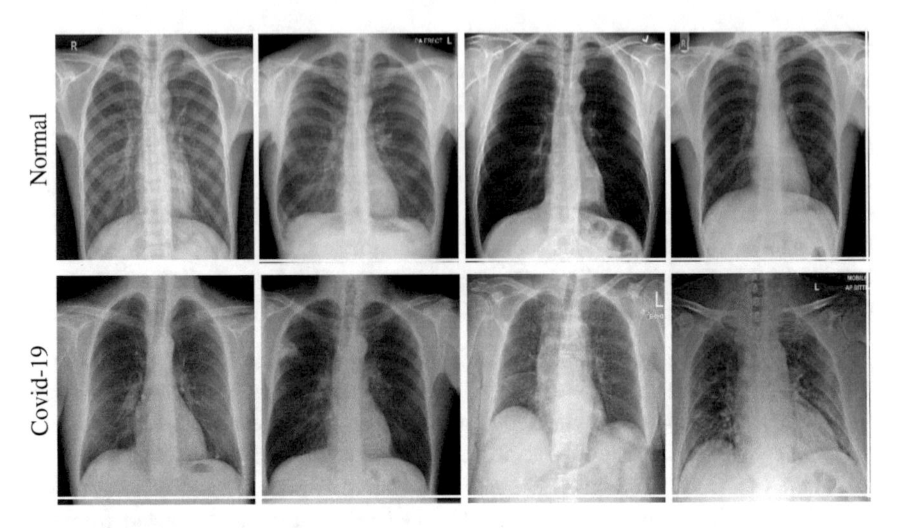


**Fig. 6** Samples of chest X-ray images of two different cases: normal healthy person (first row), and patient suffering from covid-19 (second row)

(long shares) with internal depth was observed; the level of strength was the same for each lung, so TSS was a pair of (Fig. 7b), (iii) or neither; 27-year-old male whiney of cough and fever. Initial chest X-rays revealed minor pulmonary nodules in the center of the right area (yellow arrow) and neither peripheral nor oceanic space consolidation opacity (yellow arrows) was found, and the intensity score was 1 per lung, so TSS 2 (Fig. 7c), (iv) left, left, or binary lung and up; male 23 years without symptoms. Initial chest x-rays revealed an opacity to improve the upper right area (yellow arrow).

The total severity was one (Fig. 7d), (v) and therefore the lower region; the 32-year male patient was whining of SOB and cough. Initial chest X-rays show the air space's blackout (yellow arrow) within the left and lower respiratory organ areas (with a lower area prevailing). The degree of intensity was 2 for the correct lung and a couple of on the left lung, therefore TSS four (Fig. 7e), (vi) the impact of the lung on monotheism, and the ground glass opacity (GGO); A 29-year-old feminine was affected by coughing and allergies. Initial chest x-rays showed a two-way glass opacity (yellow arrows) with a tiny low opacity within the higher left region (yellow arrow). The density was 2 per lung, so the TSS was 4 (Fig. 7f) (vii) was the reconfigured cellular thickness; the 48-year-old chest X-ray revealed the thickness of the binary biblical lung with opacity was still seen (yellow arrows) (Fig. 7g), (viii) and pulmonary nodules; the 49-year male was complaining of fever. Initial chest X-rays revealed two well-defined small pulmonary nodules in the middle lung region (yellow arrows). The degree of severity was 1 per lung, so TSS was 2 (Fig. 7h), (ix) had pleural or chest rest; the 61-year-old male complained of fever, cough, and diarrhea. Post-chest X-rays revealed a light left pleural effusion with a bilateral opacity to help the lower logical air region (yellow arrows).

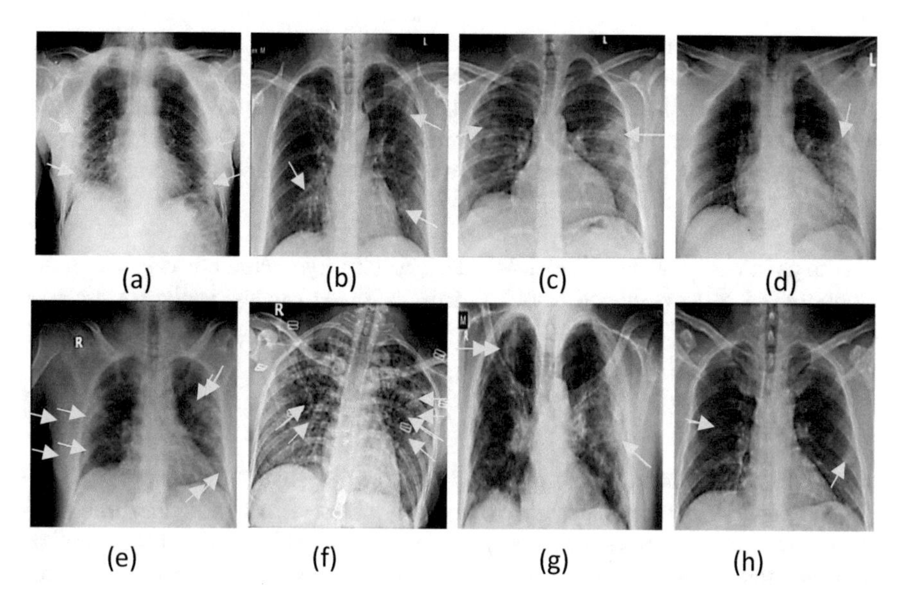


**Fig. 7** Examples of chest X-rays for different patients with positive COVID-19

## 4.2 *Performance Measures*

In order to assess the potency of the planned algorithm, varied strategies are accustomed to determine the afflicted person (+ve) and, therefore, the normal person (−ve) within the X-rays images studied. The cross-checkable was used and resulted in an uncertainty matrix, as seen in Fig. 8. The confusion matrix consists of 4 predicted findings are as follows.

Four main components are used to generate the confusion matrix True positive (TP); True negative (TN), False positive (FP), and False negatives (FN). TP may be an assortment of abnormalities that are detected with the correct diagnosis. TN is an incorrectly calculated variety of traditional ideals. FP is a set of normal cases referred to as FP for the detection of abnormalities. Finally, FN is a list of anomalies

| | | **Actual Value** (as confirmed by experiment) | |
|---|---|---|---|
| | | **Positive** | **Negative** |
| **Predicted Value** (predicted by the test) | **Positive** | True Positive (TP) | False Positive (FP) |
| | **Negative** | False Negative (FN) | True Negative (TN) |

**Fig. 8** Confusion matrix

that have been determined as an usual diagnosis. Quality metrics may be determined by measure the values of the doable outcomes within the confusion matrix, Output metrics are often measured. Check trained models are analyzed exploitation common output metrics like precision, accuracy(ACC), recall (selective), specificity, and F1 score (dice coefficient). It shall be represented as follows:

1. **Accuracy**

It is a parameter that tests the method's capacity by accurately calculating the proportion of cases predicted from all cases. ACC is mathematically expressed as follows [28]:

$$ACC = \frac{TP + TN}{TP + FP + FN + TN} \quad (8)$$

where TP (TN) is the correctly predicted positive (negative) cases and FP (FN) is the incorrectly predicted positive (negative) cases. However, exactness isn't invariably smart for evaluating every model type, notably within the case of an associate uneven knowledge set. there's conjointly a requirement to research alternative success metrics to validate the model.

2. **Precision**

Precision represents the magnitude relation of well-anticipated positive cases to the general expected positive cases. High accuracy compares to the poor, inaccurate positive rate. It's expressed as [29]:

$$\text{Precision} = \frac{TP}{TP + FP} \quad (9)$$

3. **Recall (Sensitivity)**

This is the magnitude relation of the properly expected positive notes to any or all operations within the actual class i.e. true positive rate. The confusion matrix is employed to assess sensitivity and is mathematically assessed as [30]:

$$\text{Recall } (\boldsymbol{Sensitivity}) = \frac{TP}{TP + FN} \quad (10)$$

4. **Specificity**

It is that the proportion of accurately forecast negative notes on any or more of the real negative observations [31]:

$$\boldsymbol{Specificity} = \frac{TN}{FP + TN} \quad (11)$$

5. **F1-Score**

The F1-score may be a comprehensive calculation of the accuracy of the model that blends exactness and recollection, the F1-score is double the quantitative relation between the multiplication of accuracy and also the retrieval measurement. Provides a balance between precision and recall [29]:

$$F1 - \boldsymbol{Score} = \frac{2 \times \boldsymbol{Precision} \times \boldsymbol{Recall}}{\boldsymbol{Precision} + \boldsymbol{Recall}} \quad (12)$$

## 4.3 Results and Discussion

The prompt design has been trained using the following criteria: two hidden layers, and 1000 epochs. The suggested framework has been applied employing a software package kit (MATLAB 2017b). Computer-implementation was for CPU and works during a 64-bit Windows environment. Each tests were conducted on the Intel R processor (2.60 GHz) and 4.00 GB Ram. Five collections of options are arranged: (i) morphological features only; (ii) morphological features of 4 PCs; (iii) Morphological options with eight computers; (iv) morphological features with sixteen computers; and (v) morphological features with twenty four computers. There is no reason to extract extra machines due to the information indicating no clear distinction within the utilization of over 24 PCs. Then, we tend to used NTs techniques together with the MLP classifier. NTs are additional to the three Neutrosophic parts' derived options, i.e. participation, non-identification and in organism, the foremost potent element of the three components per radiofrequency. Therefore, the machine is that the organic component that was fed to the MLP classifier. Figure 9 displays a graph of the Neutrosophic Components (NCs) graph; a sample of all MFs with machine features in the Neutrosophic environment.

The proposed algorithm's performances were measured by employing various performance metrics—confusion matrix, accuracy, precision; sensitivity, specificity, and F1-Score. Figure 10 shows the fusion matrix summary of the proposed algorithm, TP is that true positive within the case of COVID-19, and TN is the true negative in the normal state. In contrast, FP and FN are wrong model estimates for COVID-19 and also the alternative case.

Although the results are seen in Table 2 and Fig. 11 it's been found that the MLP-based compiler of MFs with options of 24 PCs features the most effective results while not exploiting NTs for all the cases tested for geared up of distinctive features, but once applied NTs to any or all cases tested for all set of features, we tend to found the results showing this using NTs of (MFs with 4PCs) in combination with MLP based classifier provides better result than using the MLP classifier with the other groups of features: ACC = 0.9846; precision = 0.9818; sensitivity = 0.9818, specificity = 0.9867 and F1-Score = 0.9818.

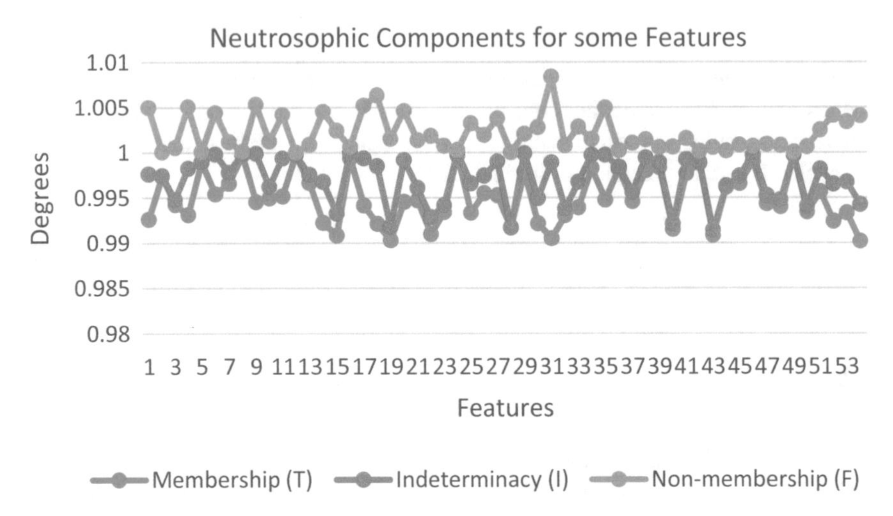


Fig. 9 Neutrosophic three components for some features

| | | Actual Covid-19 | |
|---|---|---|---|
| | | (+ve) | (−ve) |
| Predicted Covid-19 | (+ve) | True Positive 108 | False Positive 2 |
| | (−ve) | False Negative 2 | True Negative 148 |

Fig. 10 Confusion matrix analysis of the NCs for (MFs + 4 Pcs)

This means that the algorithm relies on a small number of tested features for Classification, i.e. only MFs with 4PCs, which saves time and reduces system complexity while achieving higher efficiency in the classification process. This means, that a small set of features is sufficient to classify COVID-19 images using the NTs also there no any need for much more than 4 PCs with using NTs since the results show no noticeable difference.

Table 3 represents a comparison result between the proposed architecture and the other related works which indicate that our approach has worked better for classifying the images of chest X-ray into two classes. Also achieved a high-performance, efficiency, simplicity, and speedy in Classification.

**Table 2** Performance of the proposed framework

| Performance measures | MFs | MFs + 4 PCs | MFs + 8 PCs | MFs + 16 PCs | MFs + 24 PCs | NT for MFs + 4 PCs |
|---|---|---|---|---|---|---|
| Accuracy | 0.9038 | 0.9462 | 0.9577 | 0.9692 | 0.9731 | 0.9846 |
| Precision | 0.8636 | 0.9091 | 0.9364 | 0.9545 | 0.9636 | 0.9818 |
| Sensitivity | 0.9048 | 0.9615 | 0.9626 | 0.9722 | 0.9725 | 0.9818 |
| Specificity | 0.9032 | 0.9359 | 0.9542 | 0.9671 | 0.9735 | 0.9867 |
| F1-Score | 0.8837 | 0.9346 | 0.9493 | 0.9633 | 0.9680 | 0.9818 |

Note that "MF", "PC" and "NC" stand for morphological features, principal components, and Neutrosophic technique, respectively

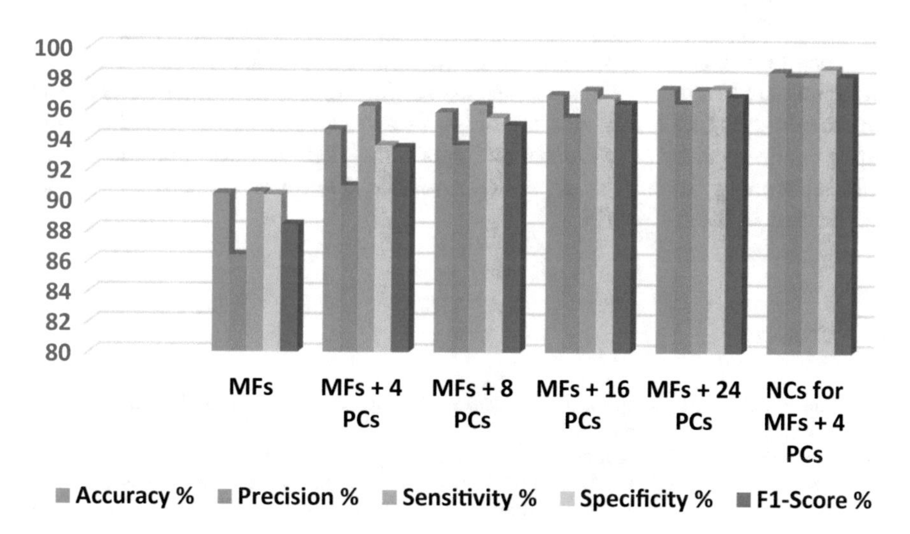


**Fig. 11** The performance metrics of the proposed pipeline for COVID-19 diagnosis

**Table 3** The proposed architecture results and the other related works

| Performance measures (%) | Accuracy | Precision | Sensitivity | Specificity | F1-Score |
|---|---|---|---|---|---|
| Apostolopoulos and Mpesiana [7] | 96.78 | NA | 98.66 | 96.46 | NA |
| Sethy and Behera [8] | 95.38 | NA | NA | NA | 95.52 |
| Afshar et al. [10] | 95.7 | NA | 90 | 95.8 | NA |
| Abbas et al. [12] | 95.12 | NA | 97.91 | 91.87 | NA |
| **Proposed** | **98.46** | **98.18** | **98.18** | **98.67** | **98.18** |

## 5 Conclusion

This paper has proposed a novel, robust, automated, and intelligent system for COVID-19 Classification using chest X-rays. The proposed pipeline is a hybrid technique that integrates machine learning techniques and Neutrosophic techniques. The testing accuracy, precision, sensitivity, specificity, and F1-Score for the scheme were 98.46%, 98.19%, 98.18%, and 98.67%, 98.17%. The obtained results show that using MFs with 4PCs for feature extraction in combination with Neutrosophic techniques and multi-layer perceptron-based classifier provides the best results compared to other ways. This will be significantly helpful during this natural event once the sickness and the want for hindrance action are in distinction with available resources. Our proposed pipeline's weakness is that it only offers a straightforward positive or negative COVID-19 observation to the physician. It will act as a screening aid to a medical imaging specialist. Our future research will be dedicated to providing staging for the severity of COVID-19 cases. This will be achieved by using more possible multi-database algorithms with clinical information for multi-modelling for improved prediction of the disease and deployment of the application in the real world.

## References

1. Chowdhury, M.E., Rahman, T., Khandakar, A., Mazhar, R., Kadir, M.A., Mahbub, Z.B., Islam, K.R., Khan, M.S., Iqbal, A., Al-Emadi, N., Reaz, M.B.I.: Can AI help in screening viral and COVID-19 pneumonia? (2020) arXiv preprint arXiv:2003.13145
2. Punn, N.S., Agarwal, S.: Automated diagnosis of COVID-19 with limited posteroanterior chest X-ray images using fine-tuned deep neural networks (2020). arXiv preprint arXiv:2004.11676
3. Wang, W., Xu, Y., Gao, R., Lu, R., Han, K., Wu, G., et al.: Detection of SARS-CoV-2 in different types of clinical specimens. Jama (2020)
4. Yang, T., Wang, Y.-C., Shen, C.-F., Cheng, C.-M.: Point-of-care RNA-based diagnostic device for COVID-19. Multidisciplinary Digital Publishing Institute (2020)
5. Huang, C., Wang, Y., Li, X., Ren, L., Zhao, J., Hu, Y., et al.: Clinical features of patients infected with 2019 novel coronavirus in Wuhan, China. Lancet **395**, 497–506 (2020)
6. Allen, J.N., Davis, W.B.: Eosinophilic lung diseases. Am. J. Respir. Crit. Care Med. **150**(5), 1423–1438 (1994)
7. Apostolopoulos, I.D., Mpesiana, T.A.: Covid-19: automatic detection from x-ray images utilizing transfer learning with convolutional neural networks. Phys. Eng. Sci. Med. **1**, 635–640 (2020). https://doi.org/10.1007/s13246-020-00865-4
8. Sethy, P.K., Behera, S.K.: Detection of coronavirus disease (covid-19) based on deep features. Preprints **2020030300**, 2020 (2020)
9. Khalifa, N.E.M., Taha, M.H.N., Hassanien, A.E., Elghamrawy, S.: Detection of coronavirus (covid-19) associated pneumonia based on generative adversarial networks and a fine-tuned deep transfer learning model using chest x-ray dataset (2020). arXiv:2004.01184
10. Afshar, P., Heidarian, S., Naderkhani, F., Oikonomou, A., Plataniotis, K.N., Mohammadi, A.: Covid-caps: a capsule network-based framework for identification of covid-19 cases from x-ray images (2020). arXiv preprint arXiv:2004.02696

11. Wang, L., Lin, Z.Q., Wong, A.: Covid-net: a tailored deep convolutional neural network design for detection of covid-19 cases from chest x-ray images. Sci. Rep. **10**(1), 1–12 (2020)
12. Abbas, A., Abdelsamea, M.M., Gaber, M.M.: Classification of COVID-19 in chest X-ray images using DeTraC deep convolutional neural network (2020). arXiv preprint arXiv:2003.13815
13. OpenCV: Image thresholding (2020). https://docs.opencv.org/master/d7/d4d/tutorialpythresholding.html. Online; Accessed 12 April 2020
14. Khalid, S., Khalil, T., Nasreen, S.: A survey of feature selection and feature extraction techniques in machine learning. In: 2014 Science and Information Conference, pp. 372–378. IEEE (2014)
15. Ruck, D., Rogers, S., Kabrishy, M.: Feature selection using a multilayer perceptron. J. Neural Netw. Comput. **2**(2), 40–48 (1990)
16. Jing, X.J., Yu, N., Shang, Y.: Image filtering based on mathematical morphology and visual perception principle. Chin. J. Electron. **13.4**, 612–616 (2004)
17. Ansari, A.Q., Biswas, R., Aggarwal, S.: Neutrosophic classifier: an extension of fuzzy classifer. Appl. Soft Comput. **13**(1), 563–573 (2013)
18. Mohammed, M.N., Syamsudin, H., Al-Zubaidi, S., AKS, R.R., Yusuf, E.: Novel COVID-19 detection and diagnosis system using IOT based smart helmet. Int. J. Psychosoc. Rehabil. **24** (7), 2296–2303 (2020)
19. Yasser, I., Twakol, A., El-Khalek, A., Samrah, A., Salama, A.A.: COVID-X: novel health-fog framework based on neutrosophic classifier for confrontation covid-19. Neutrosophic Sets Syst. **35**(1), 1 (2020)
20. Fang, Y., Zhang, H., Xie, J., Lin, M., Ying, L., Pang, P., Ji, W.: Sensitivity of chest CT for COVID-19: comparison to RT-PCR. Radiology **200432** (2020)
21. Salama, A.A., Smarandache, F., Eisa, M.: Introduction to image processing via neutrosophic techniques. Neutrosophic Sets Syst. **5**, 59–64 (2014)
22. Dey, N.: Classification Techniques for Medical Image Analysis and Computer Aided Diagnosis. Academic Press (2019)
23. Mohammadian, A., Miller, E.J.: Nested logit models and artificial neural networks for predicting household automobile choices: comparison of performance. Transp. Res. Rec. **1807**(1), 92–100 (2002)
24. Biswas, M., Adlak, R.: Classification of galaxy morphologies using artificial neural network. In: 2018 4th International Conference for Convergence in Technology (I2CT), pp. 1–4. IEEE (2018)
25. Monteral, J.C.: COVID-Chestxray Database (2020). Available: https://github.com/ieee8023/covid-chestxray-dataset
26. Mooney, P.: Chest X-Ray Images (Pneumonia) (2018). Available: https://www.kaggle.com/paultimothymooney/chest-xray-pneumonia
27. Yasin, R., Gouda, W.: Chest X-ray findings monitoring COVID-19 disease course and severity. Egypt. J. Radiol. Nucl. Med. **51**(1), 1–18 (2020)
28. Jain, G., Mittal, D., Thakur, D., Mittal, M.K.: A deep learning approach to detect covid-19 coronavirus with X-Ray images. Biocybernetics Biomed. Eng. **40**(4), 1391–1405 (2020)
29. Hemdan, E.E.-D., Shouman, M.A., Karar, M.E.: Covidx-net: a framework of deep learning classifiers to diagnose covid-19 in x-ray images (2020). arXiv preprint arXiv:2003.11055
30. Pannu, H.S., Singh, D., Malhi, A.K.: Improved particle swarm optimization based adaptive neuro-fuzzy inference system for benzene detection. CLEAN–Soil Air Water **46**(5), 1700162 (2018)
31. Kaur, M., Gianey, H.K., Singh, D., Sabharwal, M.: Multiobjective differential evolution based random forest for e-health applications. Mod. Phys. Lett. B **33**(05), 1950022 (2019)

# COVID 19 Prediction Model Using Prophet Forecasting with Solution for Controlling Cases and Economy

**Shivani Bhalerao and Pallavi Chavan**

**Abstract** Coronavirus disease outbreak (COVID-19) has threatened the entire world and has made lives difficult. It has drastically affected the way of living, working, and managing routines for the human beings, by living indoors. In a country like India, with a population of about 1.35 billion, the virus is spreading so fast that the control has become unmanageable. This paper presents COVID 19 data analysis and the prediction model that helps plan and organize things as precautionary measures. In this chapter, analysis is performed on COVID 19 data, and the prediction model is proposed for October. The analysis and prediction is performed using two methods, viz. random forest and time series. The chapter also compared the analyzed results. The idea behind analyzing the available dataset and the comparison of two prediction models is to supply some solutions to control the spreading of COVID 19. In this chapter, analysis is presented state-wise and country wise for the active number of cases and the date cases. Recovery rates are also analyzed. Gender-specific detailed analysis is also presented in this chapter with different age groups in India.

**Keywords** COVID 19 · Prediction · Prophet model · Time series analysis

## 1 Introduction

Covid-19 has sharply affected the Indian Economy. India's GDP for Q1 FY21 was contracted by 23.9% compared to the same quarter in the previous year and was the worst contraction in the Indian Economy's history [1]. Stress on the trade and supply chain and the devastating hospitality industry has affected many families [2].

S. Bhalerao
Tata Consultancy Services, Ltd., Mumbai, MH, India

P. Chavan (✉)
Department of Information Technology, Ramrao Adik Institute of Technology Nerul, Navi, Mumbai, MH, India
e-mail: pallavi.chavan@rait.ac.in



A.-E. Hassanien et al. (eds.), *Advances in Data Science and Intelligent Data Communication Technologies for COVID-19*, Studies in Systems, Decision and Control 378, https://doi.org/10.1007/978-3-030-77302-1_8

According to the ILO-ADB report, the COVID 19 pandemic resulted in the loss of jobs of 41Lakh youth in India [3], and 6.6 MN professionals with white-collar jobs, lost their job from May to August [4]. This problem has directly/indirectly affected many professionals, like movement restriction affecting the car rental business and online teaching affecting the school-uniform-making businesses, which earn good only once a year. All of these economic devastations resulted from a strict and important lockdown, from March 25 2020 to May 31 2020 in India, which was essential in order to slow down the spread of the virus.

Later, multiple unlock phases aimed at fighting the economic crisis in India. On the other hand, the current COVID status is that the number of cases each day in India is rising by around 95 K (by the mid of September 2020). The above two problems of Economy and COVID seem to be mutually exclusive to be solved. Working hard for the Economy could increase the active cases, while Lockdown or restrictions in movement could harm the Economy.

Further, this is not the case just for India. Most of the countries are suffering from the same crisis. In Japan, this outbreak resulted in the cancellation of Tokyo Olympics. In the USA, the number of unemployed persons fell by 2.8 MN [5]. In Spain, GDP contracted by 5.2% in Q1 [6] And, many more.

In this chapter, the authors analyzed COVID 19 data and proposed a prediction model using two approaches. First approach is using random forest algorithm and another approach is using time series. The idea behind analyzing the data and building a prediction model is to provide the solutions to this pandemic situation.

## 2 State of Art

As per the research work started in 'SEIR and Regression Model-based COVID-19 outbreak predictions in India', the authors used two models to predict future confirmed cases. The authors considered 70% of the population in the susceptible class. As the SEIR model also considers intervention with time, the authors stated that they used the hill decay model for the intervention. Also, due to community spreading, they added that, there will be an increase in the number of cases. Due to the insufficient data in March 2020, they could not predict death cases properly [7].

In the literature and the findings of 'Forecasting COVID-19 epidemic in India and high incidence states using SIR and logistic growth models', Malavika et al. evaluated the impact of 21 days Lockdown using different prediction models and also suggested that there is no proof that Lockdown has a positive impact to control active cases [8].

'Predictability of monthly temperature and precipitation using automatic time series forecasting methods' is the research stated by Georgia et al. which is about predicting the monthly temperature and precipitation on 98,540 year monthly temperature and 155,240 year monthly precipitation using ARFIMA, Box-Cox transformation, ARMA, BATS, simple exponential smoothing, Theta and Prophet methods [9].

The paper on SARIMA and Prophet based Time Series model for Air Pollution Forecasting, by K. Krishna et al. is an experimental analysis of air pollution forecasting to indicate the effectiveness of their proposed method capable of providing a rough approximation of the future pollution levels using SARIMA and Prophet model [10].

The paper 'Food Supply Chains during Covid 19 pandemic' by Jill E. Hobbs asses the implications of COVID 19 on the Food Supply Chain. Further, he also discussed panic buying behavior for some food items by the customers. He also furthers whether COVID would have long term effects on the food delivery sector and the study that customers would prefer local foods. Therefore, after his complete investigation, Jill concluded that they need to focus on maintaining and growing food supply chains, for which there should be robust and resilient supply chain relationships [11].

In 'COVID-19's Disruption of India's Transformed Food Supply Chains', Thomas Reardon et al. mentioned that COVID has put food security at risk as the private sector purchased 92% of food consumption, 96% food is sold by the private sector and 4% by the government. Therefore they suggested a two-pronged strategy of implementing strong health care measures and addressing food security impacts on the public income and employment [12].

In a complete study of 'The effect of COVID-19 and subsequent social distancing on travel behavior', Jonas De Vos discusses the potential impacts on the travel patterns, how people reach the out-of-home activities, demand for public transport might decrease. Also, social distancing could result in low physical activities. Therefore, public transport should look out for safer ways for people who use it for travelling [13].

According to an article in The Indian Express, the GDP of India for Q4- 2019–2020, had noticed a growth of 3.1%, and according to the government data, the gross value added at the basic price dropped by 20.6% in Q1 2020–21. According to the NSO data, construction witnessed a drop of 50.3%. The manufacturing industry decreased by 39.3%, trade, hotels, and transport by 47%.

Therefore, according to this article, GDP for Q1-2020–21 was the worst contraction by 23.9% in the Indian economy history.

According to an article in The Economic Times, there was a 26% fall in employment among industrial workers over the year. Approximately 6.6 mn engineers, teachers, and other white-collar employees lost their jobs, as number of employed white collar employees fell from 18.8 mn to 12.2 mn from May–August 2019 to May–August 2020.

According to an article in the financial express in Feb 2020, even though the Indian Economy faces a slowdown, the IT and BPM industry is a major part that provides support through jobs and revenues as it employs 4.1 Lakh professionals with 31,922 registered IT companies, with maximum number of IT firms in Delhi and Maharashtra. But geopolitical and technological, and other uncertain issues can slow down the industry if not rapidly adapted to change [14].

Also, a complete article 'Modeling Logistic Growth' from towards data science compares logistic growth and exponential growth, reason and explanation of using logistic growth, logistic growth formula using mathematics and a hypothetical case for logistic growth, steps for non-linear least squares estimation [15].

The paper on Fractionally different ARIMA models for hydrologic time-series by Alberto Montanari, Renzo Rosso, and Murad S. Taqqu provided the framework to represent hydrologic time-series for short term and long term persistence, further indicating that FARMA can improve the stochastic modeling of hydrologic time-series[16].

The news article on the COVID 19 outbreak affecting the joblessness in India from Times Now News mentions that the pandemic and lockdown measures in India lead to the spike in India's unemployment rates especially in April and May, CMIE also calculated the overall employment rate through the survey incl 174,000 households in the 4 months [17].

According to the article in The Hindu, the global Economy could contract by approx. 4.3% this year, with global output by year's end over US$6 trillion short, with the equivalent of the wipeout of Brazilian, Indian and Mexican Economies, global recession matching the great depression. Forecaster's talk of a misleading V-shaped recovery [18].

Studying from the above given references, there were some attempts for predicting COVID 19 cases and recoveries further, deaths were unable to predict due to the lack of data. Also, Prophet based predictions have been used to predict various time-series data. In the following study, authors have predicted comparably accurate COVID 19 data than the previous studies, due to the large data source. They have also used the recent prediction procedure that is the Prophet procedure to predict the time-series data.

## 3 Analysis of Random Forest Regressor and Prediction Using Prophet Time Series Model

Initially, the authors decided to use Random Forest Regressor to analyze the COVID 19 data. Therefore they used the cumulative recovery data to build a prediction model. Practically a cumulative data prediction refers to an increasing graph with no single decreasing value. But surprisingly as shown in Fig. 1, the model learned it in a wrong way and displayed a fluctuating plot, which is very far from reality. For improving the accuracy, they worked on various methods and then found a near to perfect approach of prediction-Prophet Procedure.

Therefore, the authors decided to use the Prophet procedure to build a time series prediction model because it fits the trend flexibly and is an additive regression model with four components [19].

1. It detects changes in the trends by selecting the change points from the data.
2. Yearly components using Fourier series model.
3. Weekly components using dummy variables.

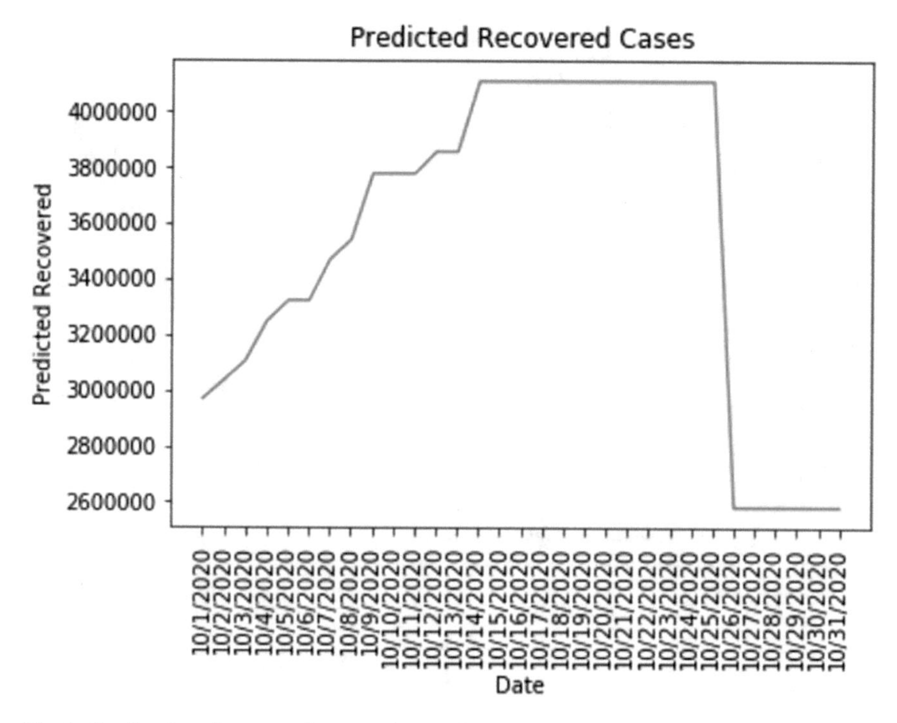


**Fig. 1** Predicted total recovered cases using random forest

4. User-provided list of holiday-dates.

The expected the prediction of total recovered cases was as shown in Fig. 2.

## *3.1 Prophet Procedure for Time-Series Forecasting*

Prediction of the Daily Recovered cases till October 31, 2020 is given in Fig. 3.

The daily recovery is forecasted to be reached to approximately 1.25 Lakhs per day by the end of October. Prediction of total deaths till October 31, 2020 is given in Fig. 4.

The total deaths are forecasted to be reached to approximately 1.25 Lakhs by the end of October. Prediction of daily deaths till October 31, 2020 is shown in Fig. 5.

Daily deaths are forecasted to be reached to approximately 1.6 K per day by the end of October. Prediction of Active cases till October 31, 2020 is shown in Fig. 6.

Active Cases are forecasted to be reached to approximately 1.4 Million by the end of October. This is a huge number and could create a bottleneck on the health

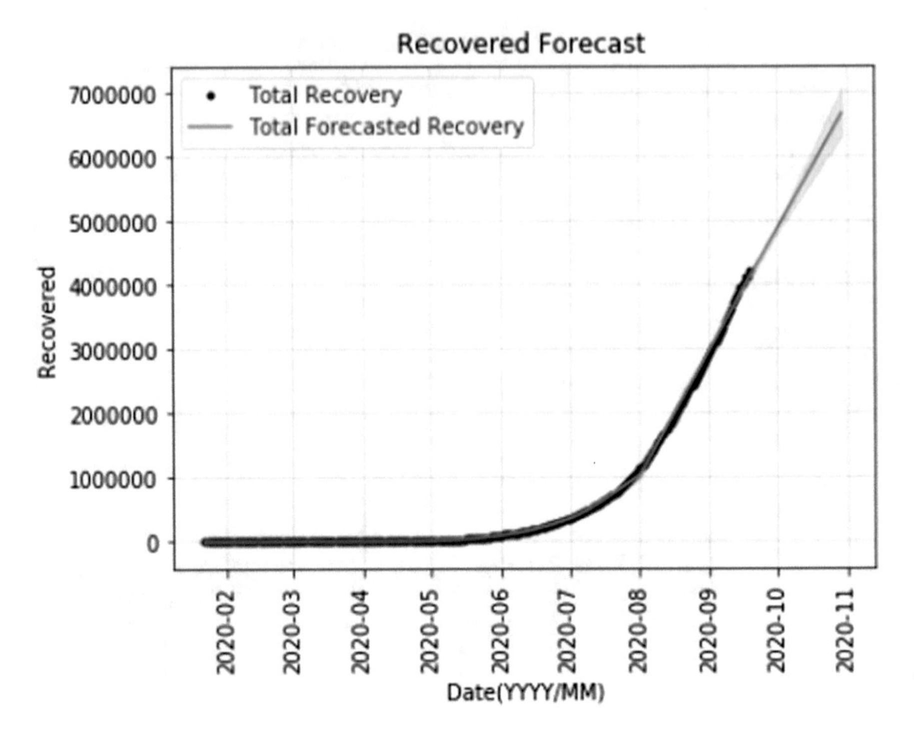


**Fig. 2** Predicted total recovered cases using prophet procedure

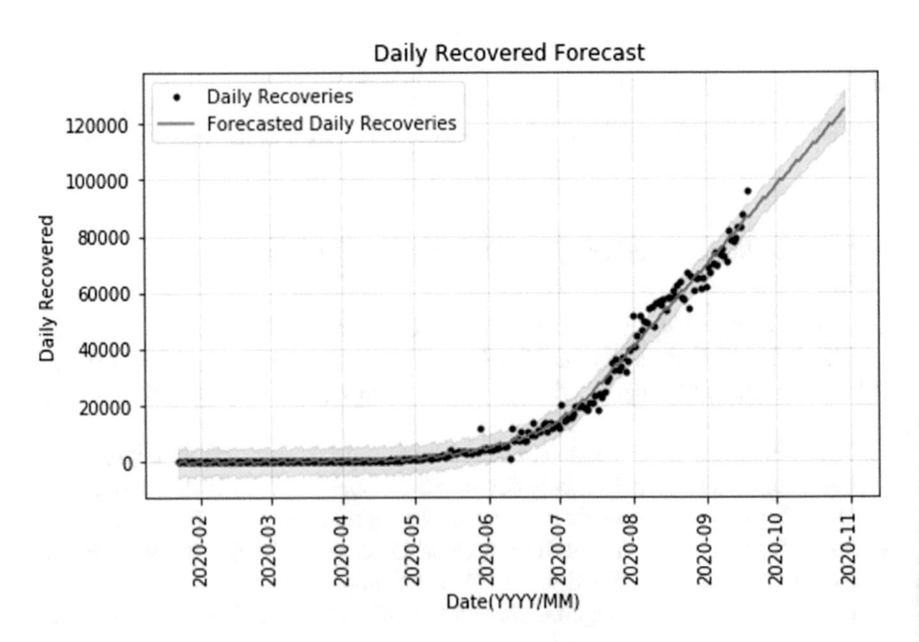


**Fig. 3** Predicted daily recovered cases using prophet procedure

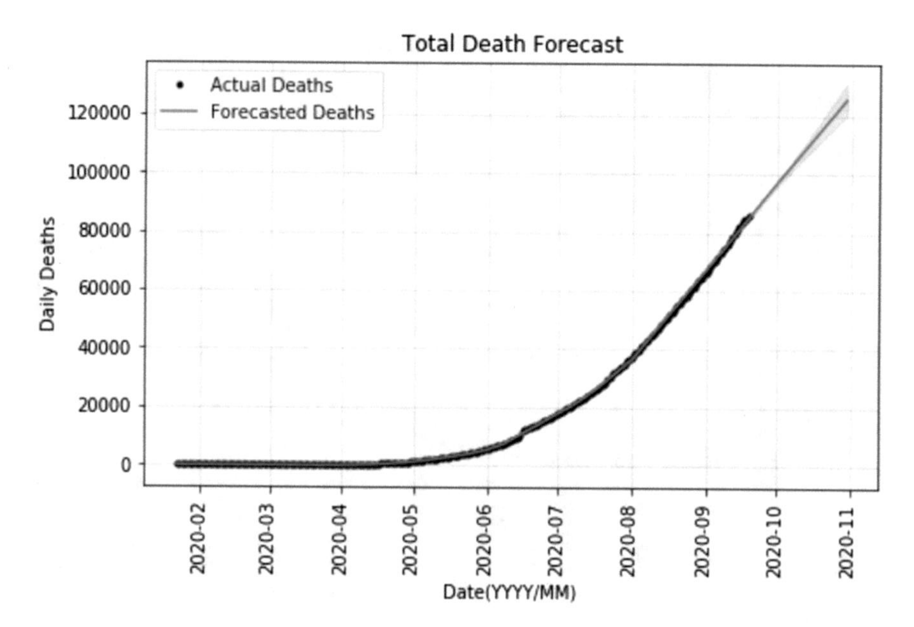


**Fig. 4** Predicted total deaths using prophet procedure

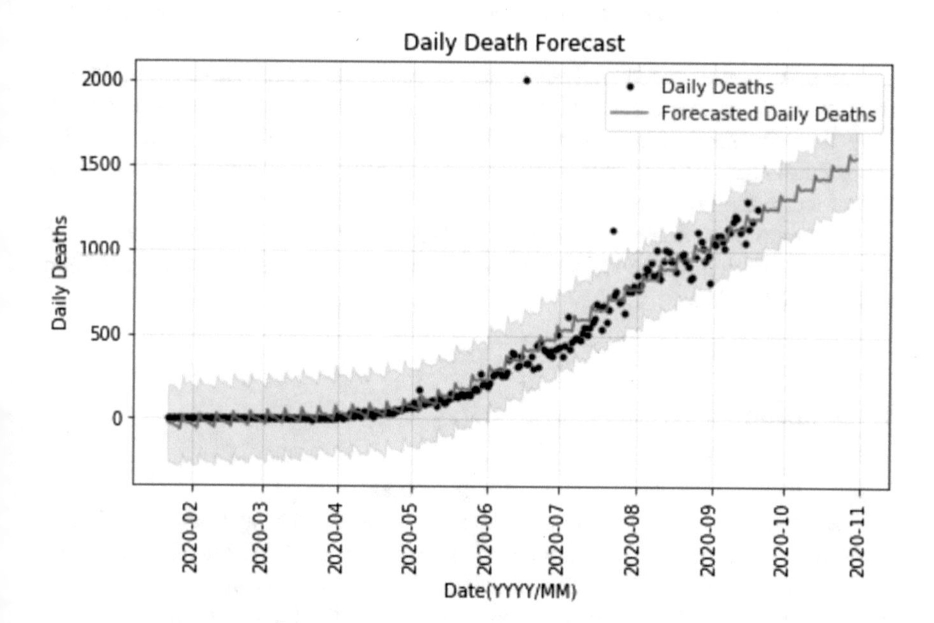


**Fig. 5** Predicted daily deaths using prophet procedure

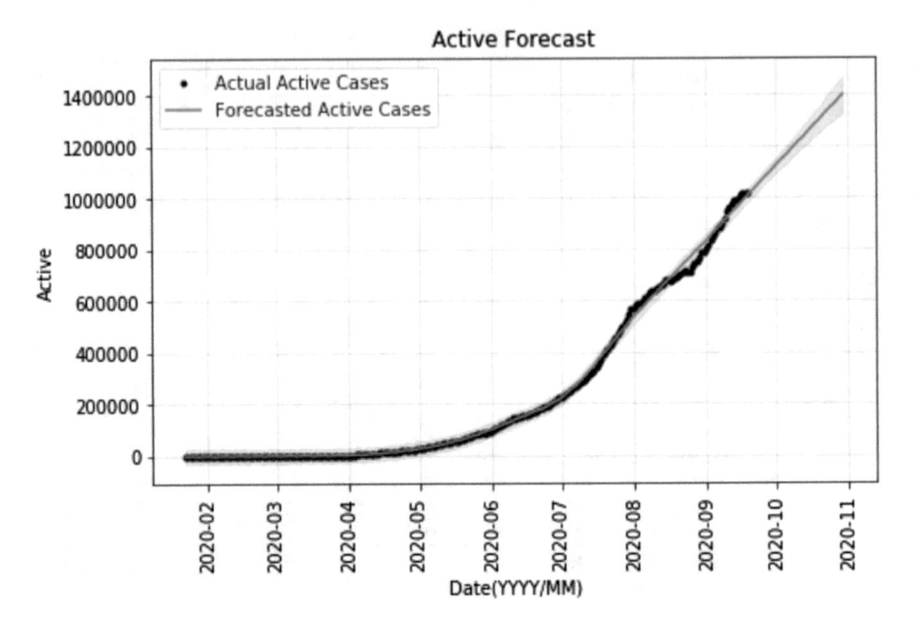


**Fig. 6** Predicted active cases using prophet procedure

and medical services. Hence, there needs to be some solution to at least reduce some cases.

Statewise gender based analysis is also shown here in Fig. 7. According to the available data, no single state in India has number of females affected by Covid 19 more than the number of males. It might be due to the: Female to male ratio: According to the 2013–15 data, there are 900 females per 1000 males in India [25], Difference in the hormones.

Age group wise analysis is shown in Fig. 8. as shown in figure, most of the population affected due to COVID 19 is between 19 to 59 years of age, according to the available data.

## 4 Solution

Figure 9 depicts the rate of growth of active cases.

The green-colored slope refers to the rate of growth during Lockdown, it was uniform and a slow growth with the slope as 1291.83, while on the other hand, the red colored slope refers to the rate of growth during unlocking phases, which has a slope of 11,167.56 (Nearly 10 times as compared to the Lockdown).

In order to control the growth of active cases, at least till people are vaccinated against the virus, the authors suggested to come to a middle ground of Lockdown

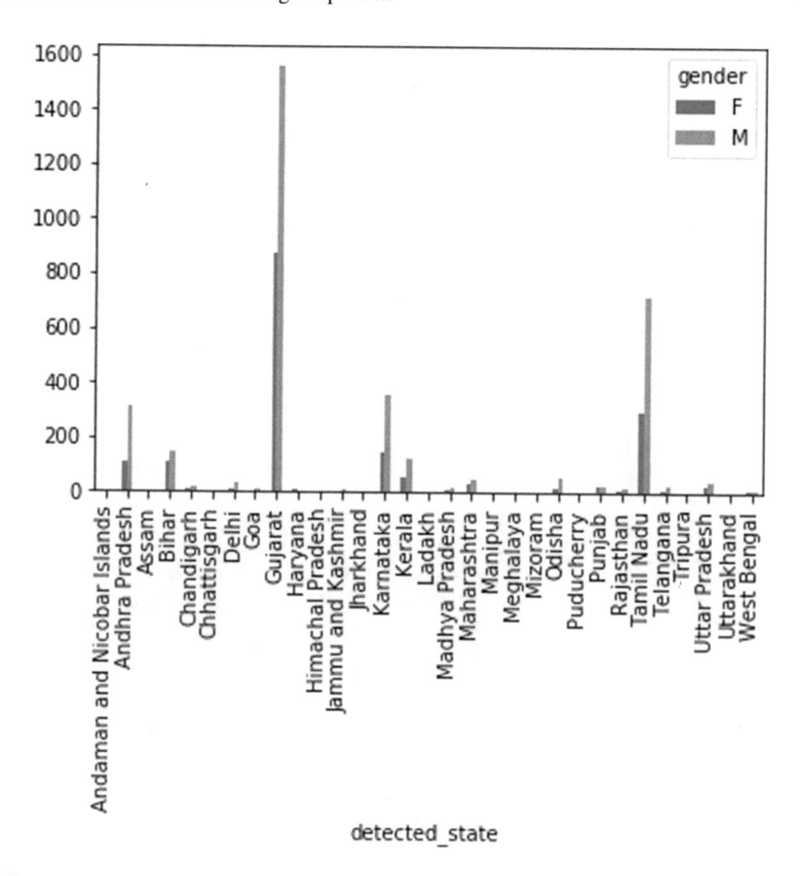


**Fig. 7** COVID distribution among the gender

and unlock, to solve economy crisis as well as avoid the bottleneck on the health and medical services.

Step 1: Work from home became a new common in India, and 70% of the companies decided to extend the work from home [20], therefore staying at home like a lockdown and at the same time working is not a problem has acted as a cushion for India's Economy. Also, the information technology and business process management industries contribute to India's GDP by about 7.9%.

Step 2: In India, 85% of the population cannot afford the cost of their personal vehicle, and, they depend largely on public transport. Therefore, the share of public transport is the largest contributor to GDP (8% share) [21]. For example, Local train is the lifeline of the financial capital of India, Mumbai, with over 75 Lakh people traveling every day [22]. Therefore, such public transports can be made safe with the increase in the number of work shifts. If the companies decided to work in three shifts of 8 h each, it can provide service for 24 h, and at the same time, the crowd in the transit could be distributed reduced to almost 34%, making it safer for the citizens during COVID times.

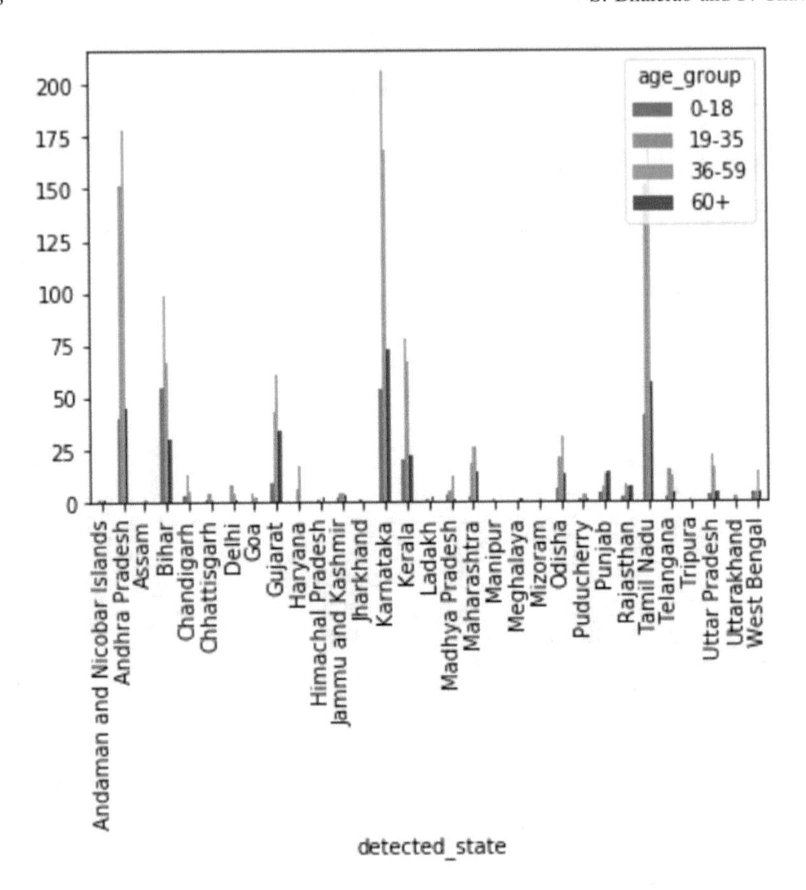


**Fig. 8** COVID distribution among the age-groups

Step 3: Indian companies mostly follow a Monday to Friday pattern, with 8–9 h of work per day [23]. But, Microsoft Japan did an interesting experiment. They tried a three day weekend and the results were: electricity consumption was reduced by 23%, productivity rose by 40% as those who worked less, worked smarted. Perpetual Guardian, a firm in NZ tried a four day work week and analyzed that the staff was more punctual, more creative and more productive [24]. Therefore, if the working pattern is changed to a four day workweek, this might benefit the Indian Economy, maintaining the work-life balance.

Full Lockdown on weekends could lower the Economy as the working crowd would hardly be able to buy anything. But, a partial lockdown could help control the rise of the cases.

Therefore, the major part is that crowded places like malls and markets can remain closed on weekends. As the weekend will be of three days, people buying essentials would maintain the flow of money, thus balancing the Economy. At the same time, it would lower the crowd at a place by segregating it in three days.

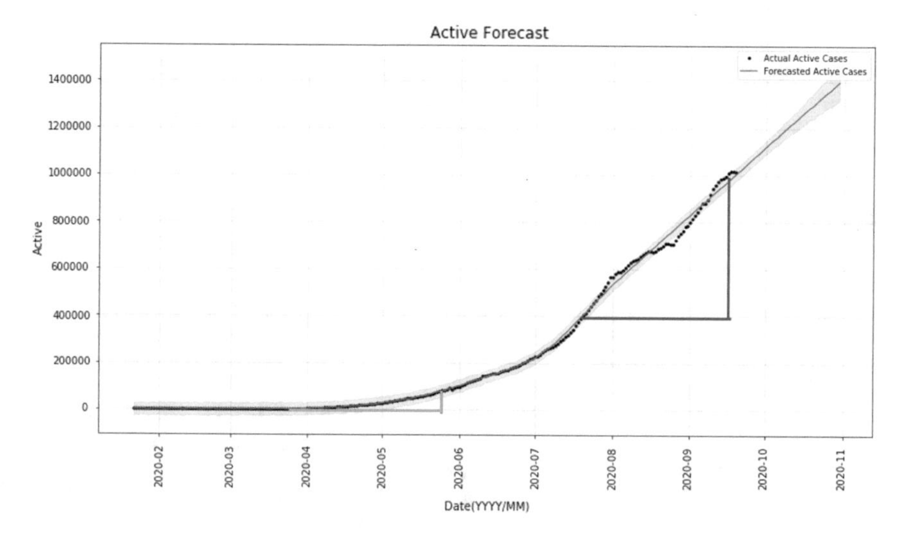


**Fig. 9** Rate of growth of active cases

Maintaining social distancing without a lockdown could be impossible for India with a very high population density. Therefore the focus should be on handling the crowd's flow to avoid gatherings at a single place.

## 5 Conclusion

The authors have analyzed and predicted future active, recovered, and death cases for India till October 31, 2020 using Prophet Procedure and suggested a solution for avoiding such large numbers of active cases, and at the same time, maintaining a good economic balance in the future. They studied that, in order to preserve social distancing in a country with a high population density, there can be a separation in the movement of the crowd by encouraging work from home as much as possible, and increasing the number of shifts for segregating the traveling and working crowd.

Further, they suggested a four day work week instead of a five day work week based on the trials of different firms in different countries, to increase the work-life balance and productivity and enhance the ideas and creativity. In this way, social distancing could be truly possible and the same time, the Economy can be boosted. Further, they also visualized the state-wise data for gender and age-group-wise covid 19 data. They found that more males than females are affected by the Corona Virus and the middle-aged group 19–59 are mostly involved in India's states.

Hospitals and other medical facilities can be kept ready for any emergency by considering these approximate figures to avoid any delay. Other parameters like number of beds, number of testing, and the probability of vaccine arrival or immunity booster are not considered in this study. One can use these parameters in the time series model, to get more accurate predictions, as accurate predictions can be closely used to create a real-time prediction system.

## References


1. indianexpress.com 2020 [online]. Available at: https://indianexpress.com/article/business/economy/gdp-1st-quarter-growth-rate-data-india-april-june-2020-6577114/. Accessed Sept 2020
2. cnbctv18.com 2020 [online]. Available at https://www.cnbctv18.com/hospitality/covid-19-impact-hospitality-sectors-revenue-loss-estimated-at-almost-rs-90000-crore-in-2020-6655261.htm. Accessed Sept 2020
3. economictimes.indiatimes.com 2020 [online]. Available at https://economictimes.indiatimes.com/news/economy/indicators/41-lakh-youth-lose-jobs-in-india-due-to-covid-19-pandemic-ilo-adb-report/articleshow/77613218.cms. Accessed Sept 2020
4. economictimes.indiatimes.com 2020 [online]. Available at https://economictimes.indiatimes.com/news/economy/indicators/6-6-mn-white-collar-professional-jobs-lost-during-may-august-cmie/articleshow/78162428.cms. Accessed Sept 2020
5. U.S. BUREAU OF LABOR STATISTICS, Washington, DC (2020). Available at https://www.bls.gov/news.release/empsit.nr0.htm. Accessed Sept 2020
6. economics.rabobank.com 2020 [online]. Available at https://economics.rabobank.com/publications/2020/may/spanish-economy-contracts-due-to-covid-19/. Accessed Sept 2020
7. Pandey, G., Chaudhari, P., Gupta, R., Pal, S.: SEIR and Regression Model based COVID-19 outbreak predictions in India (2019). https://doi.org/10.1101/2020.04.01.20049825
8. Malavika, B., Marimuthu, S., Joy, M., Nadaraj, A., Asirvatham, E.S., Jeyaseelan, L.: Forecasting COVID-19 epidemic in India and high incidence states using SIR and logistic growth models. Clin. Epidemiol. Glob. Health (2020). https://doi.org/10.1016/j.cegh.2020.06.006.Advanceonlinepublication.doi:10.1016/j.cegh.2020.06.0069
9. Georgia, P., Tyralis, H., Koutsoyiannis, D.: Predictability of monthly temperature and precipitation using automatic time series forecasting methods. Acta Geophysica **66**, 807–831 (2018). https://doi.org/10.1007/s11600-018-0120-7
10. Samal, K., Babu, K., Das, S., Acharya, A.: Time series based air pollution forecasting using SARIMA and prophet model. In: ITCC 2019: Proceedings of the 2019 International Conference on Information Technology and Computer Communications, pp 80–85 (2019). https://doi.org/10.1145/3355402.3355417
11. Hobbs, J.: Food supply chains during the COVID-19 pandemic. Can. J. Agric. Econ./Rev. Can. d'agroeconomie **68** (2020). https://doi.org/10.1111/cjag.12237
12. Reardon, T., Mishra, A., Nuthalapati, C., Bellemare, M., Zilberman, D.: COVID-19's Disruption of India's transformed food supply chains. Econ. Pol. Wkly. **55**, 18–22 (2020)
13. De Vos, J.: The effect of COVID-19 and subsequent social distancing on travel behavior. **5** (2020). https://doi.org/10.1016/j.trip.2020.100121
14. inancialexpress.com 2020 [online]. Available at https://www.financialexpress.com/industry/it-industry-may-become-lighthouse-for-indias-growth-heres-how-many-it-firms-operate-in-india/1870795/. Accessed Sept 2020
15. towardsdatascience.com [online]. Available at https://towardsdatascience.com/modeling-logistic-growth-1367dc971de2. Accessed Sept 2020
16. Montanari, A., Rosso, R., Taqqu, M.: Fractionally differenced ARIMA models applied to hydrologic time series: identification, estimation, and simulation. Water Resour. Res. **33**, 1035–1044 (1997)

17. timesnownews.com [online]. Available at https://www.timesnownews.com/business-economy/economy/article/how-the-covid-19-outbreak-has-affected-the-joblessness-rate-in-india-explained-in-4-charts/634284. Accessed Sept 2020
18. thehindu.com [online]. Available at https://www.thehindu.com/news/national/amid-covid-19-impact-indian-economy-forecast-to-contract-59-in-2020-un/article32675047.ece. Accessed Sept 2020
19. rersearch.fb.com 2017 [Online]. Available at https://research.fb.com/blog/2017/02/prophet-forecasting-at-scale/. Accessed Sept 2020
20. financialexpress.com 2020 [Online]. Available at https://www.financialexpress.com/industry/coronavirus-aftermath-over-70-companies-likely-to-continue-work-from-home-policy-for-next-6-smonth-says-knight-frank-survey/1965348/. Accessed Sept 2020
21. inancialexpress.com 2020 [Online]. Available at https://www.financialexpress.com/auto/industry/government-public-transport-indian-economy-delhi-transport-bus-delhi-metro-local-trains/2063737/. Accessed Sept 2020
22. mid-day.com 2017 [Online], Available at https://www.mid-day.com/articles/mumbai-news-metro-monorail-local-trains-overcrowded-commuters-increasing/18155192. Accessed Sept 2020
23. Dunung, D.P.: bWise: Doing Business in India-Chapter 6 (2015)
24. economictimes.indiatimes.com 2019 [Online]. Available at https://economictimes.indiatimes.com/jobs/3-day-weekend-good-for-you-and-your-employer/microsoft-japans-4-day-work-week-trial/slideshow/71920356.cms. Accessed Sept 2020
25. Goi, N.A.: “NITI Ayog,” Government of India [Online]. Available at https://niti.gov.in/content/sex-ratio-females-1000-males. Accessed Sept 2020

# Artificial Intelligence for Strengthening Administrative and Support Services in Public Sector Amid COVID-19: Challenges and Opportunities in Pakistan

**Kalsoom B. Sumra, Mehtab Alam, and Rashid Aftab**

**Abstract** The public sector establishments are increasingly interested in using artificial intelligence (AI) data science and capabilities to engender administratively and support services. The AI is identified for deepening the scientific knowledge in administration, support services, and addressing public queries. Other than the private sector, public services focus on the advancement of technology through AI's arrangement. Correspondingly, this chapter strives to view artificial intelligence within governance domains while intervening in public services and localities during COVID-19. It considers the factors that the study endorses with opportunities and challenges that disclose the prospects in the current scenario. The study aims to highlight AI's use in governance and support services through the online survey from the local departments and administrative service providers. The scope addressed in the study included the key state divisions, trailed by the National Health and Services, Science and Technology, Relief and Welfare departments, along with the key stakeholders in localities. The study examines the facilitative governance, shared objectives, imparting information, communication, socializing, AI expertise, and decision making over coronavirus governance. The results reflect a comprehensive strategy through artificial intelligence application to minimize the devastation caused by a coronavirus. It also invites public and private parties to support the Government in technological affairs of governing the crisis of COVID-19. The research contributed to further investigation in the tools that can better the state government services towards the ordinary citizen of Pakistan. Significant future implications are noted because of the challenges and opportunities received and incorporated in the study.

**Keywords** Artificial intelligence · COVID-19 · Public sector governance

---

K. B. Sumra (✉)
Center for Policy Studies, COMSATS University, Islamabad, Pakistan
e-mail: Kalsoom.sumra@comsats.edu.pk

M. Alam · R. Aftab
Riphah Institute of Public Policy, Islamabad, Pakistan



A.-E. Hassanien et al. (eds.), *Advances in Data Science and Intelligent Data Communication Technologies for COVID-19*, Studies in Systems, Decision and Control 378, https://doi.org/10.1007/978-3-030-77302-1_9

# 1 Introduction

The race for world leadership in Artificial Intelligence (AI) technologies is in progress of advancing and the governments are promoting the idea of research and development [1]. This is to encourage commercialization of AI and to catch up the leading AI nations across the world [2]. It follows swift changes and description of governance through various state analyses for enhance used of AI [3].As per the publisher standard For some states the comparison is ascertain and it received well-known response from their concern public and people in responding the challenges of coronavirus [4]. Maintaining the opportunity of social distance along with separation from the social events injected the developing countries towards vague and inconsistent governance role. For the reason of less AI awareness in these areas the services are not actively served to the victims of COVID-19 [5]. However, the case is not restricted to awareness,it is the adoption of policy planning in lieu of understanding of AI as a driving force in the digital revolution [6]. The incumbent era is inclusive of both potential and risk in terms of social, economic and, to some extent, security procedures of states [7].

Consequently, governments often leave aside any existing sector or digital strategies which is therefore be preceded by a description of AI concepts for people of shared interest [8]. The subject is not too specialized rather it is conceived of and anchored in a cross-sector approach with the lack of target systems [9]. Here the approaches are predominantly formulated in general terms and their partial and unclear objectives relate to different levels of impact on the service delivery during the COVID-19 is considerable [10]. It provides for governing bodies to overtake the more difficult areas of health sector to control the process of implementation and to measure the level of achievement [11]. For example, measuring the economic strength of the AI industry and other metrics, various states set targets for the controlling the spread of COVID-19 over the number of projects approved through the measurable targets [12].

The facilitative Government is something that needs to understand whether the COVID-19 leads to the desired effects or it requires a comprehensive mechanism that allows for the observation of global AI trends [8]. Such are the actions taken by the countries to inter-align other nations where the progress is at emerging stages. The utilizing of AI in a lower and restricted context is benefiting the social distancing and enhancing relationship among key stakeholder under single umbrella of National Command & Control Center (NCOC) [14]. The developed governments are using AI and adopted it with global integration while challenging superpowers of AI and confrontation of less-develop countries with the COVID-19. These challenges are enhancing and requiring certain implementation tools of AI. Those countries having leading position to adopt and followed the instruments of technology have strong corporate and independent sectors [15]. The dynamics of AI development are largely determined by the private sector, particularly young companies, and Internet companies with global operations [16].

The international trend in the use of AI is towards further deregulation, as it is seeing tendencies towards increased state control of the large technology companies for furthering operation under AI. This is essential during the times of crisis situation like Covid-19 which has slow down the worldwide distribution of AI instruments [17]. Same is the case with countries having more focused on science and the protection of the individual than to use other channels. Such global players partly are playing their role due to a very high degree of skepticism towards digital technologies [18]. It becomes a protective stance in governance where the support is aimed at the development of research networks along with pools of open data and technology transfer. It established manufacturing industry and SMEs which are prerogative and formative voice for coping the challenges of COVID-19 [19]. However, in divergence to the AI where the permeability between business and science has grown over the past ten years, the success of certain attempts is to achieve this permeability with limited approach. Scaling this may develop a constructive voice on the world stage, with governance which must point out ways to trigger a more-effective exchange during crisis events [20].

The separate fortresses of science and business project linked with economic powers are globally producing more AI against the production of digital society. For example, in the region of East Asia, the fourth industrial revolution in countries like South Korea and Japan which are considered as the inevitable next stages in human development [21]. Hence AI will have influence over the spheres of life that results from a constellation of specific economic growth targets and demographic limitations [22]. It is facilitating governance by a focus on research, commercialization and rapid scale-up of AJ applications in communication, decision making, socializing, and enhancing expertise of AI. The risks that are focusing too closely on manufacturing aspects, despite further specializing in basic research and the concept of industry need to be comprehended through the investment in upcoming cognitive age which could actively shaped it [23]. The digitally strong, qualified, and young population is needed specifically for countries like Pakistan as a home base where the success is to mainstream the society. The governance style must open itself toward the potential of this technology, both critically and constructively and quickly against the needs of pandemic situation [24]. The ethical and human orientation in communication and enhancing AI expertise is strengths of the developing countries [11].

These ethical issues are discussed in different social forums in with the topic is not given any high priority by the respective governments or this creates the space to take the lead on the issue of ethical AI [25]. Humans are central to world developments and it has been attempted by many with the initiative in developing an international AI governance architecture. Other independent movements and issues have weakened the country's ability to project AI developments. The challenges of COIVD-19 stems from the fragmentation of the internal data market that potentially could comprise of 200 million people [26]. The calls for infrastructural design of AI have hardly been accompanied by concrete suggestions to help shape future innovation of cognitive machines for both the developing countries digital market [27]. For other economic regions, it will be necessary to think outside the box and beyond the basic data-protection regulations the sense of responsibility can be conveyed more easily through digital economic and social control [28].

Major issue for the states like Pakistan is unavailability of computing capacities as an along with the availability of data and training of specialists. These need to be promoted by aligning social and operational roles of public sectors as a prerequisite for AI research and commercialization [29]. It must focus on the expansion of domestic computing capacities along with resolution of global trade conflicts that show that availability of powerful chips or access to cloud-based computing power. It is a strategic necessity as compared to the crisis of governance during the coronavirus where the developing countries are not competitive in this arena [30]. Further the ecosystems as a strategic asset can be based on developing higher quality AI solutions and implement them into the international dialogue. It will need experienced investors and an agile legislature in addition to its researchers, talented developers, and strategic businesspersons [31]. These parties can provide the fertile ground for successful commercialization of AI in Pakistan with significant support to the governance mechanism in times of coronavirus issues [32]. There can be the small number of AI startups which could bear the witness to the fact that such governance systems could be supported to the extent necessity of curbing coronavirus.

The public interests of support services and governance through artificial intelligence required for the achievement of following key objectives of the study. These include:

- To inspect the role of artificial intelligence for strengthening administrative and support services during COVID-19 crisis in Pakistan.
- To suggest the possibility of managing artificial intelligence for effective governance and communication over social distancing and isolation.

## 2 Literature Review and Proposed Hypotheses

Existing literature available for governance challenges during Covid-19 is researched and explained by scholars concerned to significant role of Artificial Intelligence. [33] provided that most influential role of AI, is estimated with 3,000-plus doctoral scholars in the area each year, with 1,400 AI initiatives with the ten global and largest knowledge firms. The cooperation among academies, public authorities and businesses has full-fledged during the last 40 years. Such combination of factors is the reason for global leadership in AI with increasing competition, and the strategic relevance for the economy and society. The [34] illustrate that administration must present the world's national AI strategy for governance model government which has yet to produce for challenges of COVID-19. This is in addition to funds from the private sector with elementary study that has finance through a key share of the administration's yearly AI resources. The finance from specific sections also cares the AI progress, with the spending 6.3 billion on AI. It is the Washington which supports the commercialization of AI by reducing and removing regulatory fences.

Other scholar [35] reviewed the system of governance during COVID-19 remains to be seen, whether the organizations inhibits the ability to attract AI experts or it is seen how different initiatives from the executive and legislative branches will form and shape new AI rules. As the AI comprises of three-step plan that can influence the world's leading AI nation by 2030. Certain measurable macroeconomic targets are set under AI. To achieve the purpose there are more than 700 million internet users and growing groups of high-performance hardware and technology companies are playing their role. According to the [36] there is still lags in the primary study, training of qualified specialists, and the number of AI startups and internationally enforced patents. But the developments in recent years leave no doubt that AI is catching up with advancement and promotion of instruments like the microchip industry and other used at the sub-national level.

Certain agreements according to [37] is based on the country's strengths where AI research is very influential and having more AI startups concentrated in one place than other. The governments set the foundation for the progress of proper procedures for AI, counting the establishment of a Centre for Ethics and a global AI Ethics Conference planned for 2019 with the commercialization of research, as expressed by the small number of patent publications, among other factors. This is considered a historic weakness as it is still unclear how the country will continue to attract scientists and entrepreneurs or it will compensate for its loss of influence in the region.

Further the scholar [38] reflected that the success in AI requires during COVID-19 is based on a broad spectrum of experiments in various fields of application with a striking trend is the small number of institutes and teaching staff. The actively conduct research in areas directly related to AI as well as the lack of cooperation between academia and industry requires a planned center of excellence will help bring together scientists to work with users in projects and scenarios. These areas must be extended with autonomous powers and certain new rules established for governance will allow researchers to work concurrently in the academic and private areas. The AI is examined in this review with incorporation of administration document that's specifically dedicated to work under AI domain.

For the filing of the most AI patents [39] show that other than AI there is considerable amount which have been made available to be spent on brain research. Here the findings reflect the expectations which lead to the next advance in the arena of neuronal systems and to allow quicker marketing of AI submissions. The intended companies are developing their goods to shop as administration consent measures are still in progress which could take place, for example, in the large-scale, networked like tech parks. The AI research is being conducted right next to industrial sites where the data and knowledge obtained will then be shared with startups and SMEs to help them develop new services. The technologies and promotion of governance knowledge with Institutional framework in situation of coronavirus is considered for data protection.

However, the ethical questions are gaining an importance among experts, as revealed by discussions about moral machines and stronger controls. This according to [40] corresponds to a general tendency towards centralization, as shown by the Government's attempts to control domestic technology. The spending in 2016 was only around 2.1% of GDP, in absolute terms spends the most money on R & D worldwide. The increased investment is more than three-quarters of research which are made by companies and it requires for governance role during COVID-19. It is concluded that there is there is a general ascendency of practical study where the boundaries between business and government are not altogether clear.

Further the impact of such publication's lags status as one of the "leading global AI hubs" is mainly the result of private investments in application-oriented R & D. This resulted from the government-promoted approchement of universities and companies which are as part of the reform and open-door policy allow universities to grant greater autonomy and public funding. In order to stimulate the acquisition of third-party funds, which led to a concentration on commercialization the preference of following swift financial aids, societal wealth and ranked position has condensed the inducements for investigators to start long-term elementary study.

## 2.1 Services Governance

The governance of coronavirus COVID-19 at this rule can have an influence on fiscal security for a huge number of individuals and same is case with this research. Despite available resources the public sector is reliant on these recipients, an important significance for a low-income state like Pakistan. AI is effective in raising a huge number of inhabitants out of deficiency where commended IT industry is already sensing the squeeze of robotics. This suggests that a disaster caused by COVID-19 fatalities could hit the people over the following insufficient time but it has lateral influence of AI which force to take lengthier and fail to reinforce the social distance.

## 2.2 Facilitative Administration

The understanding of administration system is reflective through a societal grading with historical origins of community. This endures to continue difference in dissident and unseen conducts, by moving salaries, service, captivity, and easy contact to praise from financial institutions. The dawn of AI has become a rising fear that data-driven procedures can prefer up prejudices from the data to facilitate, communicate, inform, and ensure the effective administration during COVID-19. [41] illustrated that technological cooperation with the various states is needed and it has to do without the centric and applicable approach for AI. It formulated a leadership in governance with the claim by striking the balance between the policies with the centralized structure and organization of the political system and especially in the areas of health, mobility and communication. The relevant ministries are using strategies and resources to stimulate AI applications which can only be questionable whether innovations can be centrally controlled in the long run or not.

*H1: Artificial Intelligence in governance is influence by the availability of health infrastructure and effective communication.*

## 2.3 Information Governance

For the assessment of information flow and exchange during COVID-19 there are suspected to show racial biases among the inter-departments of public services. The information is influenced by the businesses and it restrict the communicable exchange of information through stakeholders. The affective data-driven might be used to assess applications for patients and victims of COVID-19 where the experiment reflected that conduct of evidence of discrimination is not possible in case of applying AI applications. The researcher [42] unfolded that the network of voluntary AI experts is to direct the state on the requirement of technologies and support virtual safety as in Helsinki, it is clear that the big strings in AI development are pulled away. This is interesting feature which enable companies to purchase expertise from companies or researchers and use it to grow. The potential for AI beyond Finland's borders is obtainable in the health area, based on the Finnish genome project "FinnGen" along with the cooperation with IBM Watson Health along with other areas. This assumes as.

*It provides that H2: AI is the successful tool in executing decision making and opportunity of curbing Covid-19 Infections.*

## 2.4 Communication

The effective role of communication among provinces, health and security departments deduce that the choices were made by human assessors not learning machines. This is a reason for worry if these choices are ultimately used to guide a procedure for showing data by increasing the disparity among patients of COIVD-19 and those who are real victims. As increase in internet users and other media operators the influence of technology is central and expected to grow with easy interaction among stakeholders [43], stressed for ethical implications of AI which have so far lagged the economic considerations and the AI laws, regulations, ethical standards and security controls are going to be established in a step-by-step manner. For example, the Social Credit System planned for 2020, serves more to control society and avoid political and social conflicts than to protect data and it also reflect public opinion.

*H3: Communication among healthcare societies and entities is not effective through the AI challenges for Pakistan.*

## 2.5 Socializing

Social distancing and isolation are the pre-requisite during COVID-19 but the portable phones are the main contact point to the cyberspace, mainly in rural areas where 60% of Internet entree is through phones which support AI and seems an overall a boon for AI. The socializing remains slow if effective use of AI is applied. Here it must also be assumed that the economic objectives of the AI be achieved through government support for research and services of health care in pandemic situation. [44] reflect that in relation to the overall population, the number of scientists is comparatively low, and there's also a lack of AI researchers with 271 scholars have been performing active research in the fields of AI, computer vision, appliance knowledge, information withdrawal, usual linguistic dispensation. Such figures are still too low to fulfil the requirements to achieve leadership and governance in AI research which is measured by the number of scientific AI publications has already achieved global leadership COVID-19.

*H4: Community and public institutions are not collaborating for social distance through the AI application during COVID-19.*

## 2.6 Decision Making

Public services and governance are significantly dependent on key state decisions during crisis situation. Same is the case with AI which can support a better access and share the cyberspace with easy communication among stakeholders. The infiltration of AI informs the relevant public to identify cases of COVID-19 through which state could decide the intervention and tackle the crisis accordingly. Al et al. [45] explained that scientists, academics, and entrepreneurs living abroad and to facilitate international experts' have access to research institutions which reduced the bureaucracy and financial incentives. This has already lured many talents back and the scholars who have done their PhD gradations barely come back for governance and public sector services. Doing on the provisional basis and the agenda situations for study is more vital than the fiscal inducements like Alibaba, Sense Time and other technology companies which are responding to the dual issues. The one is the lack of able workers and inadequate basic study in a slightly pragmatic way & second is to setup up of their private billion-dollar study workshops.

*H5: Decisions are highly effective in the immunity support services and applicable to the infectious crisis situations.*

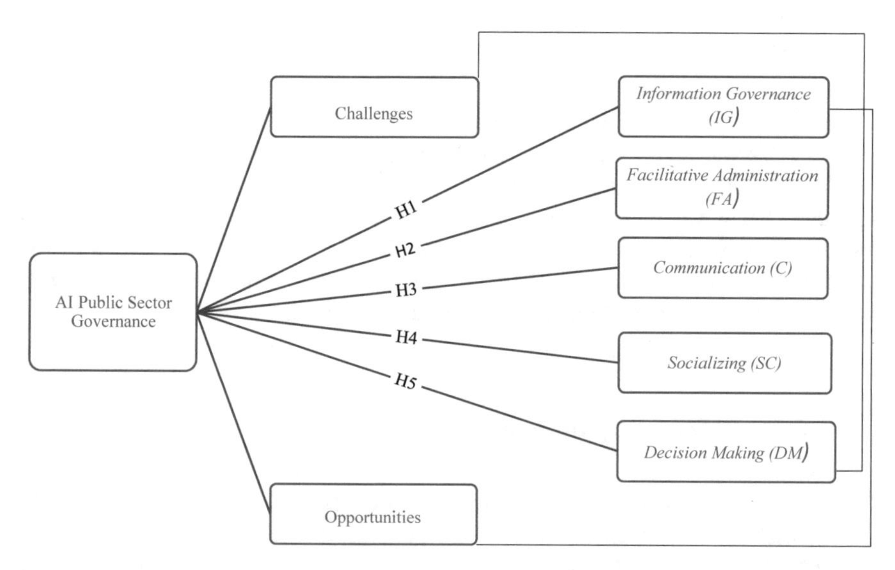


**Fig. 1** Conceptual framework. *Source* Author's Development

## 2.7 *Conceptual Framework*

The framework designed under Fig. 1 aims at the distribution of each construct, in their capacity the contribution for the investigation of AI. In public sector governance AI role is less where no facilitative governance, effective communication and easy socializing along with sustainable decision making is considered as weak areas. This framework highlights the relationship of these construct with each other and ultimate impact on public sector administration. Opportunities are presented as the way to achieve smart success and challenges are the restrictions posed by officials and key stakeholders with lack of infrastructure. Each item presented in the framework is contributing to the issues of main topic through the designed objectives.

# 3 Methodology and AI-Centric Approach

For the unconventional ways the creation of a climate of hope and optimism through artificial intelligence is significant. In this regard the conduct of survey from concerned departments and people aligned with AI is used in this research. The response has anticipated and it anticipate the possible dangers of AI-driven progress with the concerns that override AI.

### 3.1 Instrument/Scale Development

Through the online survey from the departments and sections concerned this research developed a likert scale based on the sections/categories of A, B, C, & D. Section A employ governance issues during Covid-19 and the response by state agencies. Section B considers the question relevant to cooperation and communication of provinces and key organs of state through AI. Section C increases the main concerns that arise from COVID-19 and socializing problems while section D included administration, information, and decision-making effects of AI.

### 3.2 Respondent Profile & Relevancy

Following the socio-economic context there is no exclusion to the worldwide AI wave which is commencement to evacuate governance issues through AI. Here the respondent is selected based on their choices and opinions, which follows strongly agree to strongly disagree options. Variables included are the coronavirus governance, facilitative administration, imparting information, communication, socializing, and decision making. The categories of the respondent against the variables and research objectives includes the development of likert scale questions grounded on variables followed by the research content & reflection of issues in the query's objects. There is comprehensible content for the Artificial Intelligence in governance of COVID-19 where suitable level of validity & reliability is categorized for impacts and use of AI participants. Further the data flow of general to specific questions with problem-based documentation and defendant approval and reviewed questionnaire is relevant to the study.

The participants incorporated or the online are based on the qualification and experience relevant to the part of AIi n governing COVID-19. People associated with the developments in applied knowledge and usual language boundaries and their role in public sector organizations. Against the variables the respondent participated in survey to reflect their opinion on the issues of artificial intelligence in context of governing COVID-19 challenge that requires a well-organized and advance response from public sector organizations. The profile incorporated the age, technological/AI education, involvement in science & technology, attachment with the support services and association with the theme of this research.

### 3.3 Sample & Data Collection Process

The data collected from the concern departments of National Health & Services, Science and Technology, Relief & Welfare, and Institutions related to public sector. The respondent selected under the William Gooden sample calculation formula of

2004 where the questionnaire is distributed among the 245 with return numbers of questionnaires are 210.

$$\text{Finite Population: } n = \frac{n}{\dfrac{1 + z^2 \dot{x}\, \acute{p}\,(1 - \acute{p})}{\varepsilon^2 \dot{N}}}$$

where:
n = sample size.
z = standard score,
p is population proportion,
N = population and,
€ is margin of error.

The distribution of the questionnaire is explained as workers from health care services with 50 in numbers. Same is the quantity for Relief and Welfare departments. However, there are 80 questionnaires served for skill based knowledge with precise emphasis on the role of artificial intelligence in public sector institutions. Maintaining the confidentiality of each respondent and effectively following the procedure is given due precedence. The participants have independent view over the subject and key issues of the research (Table 1).

## *3.4 Measuring and Analysis of Data*

For the investigation and gaining the outcomes following are the analytical tool used with exact and compulsive results against the responses and objects. The tools included the frequencies of demographic variables, pearson correlation matrix, regression testing model (Cronbach's Alpha) and reliability study.

The events of examining problem of artificial intelligence and its relation to the public section in combating the COVID-19 is based on the data collected through primary data collection method. The desire response from concerned stakeholders and after the gathering of data through questionnaire software and computational techniques are applied. The interpretation is explained through tested data on the SPSS software which is common and frequently used for quantitative analytical techniques. The data is verified through the correlation, regression, and examining descriptive. For predicting the effects of one variable on other there is used of

**Table 1** Data collection and sample

| Data collection process | Sample |
|---|---|
| Survey | Workers/Staff National Health Services<br>Science & Technology Department<br>Officials from Relief & Welfare Services<br>Representatives/Key Stakeholders Artificial Intelligence |

regression, while correlation is to see the nature of relationship among the constructs of artificial intelligence.

## 4 Results and Discussion

The participants responded and reflected against the questionnaire provides that AI in Government & public sector requires the architectural project, building, practice, and assessment of reasoning calculation. The mechanical knowledge here is to improve the organization of public activities, the decisions makers that project and implement the public plans (Table 2).

The related governance instruments along with public area is discovering the practice of AI with the most extensively recognized and adopted language. This replicate skills that enlarge human aptitudes and in performing both meek and multifaceted services during crisis of COVID-19 in Pakistan. Coronavirus governance is related with the terms including cognitive computing, robotic process automation and machine knowledge. Communication among provinces or key state agencies also followed through AI and it provides significant relief to the victims of disaster. The government agencies will be able to expand the abilities of their staff by dispensation (Table 3).

**Table 2** Administrative support services

| Reference | Statements | Measure Research | | Hypothesis |
|---|---|---|---|---|
| | | Dependent | Independent | |
| [12] | Artificial Intelligence in governance is influence by the availability of health infrastructure and effective communication | PSG | FA | AI (+) |
| [39] | AI is the successful tool in executing decision making and opportunity of curbing Covid-19 Infections | PSG | IG | AI (+) |
| [38] | Communication among healthcare societies and entities is not effective through the AI challenges for Pakistan | PSG | C | AI (-) |
| [7, 27] | Community and public institutions are not collaborating for social distance through the AI application during COVID-19 | PSG | SC | AI (+) |
| [25] | Decisions are highly effective in the immunity support services and applicable to the infectious crisis situations | PSG | DM | AI (−) |

*Source* Author's development

**Table 3** Analysis through Pearson correlation

| Description | Mean | S. D | AVE | CR | FA | IG | C | SC | DM |
|---|---|---|---|---|---|---|---|---|---|
| Public sector governance | 4.27 | 1.53 | 0.65 | 0.063 | | | | | |
| Facilitating admin | 3.31 | 1.64 | 0.54 | 0.052 | 1 | | | | |
| Info governance | 5.32 | 1.09 | 0.61 | 0.060 | 0.230 | 1 | | | |
| Communication | 4.01 | 1.41 | 0.53 | 0.079 | −0.064 | 0.321 | 1 | | |
| Socializing | 3.71 | 1.83 | 0.68 | 0.079 | 0.069 | 0.064 | 0.068 | 1 | |
| Decision making | 3.71 | 1.83 | 0.53 | 0.230 | 0.079 | 0.060 | 0.230 | 0.079 | 1 |

S. D = Standard Deviation, AVE = Average Variance Explained CR = Composite Reliability

Knowledge from large amounts of dissimilar facts has to be spread across varied schemes in response to administration facilitation.

*Hence the Hypothesis H1: Artificial Intelligence in governance is influence by the availability of health infrastructure and effective communication is accepted.*

The interaction with humans through natural language processing enable common man's cognitive systems at public places to allowed their staff from tasks that are suitable for mechanization due to their organized and expected landscape. Various examples are quoted that include the distribution of insurance cards, interest free loans, biometric payments, and other projects for victims through online services. These applications are designed in support and help of the citizens to navigate the easy communication and dispersion of information and updates related to the crisis situation. Majority of the queries resolved and progressed through the artificial intelligence support systems. The navigation of the governance during the crisis of COVID-19 in Pakistan is highly required for implication of the artificial intelligence. For the public sector departments there is increase interaction among pubic over the queries and issues of health care. While answering to the questionnaire few respondents reflect the less need of AI as society is not ready to accept the technological movements and advancing crisis through AI.

*This proves that Hypothesis H2: AI is the successful tool in executing decision making and opportunity of curbing Covid-19 Infections is accepted.*

The Government of Pakistan is processing millions of applications through SMS and services that requires fifth generation internet facilities with support for AI. Different types of applications are processed about with relevant the facilities of administration in dealing the travel documents, identity cards, and other subjects (Table 4).

The AI assist managements track the usefulness of policy interferences and indorse sequence alterations as required for various departments and cities in the country. The travel tracking and Government respond services are recently installed in collaboration with science and technology. The basis is to use Artificial Intelligence to transform Government, through the innovators and moving forward to implement AI in their programs and projects. It is highlighted in the promotion of healthier choices, and local and international examples of AI to recover rule-based

**Table 4** Regression analysis

| Description | Residual sum of squared | Degree of freedom | MS | Number of observation | | **210** |
|---|---|---|---|---|---|---|
| Model | 1772.373 | 12 | 443.093 | F (4, 210) | | 472.70 |
| Residual | 15,903.147 | 16,284 | 0.976 | Prob. > F | | 0.000 |
| Total | 17,675.520 | 16,288 | 1.085 | R-Squared | | 0.63 |
| | | | | Adj.- R-Squared | | 0.64 |
| | | | | Root MSE | | 0.998 |
| Public Sector Governance | Coefficient | Stander error | t-statistics | p-value | Confidence Interval [95%] | |
| Facilitating admin | 0.029 | 0.008 | 3.45 | 0.001 | 0.012 | 0.047 |
| Info governance | −0.501 | 0.014 | −1.76 | 0.000 | 0.530 | 0.472 |
| Communication | −0.008 | 0.007 | 2.09 | 0.277 | 0.022 | 0.006 |
| Socializing | 0.499 | 0.014 | 6.77 | 0.001 | 0.370 | 0.428 |
| Decision making | 0.498 | 0.015 | −7.87 | 0.001 | 0.371 | 0.429 |
| Constant | 0.199 | 0.059 | 4.35 | 0.202 | 0.826 | 0.316 |

application results for maintain social distance. The results from the responses thought leaders smearing AI in Government, demonstrating that Government can use AI to solve real subjects in bringing and contesting COVID-19 issues. The systems of AI are relying on reasoning computing competences to learn and grow even after they are organized, through absorption of additional information.

*While H3: Communication among healthcare societies and entities is not effective through the AI challenges for Pakistan is rejected.*

Public sector departments must focus on establishing information, training arrangements, and challenging performance and make instruments to collect, combine, and prepare data across manifold basis to fare. The appliances learning apparatuses and procedures can help identify designs in the dataset and build predictive models that human specialists can promote and test during national or international crisis situation. The correctness of explanations provided to known problem and solution in cases where the system gets the answer wrong, experts will provide the precise response, which will then be used by the structure in imminent restatements. The health services collaboration with public services can experiment the problem of patient care and future identification of virus through AI.

*The Hypothesis H4: Community and public institutions are not collaborating for social distance through the AI application during COVID-19 is true and acceptable.*

The effective use of limited health care resources through AI is vital to develop the system adequate and good-quality information that has balanced representation on the various elements of COVID-19. Main specialists to implant practiced data which normally happens by reviewing health cases where the system has gotten

**Table 5** Opportunities to help government move forward in critical arena of COVID-19

| AI in the public sector | Opportunities |
|---|---|
| Coronavirus governance | Investment in upgrading and modernizing the AI set-up |
| Facilitative administration | Inter-connected computing and clear cause as options to decisions/ management |
| Imparting information | Active power and acquired practices for IT |
| Communication | Recognize information rigorous issues and advantage of machine learning and cognitive capabilities, invest in data governance & to advance provision |
| Socializing | Change the public staff to take benefit of AI involve human experts in scheming, challenging, and assessing AI |
| Decision making | Progress cooperative companies with academia to initiate AI Plans corporation with public–private to plan, organize, and assess AI |
| Governance | Proactive monitor systems to track unexpected outcomes during COVID-19 with robust audit and inspection mechanisms |

decisions wrong or has given the impractical references. Further the AI-based instrument provide health policymaking or support with law enforcement calls for a much higher presentation threshold due to the potential for harm caused by a defective system during the pandemic. Such automated warning systems can attentive a trial about possible hazards and subjects (Table 5).

*The assumption H5: Decisions are highly effective in the immunity support services and applicable to the infectious crisis situations is rejected.*

## 5 Conclusion and Future Implications

The analysis provided over the governance organizations response through AI to events of coronavirus is based on the principles of protecting the patients by using advance technology. There are certain supervisory values for the construction of a healthy AI in Pakistan. The governments digital ideas are in progress and these are already effective for fulfilling common problems of coronavirus. However, the initiatives in lieu of combating COVID-19 requires enhance role of public sector departments through AI. Safety of the community information from governmental sectors is effective and be useful to create locally relevant public open sets pertaining to language, health, market places, and social distancing. For issues of effective communication AI skills such as processer system and crowdsourcing could themselves be deployed to seed the effort and it would neither be real nor maintainable if the activity of developing AI-based solutions is confined to a small individual. For the national response or wider unit of the people particularly subgroups, and rural communities be addressed by training to create and uphold AI systems for public requirements.

The instructive and clear errors can be corrected more quickly and efficiently by resident administrations and those associated with health care departments. For the information and the practical detail, the development and maintaining of communication, coronavirus governance needs an actively complement of top-down source of knowledge group and propagation within the context of AI. This is relevant to the progress of the AI facts and through the open source movement there has been reasonably successful in expanding the growth of AI reading room, values, and public sector training program like Seylani Welfare Organization has adopted. The compensations of consuming a recognized scheme and a well-trained workforce with the supply knowledge and skill are match for the demand of governance system during COVID-19. In context of this AI could stimulate new attentions to pursue occupations and is also able to the publication of stimulating facts and the government organizations of competitions. The centers of excellence in research could provide leadership not just in core AI technologies, but also in interdisciplinary areas where socializing, imparting information and decision making.

The recommendations and the future implications of the AI measures are to guide a big staff to build applications using image, language, and staff for public services. There would be an optimistic stage which may also help by absorbing some of the shock created by job fatalities or business. However, the new entrants will play a vital role in classifying and understanding the benefits of AI across diverse sectors as Pakistan has a flourishing tech free enterprise environment. There is entree to capacity, capital, and large markets with more than 100 + startups are focusing on AI. The presidential initiative with welfare organization for AI projection is reflecting the state response to advance technology. However, there is low services in comparison to response ratio of other countries. The investments in the AI requires enhancement with active system of operations. The facts and flair are both challenges that governments in developing countries will have to exchange. The earlier teamwork with universities could help in the latter aspect and it can be forced to retain danger low can focus on high-volume, low margin public entities. For the growth of AI in times of COVID-19, Pakistan needs to have supervisory machines with safety and quality standards; legal frameworks addressing data security, privacy, and obligation; and beliefs with governance evaluation commissions.

## 6 Annexure

Survey questionnaire
AI public sector governance

| Questions | Strongly disagree | Disagree | Neutral | Agree | Strongly agree |
|---|---|---|---|---|---|
| Do you consider satisfaction of effective AI implication is significant in curbing the challenge of COVID-19 for Pakistan? | | | | | |
| Did you expect the positive results of cooperation among federal and provincial sectors through AI management? | | | | | |
| People are satisfied in performing their activities during the COVID-19 | | | | | |
| Is there any Impact of improved coordination on Public Sector departments? | | | | | |

### Facilitative administration

| | | | | | |
|---|---|---|---|---|---|
| The reflection of positive administration through AI is significant | | | | | |
| Do you agree that AI effective communication can improve services during COVID-19? | | | | | |
| The affective AI is influence by facilities provided to public sector departments | | | | | |
| The standardized transport services can attract services to combat COVID-19 in Pakistan | | | | | |

### Information governance

| | | | | | |
|---|---|---|---|---|---|
| Is the implementation of AI impart information for effective services of Covid-19 | | | | | |
| Communication can be facilitated by support of AI | | | | | |
| The standards of health services are relevant with various AI tools | | | | | |
| Pakistan can develop tools to promote Ai in public sectors for better public management | | | | | |
| The use of social media can be effective in presenting AI awareness | | | | | |

### Communication

| | | | | | |
|---|---|---|---|---|---|
| The implementation of AI communication contributed for positive cognition of public service? | | | | | |
| Communication can be facilitated by support of AI public service organization | | | | | |
| The standards on of AI is relevant with current crisis situation | | | | | |
| Pakistan can promote instruments of AI in enhancing the affective management of patient | | | | | |
| Social media is effective in presenting AI tools to use in governance | | | | | |

### Socializing

| | | | | | |
|---|---|---|---|---|---|
| The social distancing implementation is easy through the tools of AI | | | | | |
| Social communication can be facilitated by support of AI | | | | | |
| The values on social isolation is relevant with various AI tools | | | | | |
| Public Service departments can promote AI tools through affective cognitive and public approach | | | | | |
| The use of social media can be effective in presenting AI socializing models | | | | | |

### Decision making

| | | | | | |
|---|---|---|---|---|---|
| Effective implementation of AI contributed for positive decision making | | | | | |
| Communication can be facilitated by support of AI for effective decisions | | | | | |
| The ethics of AI are helpful in determining the decisions right or wrongness | | | | | |
| There is immense potential of AI for effective decisions to curb Covid-19 management | | | | | |
| Social media can be active in presenting decisions for victims and effects of Covid-19 | | | | | |

## References


1. Renda, A.: Artificial intelligence: Ethics, governance, and policy challenges. CEPS Task Force Report (2019)
2. Winfield, A.F., Jirotka, M.: Ethical governance is essential to building trust in robotics and artificial intelligence systems. Philos. Trans. R. Soc. A: Math. Phys. Eng. Sci. **376**(2133), 20180085 (2018)
3. Cotter, T.S.: Research agenda into human-intelligence/machine-intelligence Governance. In: Proceedings of the International Annual Conference of the American Society for Engineering Management, p. 1. American Society for Engineering Management (ASEM) (2015)
4. Omar, S.A., Hasbolah, F., Zainudin, U.M.: The diffusion of artificial intelligence in governance of public listed companies in Malaysia. Int. J. Bus. Econ. Law **14**(2), 1–9 (2017)
5. Mahmud, F., Cotter, T.S.: Human-intelligence/machine-intelligence decision governance: an analysis from ontological point of view. In: Proceedings of the International Annual Conference of the American Society for Engineering Management, pp. 1–8. American Society for Engineering Management (ASEM) (2017)
6. Lauterbach, B.A., Bonim, A.: Artificial intelligence: a strategic business and governance imperative. NACD Directorship, September/October, pp. 54–57 (2016)
7. Zhang, B., Dafoe, A.: Artificial intelligence: American attitudes and trends. Available at SSRN 3312874 (2019)
8. Yu, H., Shen, Z., Miao, C., Leung, C., Lesser, V.R., Yang, Q.: Building ethics into artificial intelligence (2018). arXiv preprint arXiv:1812.02953
9. Fenwick, M., Vermeulen, E.P.: Technology and corporate governance: blockchain, crypto, and artificial intelligence. Tex. J. Bus. L. **48**, 1 (2019)
10. Casares, A.P.: The brain of the future and the viability of democratic governance: the role of artificial intelligence, cognitive machines, and viable systems. Futures **103**, 5–16 (2018)

11. Adams, W.A.: A transdisciplinary ontology of innovation governance. Artif. Intell. Law **16**(2), 147–174 (2008)
12. Sukhadia, A., Upadhyay, K., Gundeti, M., Shah, S., Shah, M.: Optimization of smart traffic governance system using artificial intelligence. Augmented Hum. Res. **5**, 1–14 (2020)
13. Board, S.A.: Law, Governance and Technology Series. Springer (2011)
14. Gasser, U., Almeida, V.A.: A layered model for AI governance. IEEE Internet Comput. **21**(6), 58–62 (2017)
15. Wang, W., Siau, K.: Artificial intelligence: a study on governance, policies, and regulations. In: Thirteenth Annual Midwest Association for Information Systems Conference (MWAIS) (2018)
16. Arnold, R., Buenfil, J., Abruzzo, B., Korpela, C.: Artificial Intelligence Ethics: Governance through Social Media (2020)
17. Dillon, S., Dillon, M.: Artificial intelligence and the sovereign-governance game. AI Narratives: Hist. Imaginative Thinking Intell. Mach. **333** (2020)
18. Cao, Y.: Research on the application of artificial intelligence in administrative governance. In: The 4th International Conference on Economy, Judicature, Administration and Humanitarian Projects (JAHP 2019). Atlantis Press (2019, September)
19. Almeida, P., Santos, C., Farias, J.S.: Artificial intelligence regulation: a meta-framework for formulation and governance. In: Proceedings of the 53rd Hawaii International Conference on System Sciences (2020, January)
20. Bhathela, N., Bennett Moses, L., Zalnieriute, M., Clarke, R., Manwaring, K., Bowrey, K., et al.: Response to 'Artificial intelligence: governance and leadership'white paper consultation. UNSW Law Res. Pap., 19–15 (2019)
21. Kerr, A., Barry, M., Kelleher, J.D.: Expectations of artificial intelligence and the performativity of ethics: implications for communication governance. Big Data Soc. **7**(1), 2053951720915939 (2020)
22. Mannes, A.: Governance, risk, and artificial intelligence. AI Mag. **41**(1), 61–69 (2020)
23. Winter, J.S., Davidson, E.: Governance of artificial intelligence and personal health information. Digital Policy, Regulation and Governance
24. Wirtz, B.W., Weyerer, J.C., Sturm, B.J.: The dark sides of artificial intelligence: an integrated ai governance framework for public administration. Int. J. Public Adm. **43**(9), 818–829 (2020)
25. Butcher, J., Beridze, I.: What is the state of artificial intelligence governance globally? RUSI J. **164**(5–6), 88–96 (2019)
26. McGregor, L.: Accountability for governance choices in artificial intelligence: afterword to Eyal Benvenisti's foreword. Eur. J. Int. Law **29**(4), 1079–1085 (2018)
27. Zhang, B., Dafoe, A.: US public opinion on the governance of artificial intelligence (2019). arXiv preprint arXiv:1912.12835
28. Reddy, S., Allan, S., Coghlan, S., Cooper, P.: AI in global healthcare: need for robust governance frameworks (2020)
29. Mahmud F., Cotter, T.S.: A systematic approach in decision governance: knowing from the existence. In: Proceedings of the International Annual Conference of the American Society for Engineering Management, pp. 1–6. American Society for Engineering Management (ASEM) (2019)
30. Marchant, G.: "Soft Law" Governance of Artificial Intelligence (2019)
31. Pagallo, U., Casanovas, P., Madelin, R.: The middle-out approach: assessing models of legal governance in data protection, artificial intelligence, and the Web of Data. Theory Pract. Legislation **7**(1), 1–25 (2019)
32. Guan, J.: Artificial intelligence in healthcare and medicine: promises, ethical challenges and governance. Chin. Med. Sci. J. **34**(2), 76–83 (2019)
33. Feldstein, S.: Artificial Intelligence and Digital Repression: Global Challenges to Governance (2019). Available at SSRN 3374575
34. Jackson, B.W.: Artificial intelligence and the fog of innovation: a deep-dive on governance and the liability of autonomous systems. Santa Clara High Tech. LJ **35**, 35 (2018)

35. Mazzini, G.: A system of governance for artificial intelligence through the lens of emerging intersections between AI and EU law. Digital Revolution–New challenges for Law (2019)
36. Sætra, H.S.: A shallow defence of a technocracy of artificial intelligence: examining the political harms of algorithmic governance in the domain of Government. Technol. Soc. 101283 (2020)
37. Maas, M.M.: Innovation-proof global governance for military artificial intelligence? How I learned to stop worrying, and love the bot. J. Int. Humanitarian Legal Stud. **10**(1), 129–157 (2019)
38. Dignam, A.J.: Artificial intelligence: the very human dangers of dysfunctional design and autocratic corporate Governance. Queen Mary School Law Legal Stud. Res. Pap. (314) (2019)
39. Walz, A., Firth-Butterfield, K.: Implementing ethics into artificial intelligence: a contribution, from a legal perspective, to the development of an AI governance regime. Duke L. Tech. Rev. **17**, i (2018)
40. De Stefano, V.: Negotiating governance and control over AI-bosses at work-Position paper presented at OECD Network of Experts on Artificial Intelligence (ONE AI) 26–27 February 2020 (2020). Available at SSRN
41. Li, G., Deng, X., Gao, Z., Chen, F.: Analysis on ethical problems of artificial intelligence technology. In: Proceedings of the 2019 International Conference on Modern Educational Technology, pp. 101–105 (2019, June)
42. Pupillo, L., Ferreira, A., Fantin, S.: CEPS Task Force on Artificial Intelligence and Cybersecurity Technology, Governance and Policy Challenges Task Force Evaluation of the HLEG Trustworthy AI Assessment List (Pilot Version). CEPS Task Force Report 22 Jan 2020
43. Sharma, G.D., Yadav, A., Chopra, R.: Artificial intelligence and effective governance: a review, critique and research agenda. Sustain. Futures, 100004 (2020)
44. Zeng, Y., Lu, E., Huangfu, C.: Linking artificial intelligence principles (2018). arXiv preprint arXiv:1812.04814
45. Al Khamisi, Y.N., Hernandez, J.E.M., Khan, M.K.: Measuring the impact of healthcare governance on its quality management using artificial intelligence. In: The World Congress on Engineering, pp. 307–315. Springer, Singapore (2018, July)

# Artificial Intelligence in Healthcare and Medical Imaging: Role in Fighting the Spread of COVID-19

**Maryam Mohamed Zainal and Allam Hamdan**

**Abstract** The research aims to define the need and importance of artificial intelligence in the healthcare sector in general and the medical imaging and radiology procedures in specific. This research is based on numbers and facts taking from other investigations about artificial intelligence in general and the present and studies of the healthcare sector's current Artificial Intelligence (AI) role. To simplify, the chapter will focus on the AI in healthcare facilities and how it can help solve problems and make the best decisions for the organization and public health by reducing human errors and increasing the perfection of discovering some diseases. Moreover, to achieve the medical providers' goals that aim to achieve the quality of care, we are going to talk about applying the AI in the radiology department. It's a role to help the radiologists in better diagnosis and how it could increase the efficiency of the operations by using the database of information gathered by the modern techniques. On the other hand, this article and according to the current global situation will talk about fighting the spread of Coronavirus with the new Medical Imaging technology in general and in Bahrain's society in specific.

**Keywords** Artificial intelligent · Healthcare · Technology · Radiology · Coronavirus · Public health · Patients · Medical imaging · COVID-19

## 1 Introduction

Up to the 1950s, artificial intelligence was still unfamiliar and unknown until 1956, when a classical philosopher adopted a Dartmouth College conference in Hanover, New Hampshire. When we ask about the definition of artificial intelligent future, it comes to mind that the robots and machines will take place in our lives, where humans will sit comfortably watching it doing their jobs and taking care of their

M. M. Zainal
Al Hilal Hospital, Manama, Bahrain

A. Hamdan (✉)
Ahlia University, Manama, Bahrain



A.-E. Hassanien et al. (eds.), *Advances in Data Science and Intelligent Data Communication Technologies for COVID-19*, Studies in Systems, Decision and Control 378, https://doi.org/10.1007/978-3-030-77302-1_10

responsibilities in different categories. Still, could robots take place in the hospitals and physician's lives to take responsibility for operating and managing the medical practice? [1, 2].

Between all definitions of artificial intelligence, the medical dictionary and environment gave this simple one. It defined AI as a technology that enables the database system and software to decide and participate in healthcare decisions and opinions to reach the perfection beside the physician input and judgments. It's a simulation of intelligent human behavior in Healthcare facilities [3, 4].

As all definitions in the medical dictionaries can conclude that the AI in healthcare is all about the remarkable ability of the modern technology and computers in our lives and health sector.

Some of the research goes beyond asking if artificial intelligence could create an information base with all patient's health experiences to serve others in a better way without human errors [5].

When we hear the word Artificial intelligence, it comes to our minds a very futuristic thing in a very modern and technologized city that can be watched only in scientific films. After some time of thinking, we come back to our lives as the AI future opportunity in the medical sector still an unanswered question that needs to be more asked and search for it "Will, it's still in the future" [6].

As per some previous research, the database and gathering of the AI information in healthcare can be applied successfully on pathology, dermatology, and image analysis (medical imaging) in Radiology, which will focus on it more deeply in this chapter. All these practices will be affected by the AI with a diagnostic speed exceeding and medical experts to reach the 100% perfection system performance besides the physician knowledge and the knowledge management system applied in the healthcare sector [7].

As has been mentioned above, in this research, the main focus will be on medical imaging and the effect of artificial intelligence on the Radiology services improvement in the healthcare sector and how it can help the doctors and the patients to discover and define the early stage of any diseases that could be missed during the human practice and the presence of human errors [8, 9].

AI and machines will not replace the physicians; oppositely, it will help them make better medical decisions and replace human error judgments in some functional regions in healthcare by unlocking the clinical information database, which will help assist the medical practice. This research will go through the future and present of the AI in healthcare and its future.

Will first briefly review the four main relevant aspects for the medical investigation and the AI usage [10].

1. The impact of AI in healthcare motivation by providing the physician with up-to-date information and feedback from a big base of papulation and from all over aspects that can encourage and motivate them in taking a better decision for them and for the patients.

2. Healthcare data can be available worldwide for all medical practitioners, especially for the diagnostic procedures done by the radiology department in different kinds of modalities and machines.
3. AI devices, according to a previous study, can be divided into two main categories (ML and NLP) that will be discussed more it later.
4. Diseases types that the AI can help in defining and discover like cancer, nervous system diseases, and cardiovascular system disease (all will be discussed later in more details).

Still, the raised question is, what is the difference between artificial intelligence in healthcare practice and normal technology?

We can briefly say that artificial intelligence is all about gaining the information, re-processing it, and giving back in the end-user to be used in decision making.

Moreover and during the global situation that more than 130 countries, including the Kingdom of Bahrain, are facing currently with the rapid spread of the Coronavirus all over the cities and between people, we are going to add to the research the impact of Al in fighting against the COVID-19 virus and the impact of using the Medical Imaging in fasting the discovery of the virus by using the most familiar imaging technique between people (the Chest X-ray) and other techniques.

## 2 Literature Review

As part of the global crisis we are facing, the Kingdom of Bahrain is trying to be up to date with the new technologies and Artificial intelligence tools used in the war against the Coronavirus. The most shown Implementation in Bahrain is the websites and mobile applications that have been created by the government to keep all the society aware of whatever we are facing.

In this chapter, we will focus on Artificial intelligence in Healthcare in the first part, then will go through the Medical imaging specific in the second part, ending the research with the role of Al in fighting the Coronavirus.

### *2.1 Artificial Intelligence in Healthcare*

To define Artificial intelligence, we should know that the physicians use the growing body of knowledge to say which diagnosis and treatment to be considered in dealing with the patient.

Artificial intelligence can be concluded as intelligent aspects and agents taking the surrounding situations to take actions and decisions to maximize the advantages and increase efficiency in reaching the goals and objectives. As a fundamental science, artificial intelligence is used to enhance modern technology and computer systems and science to utilize the knowledge for best use.

## 2.2 The Current State of AI in Healthcare

Artificial intelligence, with its capability and advantages to analyzing the significant amount of data to small information and easy to access, is one of the essential processes are being used in medicine in general and the healthcare sector in particular, especially when there are scenarios and situation that the medical practitioner cannot take a discussion in it without the reference to the big database of information. They are putting in mind that the pressure and timelessness in the medical practitioner's work life, especially during the catastrophes happening in the world just like what the essay will talk about it (COVID-19) [11].

Important figures can be used in healthcare to switch the artificial intelligence key to a useful health sector tool. Some of the new systems and programs offer a question–answer tool for all public based on hypothesis generations and evidence taken from medical professionals and decision-making in the health sector. The two main goals to be reached while converting AI to HC tool is to reduces the pressure on medical practitioners and to increase the people awareness and knowledge about the different type of diseases and keep them up to date, in Coronavirus's situation now, the advanced technology and AI use in the health sector reduce the healthcare facilities visits to more than half of the numbers, which will automatically reduce the number of infected people by lowering the public's admixture [12, 13].

As a base of professional information that can be allowed for doctors and patients, this will help both people by using the telemedicine and trustable applications.

## 2.3 Point of View of AI in Healthcare

***Optimizing the architecture***

According to [12] AI usage and the new technology will be part of the routine of each healthcare facility shortly, starting from answering the medical questions for patients going through suggestions and ending with giving treatments based on the explanation and negotiations will be taking place between the system and the patient [14, 15].

As part of global development, according to [16], companies are facing the problem of diversity of the base of the artificial intelligence that is being used in the companies, which cannot fit the traditional and culture of people in the companies.

Another problem facing the AI in this time, especially for the large companies with a large number of accesses and servers, is that many databases and information and the way to store it in safe storage from any changes and reachable at the same time.

As part of this procedure, a website has been distributed among the public since the start of the Coronavirus crisis in the world, the website is based on some questions for the people, and according to it, they suggest to take advance concern and contact the responsible governmental side in the country, or consider it as the normal flu or any other disease that can be treated by normal medications at home. Still, the raised question is it affected and accurate to trust? [17, 18].

***Social and Safety aspects of using Al in Healthcare***
As a new technology being used by people, defiantly there will be some problems, and a reduction of people responded. As the main problem facing the global development (Al dangerous) word still in mind especially for non-IT and techniques people, this problem has been suggested to be solved by raising public awareness and targeting the development and learning of the governments to help in awareness and train the public later.

Another problem facing the development of Al in modern countries is the privacy and safety of information are being shared in the servers and networks and the limitation of the access to these details [19].

As a healthcare user, saving the documents related to the history of diseases is a very sensitive goal, saving these data from any misusing or any changes to make sure that all information related to treating the new cases will get to the right person with every single letter as per original basis. Al in healthcare is directly related to the lives and health of the public that cannot play in it in any way, therefore, devices must be protected from external reliability.

Still, the raised question is how Al can be helpful in the healthcare sector?

To answer this question, Al is being used by the doctors and patients to help them in reaching the related case data and information. On the other hand, some are expecting to replace doctors with technology soon. Still, replacing the human with technology is way complicated to be a concern, the human body and the sense of feeling between the doctor and the patient are too complicated to be known and practice by a machine [20].

## *2.4 Current Researches*

The difference between the Al technology from the traditional technology in the healthcare sector that Al can gain, process, and give the user the final discussion according to the information stored in it by doctors and professional people. Still the problem here is that Al and the modern systems cannot take any urgent situation from outside the database stored in it, this will put the users in a position to switch off the technology and get back to use the human mind to deal with the urgent threatening of the human lives.

Here we will discuss some of the useful uses for Al in the healthcare sector.

**Radiology**: A study of [21] developed an algorithm that could find pneumonia in the Chest X-ray at any specific site of it in the chest. This new technology in detecting the simple diseases in one of the medical imaging modality can reduce the timing being used in checking each image, reduce the missing errors can be by the specialist Radiologist between the large number of images being seen every day, increasing the accuracy and efficiency. All these will guide the most important goal to reduce the recall cases in the hospitals and clinics, which will reduce the radiation exposure to the patients and medical staff. Still, according to the annual meeting held in North America, some of the specialists are concerned about replacing them with uncompleted technology.

**Imaging**: In 2018 a paper published by [22] stated that skin cancer can be 95% more accurate than dermatologists with 86.6% accuracy. This can approve that some of the scanning and imagining are more accurate with the help of artificial intelligent technology, using these technologies cannot replace the human being in the health sector, but can help them to reach perfection in determining the diseases and taking care of human health and lives.

**Diseases Diagnosis**

Being accurate in the diagnosis of diabetes and cardiovascular disease has taken place in many recent kinds of research. Modern technologies are being developed to help find the causes of these diseases as they are categorized as two of the ten top diseases that lead to death in the world.

An article has been published by [10] about the multiple types of Al being used in the Healthcare sector.

Over the years between 2008 and 2017 several types of research and studies took place to determine the best technology in Al to define and used in Diseases Diagnosis, still, the field is too big to end it and all studies are being re-concern every day to conclude it and find the solution.

**Telehealth**

Telehealth or Telemedicine started taking place all over the world. In the Kingdom of Bahrain and as a small country, the only one Radiology Centre in the Kingdom it's working is based on the Telemedicine used by, first outsourcing the Radiology departments in the small clinics and hospital to reduce the expenses and cutting the cost of the operations cost and the salaries of Radiology staff. Secondly, they are putting their Telemedicine devices instead of doctors all over the branches and start reading all types of medical imaging examinations are being done from the main branch, so the idea is instead of putting one doctor in each branch, a bulk of doctors are setting in the reporting center, reporting for more than 20 centers and hospital in the same time.

This idea is a very modern application of Artificial intelligence in Bahrain and in medical imaging, specifically agreed and approved by the governmental sides [23].

**Drugs Interactions**

As chronic disease patients, many people aged more than 50 years are talking more than one medication type. To avoid interactions of medicines, they created an Artificial intelligence system that can collect the information about the drugs and warn any medication's user to not fall in the error of interaction of drugs. This can be more accurate and efficient than depending on the human mind in memorizing each medication's consequences.

**Creation of new drugs**
In Jan 2020 BBC News talked about a medication for (obsessive–compulsive disorder) that has been developed by artificial intelligence machines within one year, and it's ready to be used by humans. Normally the companies and pharmaceuticals take around five years to produce such the same drug before publishing it and starting the human trail.

**Others**
As part of the COVID-19 war in the Kingdom of Bahrain, some physicians switched their actual clinics to digital consultations. This is an ideal application of Artificial Intelligence used and talked about by [21]. This is to reduce the number of visitors in the clinics and prioritize the appointments to urgent and chronic cases. The Application is based on entering the patient medical history and syndromes. Using the database and online consulting for the doctor's treatment is being given and rescheduling an actual appointment for the sever cases.

## 2.5 *Implication*

As part of modern technology and development in the healthcare sector, the raised question is how Artificial intelligence use in the healthcare facilities and application will help human health and lives?

To answer this question, first, we have to know that any use of modern and developed technology in any field in life is focused on reducing the cost and time wasted. More accurate prediction and treatment of any diagnosis will automatically reduce the timings of the patients' visit. They will reduce the cost of trying a different type of treatment every time.

Another side of using the Al in healthcare is that scientists expect to develop a modern Al technology to help the patients with talking and moving problems (Brain diseases) by preventing the causes and developing the Bain computer interfaces. The technology will help this category of people to talk and communicate by using the neural activities and converting it into useful activities [22].

The U.S. News staff wrote about the fair in replacing the doctors and human being with robots and machine, as the idea still not clear for public, the Al applications will help in the first step of diagnosis the patients to reduce the pressure on doctors, in the other hand they will focus more one-bed step and following up with the patients.

## 2.6 *Expanding Care to Developing Nations*

The developing nations are the countries will behind technology and fewer opportunities in different aspects of life. Focusing on the healthcare and medical sector in these countries can lead us to use Al in the world to help them improve and learn remotely. By treating and teaching them, doctors can diagnose the diseases faster and in more comprehensive ranges besides improving patient care [24].

## 2.7 *Disease Diagnostic and Prediction*

The rapid development in Artificial intelligence technology reflects the need for urgent techniques that can help in the diagnostic of different types of chronic and dangerous diseases. A variety of essays and research papers took place in this area to give accurate and early diagnostic of the different types of diseases [25].

Al can help in preventing and discover a different type of diseases such as.

- (Gene Expression)—in this century and with the technology and developed tools are being used in the healthcare, Al plays a very important role in interrupting the microarray of the Gens to detect any abnormalities any example that can be found in most of the modern countries is the IVF, where doctors are interfering in Industrial insemination to change the abnormalities in the gens like the sickle cell disease or just to change the color of the fetus eyes.
- (Cancer Diagnosis)—the cancer cells are about a tortuous cell with extra pulps and nodulous that can be seen and defined by different medical imaging types. With the modern techniques of Artificial intelligence, these cells can be determined by the advance technology which can be missed by the human due to human errors [20].
- (Cardiovascular diseases)—again the artificial technology can help in preventing and reducing the heart attacks be defining the places and sites of the thrombus in the blood before reaching blocked veins in the human body
- (Medical Imaging)—instead of wasting half an hour of the doctor's time, the artificial intelligence applications can do the (ECG) and determine the sites and any abnormalities in the reports [25].

Overall, Artificial intelligence has been recently involved in the management and development of a different type of the ten top causes disease and classified as the basic element in helping the physician and patients in a different type of applications (like the mobile phone application being used now in the Kingdom of Bahrain during the (COVID-19 war).

## 2.8 *Financing the Healthcare Application of Al*

As one of the most important technologies has been developed and used in the healthcare sector and concerning human health and lives, investors are competing for each other to invest in the development of diverse applications in healthcare.

In association with encouraging the use of artificial intelligence in medical practice, the [24] published a white paper to explain the impact, usage, policies, and the application that can be used to improve the impact of Al in medical imaging.

As per [24], many Al applications have been discussed, including the nursing assistance inpatient care and surgeries, helping in the administrative work, medical records, improving the diagnostic accuracy, and dosage error reduction by reducing medical imaging recall. Much more applications can be applied in healthcare depending on developing these technologies and medical staff training.

This new field in medical sector created a market and competition between the investors from entrepreneurial and corporate institutions 'recent research estimated that over 50$ billion USD has been invested in Al start-ups since 2011 till the first half of 2018, in a percent of 12% attacking the worldwide investments [26].

## 2.9 *Sustainability*

Sustainability in technology is difficult to control, especially in the healthcare sector. Whatever data, information, and technology will be saved and developed today will not be useful within 20 years from now, countries' investments are needed in this sector.

as one of the most critical moments in Artificial intelligence history in healthcare services, development of technology still going on, and trying to put the steps in the right path to design a powerful piece that can depend on it in treating and dealing with human health and lives [27].

Still, sustainability is ongoing during the changes in global situations and the patient's needs.

## 2.10 *Emphasizing Diversity*

Emphasizing diversity in public worldwide should take in mind while developing artificial intelligence systems in healthcare. The nature of the human being and the environment in Asia from example is different from the one in South Africa. The type of disease, treatment types, education, investments, and the technology's availability will be diverting from side to side. According to a study adopted in 2019 (Medical innovation and developing of heart attacks diseases discovery has been focused on male patients only, in fact recently has been approved that females are getting the same heart attack but with different symptoms. All these should be taken in mind while developing the techniques of artificial intelligence [28].

## 2.11 Limitation of Al

Still, the medical staff and healthcare facilities managements have the fair of replacing them with robots and machines. This thought and assumptions should be discussed with public. Artificial intelligence is still a question mark for most of the concerned people.

The quick development and moving in the Al technology will question the public if they are being ignored and replaced in the medical facilities. Are we going to be examined and treated by machines and computers?

As [29] said 'this emotional cost of moving forward technologically should not be underestimated. Yet the limitation of Al is still not underestimated and control and all healthcare workers that will be involved in this should be significant with the changes and well trained. It should be clear for medical staff and the public that this tool is a useful technology in different aspects starting from hospital managements, diversity in information and data, reachable details all over the world and ending with patients' advantages in fasting the diagnostic procedures and verity of options in treating them [3].

## 2.12 The Physician–Patient Relationship

The physician–patient relationship based on ethics and trust is the major decision in choosing the right doctor or the specific hospital/Medical center. How could we save this trust and relationship between humans and machines developed by Artificial intelligence?

To answer this, the patient is searching first for the doctor's Serenity before starting the physical treatment and diagnosis. Replacing the medical workers with machines will destroy this relationship.

Doctors usually and by the social life refereeing the patients from one specialty to another in the diagnosis process, with the use of applications and machines this advantage will be deleted from the process where the patient should ask and try by themselves.

In the hand, the medical issues and management in the healthcare facilities will be partially solved, but yet the medicolegal issues will increase, who? An error in diagnosis or prospection of medication will cost the health of the public and lives, who will be responsible for it if the machines were the only involved part?

## 3 Artificial Intelligent in Medical Imaging

As a high-speed developing path in medical care, Radiology has been taking a very advance stage in rapidly growing and increasing the data basis, with the fast developed, Artificial intelligence took a place in the medical imaging to catch the growth during the lack of the number of readers, radiologists, and medical staff that are specialist in radiology and imaging reading.

With the force to increase the productivity in the healthcare, the pressure on Radiologist has been increased, the average has been reduced to 3–4 s per image to interpret in it, with 8 h duty for the radiologist. In compensating with visual perception and decision making discussions and under this uncertain pressure the human errors percent is increased in a very notable way [24, 28, 30].

By Compensating the Artificial intelligence with the human effort, it's obvious that we can note the efficiency increment, errors reduction and achieve objectives with minimal interrupt by the human, to give the chance for the radiologist for more focus and interfere in the image reading in a way that will help the patients and the human health.

Two methods are being used in radiology to utilize the use of Artificial intelligence in medical imaging, first one is the term of mathematical equations usage (Ex. Determine the tumor texture and feature being shown in the images) [29] the computers can automatically use this with the bulk of images to give the reader or the radiologist more chance for more accrue diagnosis and decision making.

The second method has been focused more on it in recent years. The deep learning method is a way used by an automatic configuration of the algorithm being done without any human interruption and experts. By directly noticing the changes of the human tissue the errors percent of mess diagnosis or any unclear features for the radiologists will be reduced [31]. Putting in mind with the rapid increase of the methods and the usage of Artificial intelligence in medical imaging, we can see that the radiologist and specialist technologist can be equalized in some of the modalities images reading, costing a reduction in time-consuming and giving more chance to avoid errors and to see more patients [32, 33].

### *3.1 Impact on Oncology Imaging*

To be able to fill the gap of determining the oncology diseases (especially the most common between people the Cancer) you have to be familiar with three main topics Abnormality detection, Characterization and subsequence monitoring of change or to bring a team of people with all knowledge related to it, to can determine the diseases accurately.

1. **Abnormality detection**: in the normal steps of detecting the abnormalities and findings, the radiologist is using the normal steps they learned and dictionaries to start finding and either confirming or rejecting according to their skills that are

relying on education, experience and understanding of the patients' health and the environment and society with lifestyle coming from it and then make the decisions that are usually made by the specialist or consultants based on the appearance of any abnormalities and unusual patterns [25].

2. **Characterization**: categorizes the stage and age of the diseases and tumors by determining the segmentation and diagnosis (such as size, extent, and internal texture). Humans cannot count or find some of the characteristics and changes in the tissue due to the limitation of brain work and the number of cases seen by each reader every day. For example, it's different for the human being to define the difference between malignant and benign tumor seen by CT scan, and preferably to take a biopsy for more accurate results, this can be cost more money, time and consume the patient power, all this can be solved with modern technologies and usage of Artificial intelligence [32].
3. **Monitoring**: As a tumor monitoring to define the stages and developing of diseases like Cancer, any small error can cost the human life and public health, the Artificial intelligence usage in these cases can put a comparable image between each period and give a specific diagnosis and measuring in a way that will save the patient health, and to be ready in case of any urgent response needed.

## *3.2 AI Challenges in Medical Imaging*

Due to the fast development in public health in general and Radiology imaging in specific and in the types of a new disease introduced every period (Ex. The Coronavirus), it's difficult to catch up with the updates keep the artificial intelligence technology up to date. Now the world is facing the Coronavirus with no ability to find any medication or solution to kill this disease, still, the studies are slower than the virus development [32, 34]

Another problem is facing the artificial intelligence in healthcare. Some of the techniques require the interfere of human expertise, which will not be available full time to attend the start of technology practice [35, 36].

In the case of multiple tumors or mass, the technology cannot mark all the normal up findings in one time, which will cause a technology error that will affect the disease's discovery percent. Using the (PACS) Picture Achieving and Communications system ensured that medical imaging is electronically organized in a system that can be destroyed by the loss of internet connection and internet hackers, and result in losing all the images if no proper storage has been done [22].

### 3.3 Future Perspectives

From the early days of discovering the X-rays imaging in the 1890s to the more recent modalities are used in healthcare such as CT, MRI, and PET CT, medical imaging is developing to keep as a pillar for the medical treatment. It is a basic advantage, enabling the differentiating between the densities in the soft tissue and discovering the up normality that the trained eye cannot recognize. With the new algorithm scales and recent development, the use of artificial intelligence promised to be more accurate and see a relative improvement in performance that will increase the health services given to the patients.

## 4 Role of Medical Imaging in Fighting the Coronavirus

COVID-19 or its known Coronavirus disease is infectious that attacks the lungs or the respiratory system directly, were it cause a severe acute respiratory syndrome. It has been announced as a global pandemic in 11th March 2020, after reporting the first case on late December 2019 [31].

Since the first report of the COVID-19 case in Wuhan, China, the competition has been started between the technology companies to build teams with the clinicians, academics and government entities to fight together in the war against this new virus and to stop it from spreading after reaching more than 100 countries with millions of cases all around the world.

In the last part of this research, we will talk about the use of artificial intelligence in this war and some of the ways they are using in managing.

- AI is used to identifying, track and forecast outbreaks by using a new application has been built by Kingdom of Bahrain government, they are tracking the suspected cases by allowing a full location share from each citizen, and give a warning for any of the public that has been in contact with the positive cases in the last 14 days.
- AI can help in diagnosis the virus. In the second part of the lecture review in this research, artificial intelligence is being used in diagnosis the COVID-19 symptoms in medical imaging due to the overload on the Radiology departments in this crisis. Still, this has not been applied in Bahrain, and a study is going on it at Gulf Medical University to start using the portable CT scan to scan the patients faster than waiting for the Lab results [30, 37–39].
- Processing healthcare claims, the new technologies can be offered to the administrative parts of the hospitals and healthcare facilities and the medical staff. New insurance companies in Bahrain start using the E-cards and E-claims to reduce the direct contact between the public or the patients and the staff in the back offices. As part of reducing direct contact, some healthcare facilities and business centres, in general, offered the non-wallet payment method, which can

give the customer the chance to pay directly from their mobile phones without using the cards or tangible money [38, 40].

- Using Drone in delivering the medical supplies and to keep an eye on the streets and gatherings more than 5 as per the Bahraini's government put the new rules. Drones can be the fastest and easiest way to deliver the medical supplies without the need for direct contact between the sender and receiver. It can be a good option in countries that apply the quarantine to the public during the day-hour technique applied in Bahrain [3].
- Using robots in cleaning and sterilizing the streets and suspected places with the spread of the virus as a precaution from more spreading to the virus, as the robots are not containing soft tissue, can be the perfect place for the virus to grow.
- Developing the drugs can be done using the algorithm scales to discover the protein type and whatever is best to destroy it without the risky part of dealing with the virus directly by the lab technologist and the scientists.
- Temperature detection technology is used to detect the human temperature by capturing it in seconds with the need for direct contact or reducing distance to measure it. Bahrain's government applied this technique months ago in the entrance of the Kingdom on King Fahad causeway and Bahrain international airport as the first precaution system for the new coming travelers that entering the Kingdom.
- Online consultations have been approved in Bahrain by the national health regulatory authority, to reduce the working hours in the private medical centres and decrease face- to-face contact especially for the front-line medical staff that is the government in need to their effort in this war [33, 41].

Yet, the CT scan medical imaging is not approved as a way of COVID-19 diagnosis, as the images' findings can be related to another respiratory system disease. In general, during this crisis, the governments and public noticed the importance of using the new technology rather than healthcare technology only to fight against it and protect public health [42].

Still, the question is being raised, could artificial intelligence technology fight against the spread of COVID-19 and win the war?

## 5 Descriptive Analysis

This section will analyze demographic data collected via a questionnaire survey (which is the primary information and the main tool in the study), to collect the bulk of information such as age, gender, education, etc. The survey is consisting of multiple-choice questions to a random group of people that have received the survey by social media channels like (WhatsApp and Instagram).

To address the key research objectives, multiple choice has been divided into mandatory questions for all participates talking generally about the population knowledge and awareness about the COVID-19 in Kingdom of Bahrain, and specific questions targeting the medical students and staff only with medical backgrounds.

The study population consisted of random people from Bahrain's society with different backgrounds and education levels as the questioner was targeting the COVID-19 updates in Bahraini's society which all type of people are leaving it in the current pandemic, some of the sample were not Bahraini, but according to the research needs, the study was targeting the Bahrain's citizens in general regardless of the nationality or the gender. A total of 60 sample has been collected with random size of each industry. The targeted sample was mostly between medical and staff and students, but the sample contained only 25% of them.

According to (Table 1) the results of the first question which is asking about participants' age, shows that most of the survey participants are aged between 30 and 40 years old (33.9%) of the total sample population, the second majority are participants aged between 25 and 29 years old (32.1%), then above 40 years old (19.6%), between 20 and 24 (8.9%), and finally between 15 and 19 (5.4%). The majority of participants' gender were females (66.1%) who were higher than males (33.9%). This will provide the chance unequally between both genders to express their perceptions toward the role of Artificial Intelligence in fighting during the crisis of COVID-19, this unequally between the gender is because most interested and workers in the medical field are females.

**Table 1** Participants' responses to demographic questions

| Demographic | Answer | Frequency | Percent % |
|---|---|---|---|
| Age | 15–19 | 3 | 5.4 |
| | 20–24 | 5 | 8.9 |
| | 25–29 | 18 | 32.1 |
| | 30–40 | 19 | 33.9 |
| | Above > 40 | 11 | 19.6 |
| | Total | 56 | 100 |
| Gender | Male | 19 | 33.9 |
| | Female | 37 | 66.1 |
| | Total | 56 | 100 |
| Education | Under graduate | 2 | 3.6 |
| | High school Diploma | 8 | 14.3 |
| | Bachelor Degree | 37 | 66.1 |
| | Master Degree | 6 | 10.7 |
| | PhD | 3 | 5.4 |
| | Total | 56 | 100 |
| Are you Medical Staff/Student | Yes | 14 | 25 |
| | No | 42 | 75 |
| | Total | 56 | 100 |

The findings showed the majority of participants were Bachelor's Degree holders (66.1%), then High school Diploma degree with (14.3%), after that Master degree with (10.7%), Ph.D. degree with (5.4%) and others (Undergraduates) with (3.6%). The findings show that High School degree holders are close to Master Degree holders. Therefore, different perceptions and background may not affect that much because such topic is being discussed according to the global sorting as an international crisis and being talked about it more in details in all the social media and the news's channels (i.e. Instagram, Google, Social Groups) rather than the educational institutions.

The survey showed that most participants are Medical Staff/Students with (75%), however, non-medical Staff/Students participants came in the second and last place with (25%). We can notice that the majority of participants have Medical Background and can see clearly from their working places the effect of using the Artificial Intelligence in Healthcare sector in General and in fighting the COVID-19 during this crisis in specific.

## 5.1 Descriptive of Variables

This section and based on (Table 2) will discuss and explain different factors that impact using Artificial intelligence in the Healthcare sector in Bahrain and how it can be helpful in quickly discover and reduce the dangers of the virus. There are three main factors that needed to be addressed in this study, awareness, and Knowledge, Using the technology and the expectations for the crisis future. There are other factors that may impact participants' intention to be involved in using Artificial intelligence channels.

The above-mentioned factors considered as a major factor that impacts consumers' decision to purchase any kind of product. People's perception is built based on their knowledge and background of any new technology pop-up in the social media and between the population in Bahrain. Therefore, a source of knowledge and communication channels of certain awareness is very important as it is going to impact people's perception.

From the above table, we can see that most of the participants are following the updated of COVID-19 news and Bahrain and the new ways that are being used to speed up the discovery of the virus with the location used to alert anyone is going near an active case in the last 14 days. Of course, with multi Social media, applications, and news channels in Bahrain that already launched and developed by using the new technology and artificial intelligence, it was expected that the majority had heard about the virus and being updated. In spite that the second-highest percentage of participants' education level is high school Diploma, the majority have been part of the social media life. This is an indicator that awareness is not related to educational level especially (69.6%) of people's asses. Their knowledge about the topic as average and (30.4%) as good. Social media, applications, news channels, and other sources of knowledge may have an impact

**Table 2** Awareness and knowledge, using the technology and the expectations for the crisis future

| Questions | Answer | Frequency % | Mean |
|---|---|---|---|
| Due to Artificial Intelligence awareness in Bahrain, the number of COVID-19 cases in Bahrain is under control | Strongly agree | 16<br>28.6% | 0.29 |
| | Agree | 23<br>41.1% | 0.41 |
| | Natural | 15<br>26.8% | 0.27 |
| | Disagree | 1<br>1.8% | 0.02 |
| | Strongly disagree | 1<br>1.8% | 0.02 |
| Using Artificial Intelligence in Healthcare can reduce the diagnosis errors in finding the COVID-19 and spread of the virus | Agree | 28<br>50% | 0.61 |
| | Natural | 16<br>28.6% | 0.35 |
| | Disagree | 2<br>3.6% | 0.04 |
| Have you been following the updates of COVID-19 in Bahrain | No | 1<br>1.8% | 0.02 |
| | Yes | 55<br>98.2% | 0.98 |
| How do you assess your knowledge you gained from Artificial intelligence channels about the coronavirus | Good | 39<br>30.4% | 0.7 |
| | Average | 17<br>69.6% | 0.3 |
| Do you spread any instructions and directions you are getting through Artificial Intelligence channels | Yes | 22<br>39.3% | 0.39 |
| | Sometimes | 25<br>44.6% | 0.45 |
| | No | 9<br>16.1% | 0.16 |
| Can we use other modalities to diagnose the COVID-19 (Ex. X-ray Chest? CT Chest?) | Yes | 16<br>28.6% | 0.41 |
| | Maybe | 19<br>33.9% | 0.49 |
| | No | 4<br>7.1% | 0.10 |

more than educational entities. The findings showed that the majority of participants with (25%) with yes answers and (22%) are involved in Bahrain Team goals in participating in the campaign against COVID-19 by helping in spreading the updated news and instructions to their contacts, friends, and families.

In related to multichoice questions, people went for Mobile applications mostly, google search and online consultation with some physicians as a basic knowledge to know more about whatever is related to the virus symptoms and the global news related to any reductions/increased in the number of cases, any vaccination or medications are being developed and tried.

## 6 Conclusion

The research was done according to the new technology in general and COVID-19 in specific and based on the research and articles have been found from different backgrounds and dates of an issue without considering any other financial or country developments and backgrounds.

In the research, we started talking about Artificial intelligence in general in the healthcare facilities, how the definition has been started to be used and catch up the development up to date, use of AI with some applications from different countries, going through the limitations and the question of replacing the human with robots. In the second part of the lecture review, we went through Artificial intelligence in medical imaging and the future expectations, advantages, and disadvantages of using it.

In the last part of the research, we talked about using Artificial intelligence in the current situation during the COVID-19 spread in the world. A questioner has been distributed to a different type of people in Bahraini society and has discussed and analyzed.

according to questions has been raised about the impact of using the Artificial intelligence in fighting the spread of Coronavirus.

After more than 60 years of research, we can say that the artificial intelligence tools are being considered and used in best practice these days, after noticing its importance in our lives, especially with the last update of the world crisis of COVID-19 [40]. The crisis showed how AI has been utilized to cover most of lives needs, starting from working at home for almost 70% of Bahraini's governmental employees, passing through the online school and universities class, going to online consultations in the healthcare facilities and ending with the online shopping for all public needs.

Ultimately, there has never been a better time to start practicing this type of technology in our lives.

As to the main question raised at the beginning of this research could robots take a place in the hospitals and physician's lives to take the responsibility of operating and managing the medical practice? In the conclusion, we can say that the artificial intelligence can play an important role in fighting the diseases and an essential tool in the diagnosis and reducing the human errors, yet the human cannot be replaced or compared with the technology and machines. Human physicians and medical staff are not a tool to be replaced or changed. Choosing the right person at the right time for the right job beside the artificial intelligence tools can generate a powerful team with almost the perfection percent of reducing the errors.

Artificial intelligence tools as per [26] can be an assist for the radiologists or readers in the medical imaging reading and cannot defiantly replacing them, be highlighting the findings of the advanced cases, to be seen by the human team and mark the other normal images as normal and pass it, another advantage can be the memorization and linking between the patients' history and the diagnosis and complains to achieve a better understanding of the patient condition. This will help them reduce the time of reading and increase the radiology departments' quality and efficiency all over the healthcare facilities.

Artificial intelligence will play a more important role in the coming years. All medical practitioners should be aware and well trained to use this technology they are saying that the technology will mostly affect the radiology department before any other facility [37].

In fact, the most advanced and technological airplanes with the developed system and all artificial intelligence technology still need at least two human pilots on board. Simply anything deals with human lives should be trusted to be handled for human only.

## References


1. Alves, G.: The impact of culture and relational quality in the cooperation between export companies and local distributors. Int. J. Bus. Ethics. Gov. **1**(2), 1–19 (2018). https://doi.org/10.51325/ijbeg.v1i2.13
2. Al-Taee, H., Al-Khawaldeh, K.: Impact of health marketing mix on competitive advantage: the case of King Hussein Cancer Center. Jordan J. Bus. Adm. **16**(1), 125–152 (2020)
3. Nassar, S.: The impact of intellectual capital on corporate performance of IT Xompanies: evidence from Bursa Istanbul. Int. J. Bus. Ethics Gov. **1**(3), 1–10 (2018). https://doi.org/10.51325/ijbeg.v1i3.17
4. Sisaye, S.: The influence of non-governmental organizations (NGOs) on the development of voluntary sustainability accounting reporting rules. J. Bus. Socio-econ. Dev. **1**(1), 5–23 (2021). https://doi.org/10.1108/JBSED-02-2021-0017
5. Ramaano, A.I.: Potential of ecotourism as a mechanism to buoy community livelihoods: the case of Musina Municipality, Limpopo, South Africa. J. Bus. Socio-econ. Dev. **1**(1), 47–70 (2021). https://doi.org/10.1108/JBSED-02-2021-0020
6. Adnan, S.M., Hamdan, A., Alareeni, B.: Artificial intelligence for public sector: chatbots as a customer service representative. In: Lecture Notes in Networks and Systems (LNNS), vol. 194, pp. 164–173 (2021)
7. Awad, I.M., Al-Jerashi, G.K., Alabaddi, Z.A.: Determinants of private domestic investment in Palestine: time series analysis. J. Bus. Socio-econ. Dev. **1**(1), 71–86 (2021). https://doi.org/10.1108/JBSED-04-2021-0038
8. Elali, W.: The importance of strategic agility to business survival during corona crisis and beyond. Int. J. Bus. Ethics Gov. **4**(2), 1–8 (2021). https://doi.org/10.51325/ijbeg.v4i2.64
9. Albinali, E.A., Hamdan, A.: The implementation of artificial intelligence in social media marketing and its impact on consumer behavior: evidence from Bahrain. In: Lecture Notes in Networks and Systems (LNNS), vol. 194, pp. 767–774 (2021)

10. Jiang, F., Jiang, Y., Zhi, H., Dong, Y., Li, H., Ma, S., Wang, Y., Dong, Q., Shen, H., Wang, Y.: Artificial intelligence in healthcare: past, present and future. Stroke Vasc. Neurol. **2**(4), 230–243 (2017). https://doi.org/10.1136/svn-2017-000101
11. Ali, M.H., Hamdan, A., Alareeni, B.: The implementation of artificial intelligence in organizations' systems: opportunities and challenges. In: Lecture Notes in Networks and Systems (LNNS), vol. 194, pp. 153–163 (2021)
12. Walsh, K.: Artificial intelligence and healthcare professional education: superhuman resources for health? Postgrad. Med. J. **96**(1133), 121–122 (2019). https://doi.org/10.1136/postgradmedj-2019-137132
13. Alrabba, H., Almahameed, T.: The impact of board's characteristics on risk disclosure in Jordanian industrial corporations. Jordan J. Bus. Adm. **16**(4):790–811 (2020)
14. Al Kurdi, O.F.: A critical comparative review of emergency and disaster management in the Arab world. J. Bus. Socio-econ. Dev. **1**(1), 24–46 (2021). https://doi.org/10.1108/JBSED-02-2021-0021
15. Youssef, J., Diab, S.: Does quality of governance contribute to the heterogeneity in happiness levels across MENA countries? J. Bus. Socio-econ. Dev. **1**(1), 87–101 (2021). https://doi.org/10.1108/JBSED-03-2021-0027
16. Shinners, L., Aggar, C., Grace, S., Smith, S.: Exploring healthcare professionals' understanding and experiences of artificial intelligence technology use in the delivery of healthcare: an integrative review. Health Informatics J. **26**(2), 1225–1236 (2019). https://doi.org/10.1177/1460458219874641
17. Al Azemi, M., Al Omari, A.M., Al Omrani, T.: The reality of financial corruption in Kuwait: a procedure research according to corruption perception index & related rules. Int. J. Bus. Ethics Gov. **2**(2), 64–86 (2019). https://doi.org/10.51325/ijbeg.v2i2.52
18. Ahmed, N., Hamdan, A., Alareeni, B.: The contribution of healthcare middle managers as change agents in the era of Covid-19: critical review. In: Lecture Notes in Networks and Systems (LNNS), vol. 194, pp. 670–678 (2021)
19. AL-Hashimi, M., Hamdan, A.: Artificial intelligence and coronavirus COVID-19: applications, impact and future implications. In: Lecture Notes in Networks and Systems (LNNS), vol. 194, pp. 830–843 (2021)
20. Emanuel, E.J., Wachter, R.M.: Artificial intelligence in health care. JAMA **321**(23), 2281 (2019). https://doi.org/10.1001/jama.2019.4914
21. Ellis, L.: Artificial intelligence for precision education in radiology—experiences in radiology teaching from a UK foundation doctor. Br. J. Radiol. **92**(1104), 20190779 (2019). https://doi.org/10.1259/bjr.20190779
22. Tekkeşin, A.I.: Artificial intelligence in healthcare: past, present and future. Anatolian J. Cardiol. (2019). https://doi.org/10.14744/anatoljcardiol.2019.28661
23. Al-Mohaisen, F., Al-Kasasbeh, M.: Impact of succession planning on talent retention at Orange – Jordan. Jordan J. Bus. Adm. **17**(1), 126–146 (2021)
24. Racine, E., Boehlen, W., Sample, M.: Healthcare uses of artificial intelligence: challenges and opportunities for growth. Healthc. Manage. Forum **32**(5), 272–275 (2019). https://doi.org/10.1177/0840470419843831
25. Meyer, A.: Machine-learning based wind turbine operating state detection and diagnosis (2020). https://doi.org/10.5194/egusphere-egu2020-6532
26. Martinez-Vernon, A.S., Covington, J.A., Arasaradnam, R.P., Esfahani, S., O'Connell, N., Kyrou, I., Savage, R.S.: An improved machine learning pipeline for urinary volatiles disease detection: diagnosing diabetes. Plos One **13**(9) (2018). https://doi.org/10.1371/journal.pone.0204425
27. Areiqat, A.Y., Hamdan, A., Alheet, A.F., Alareeni, B.: Impact of artificial intelligence on E-commerce development. In: Lecture Notes in Networks and Systems (LNNS), vol. 194, pp. 571–578 (2021)
28. Yu, K., Beam, A.L., Kohane, I.S.: Artificial intelligence in healthcare. Nat. Biomed. Eng. **2**(10), 719 (2018)

29. Greenspan, H., San José Estépar, R., Niessen, W.J., Siegel, E., Nielsen, M.: Position paper on COVID-19 imaging and AI: from the clinical needs and technological challenges to initial AI solutions at the lab and national level towards a new era for AI in healthcare. Med. Image Anal. **66**, 101800–101800 (2020). https://doi.org/10.1016/j.media.2020.101800
30. Nassar, R.M., Battour, M.: The impact of marketing ethics on customer loyalty: a conceptual framework. Int. J. Bus. Ethics Governance **3**(2), 1–12 (2020)
31. Dilsizian, M.E., Siegel, E.L.: Machine meets biology: a primer on artificial intelligence in cardiology and cardiac imaging. Curr. Cardiol. Rep. **20**(12), 1–7 (2018). https://doi.org/10.1007/s11886-018-1074-8
32. Martinez-Vernon, A.S., Covington, J.A., Arasaradnam, R.P., Esfahani, S., O'Connell, N., Kyrou, I., Savage, R.S.: An improved machine learning pipeline for urinary volatiles disease detection: diagnosing diabetes (2018). https://doi.org/10.5281/zenodo.1419251
33. Emile, S.H., Hamid, H.K.S.: Fighting COVID-19, a place for artificial intelligence. Transbound. Emerg. Dis. **67**(5), 1754–1755 (2020). https://doi.org/10.1111/tbed.13648
34. Yang, H., Kundakcioglu, E., Li, J., Wu, T., Mitchell, J.R., Hara, A.K., et al.: Healthcare intelligence: turning data into knowledge. IEEE Intell. Syst. **29**(3), 54–68 (2014). https://doi.org/10.1109/MIS.2014.45
35. Gandhi, S., Mosleh, W., Shen, J., Chow, C.: Automation, machine learning, and artificial intelligence in echocardiography: a brave new world. Echocardiography (Mount Kisco, N.Y.), **35**(9), 1402–1418 (2018). https://doi.org/10.1111/echo.14086
36. Razzaque, A., Hamdan, A.: Artificial intelligence based multinational corporate model for EHR interoperability on an E-health platform. Artif. Intell. Sustain. Dev. Theory Pract. Future Appl. Stud. Comput. Intell. 71–81 (2020). https://doi.org/10.1007/978-3-030-51920-9_5
37. Unberath, M., Ghobadi, K., Levin, S., Hinson, J., Hager, G.D.: Artificial intelligence-based clinical decision support for COVID-19—where art thou? Adv. Intell. Syst. **2**(9), 2000104-n/a (2020). https://doi.org/10.1002/aisy.202000104
38. Dananjayan, S., Raj, G.M.: Artificial intelligence during a pandemic: the COVID-19 example. Int. J. Health Plann. Manage. **35**(5), 1260–1262 (2020). https://doi.org/10.1002/hpm.2987
39. Stebbing, J., Krishnan, V., Bono, S., Ottaviani, S., Casalini, G., Richardson, P.J., et al., the Sacco Baricitinib Study Group.: mechanism of baricitinib supports artificial intelligence-predicted testing in COVID-19 patients. EMBO Mol. Med. **12**(8), e12697-n/a (2020). https://doi.org/10.15252/emmm.202012697
40. Kaushik, A., Patel, S., Dubey, K.: Digital cardiovascular care in COVID-19 pandemic: a potential alternative? J. Card. Surg. **35**(12), 3545–3550 (2020). https://doi.org/10.1111/jocs.15094
41. Ali Saad, A.Z., Mohd Noor, A.B., Sharofiddin, A.: Effect of applying total quality management in improving the performance of Al-Waqf of Albr societies in Saudi Arabia: a theoretical framework for "Deming's Model". Int. J. Bus. Ethics Governance **3**(2), 13–33 (2020)
42. Tsiknakis, N., Trivizakis, E., Vassalou, E., Papadakis, G., Spandidos, D., Tsatsakis, A., et al.: Interpretable artificial intelligence framework for COVID-19 screening on chest X-rays. Exp. Ther. Med. **20**(2), 727–735 (2020). https://doi.org/10.3892/etm.2020.8797

# Intelligent Data Communication Technologies Against COVID-19

# An Intelligent Cloud Computing Context-Aware Model for Remote Monitoring COVID-19 Patients Using IoT Technology

**A. Waleed and Sally M. Elghamrawy**

**Abstract** This chapter proposes and investigates an intelligent context-aware model that adopts a hybrid architecture with both local and cloud-based components to monitor patients, particularly COVID-19 patients with mild symptoms while they are in their homes. The cloud-based part of the system makes storing and processing easier, especially that the data generated by ambient assisted living systems are huge, particularly with patients suffering from chronic diseases and require more frequent readings. On the other hand, the local part of the system monitors the patients in case of any sudden interruptions on the internet or any failure in the cloud system. The proposed model uses context-aware techniques by monitoring different physiological signals, surrounding ambient conditions, along with the patient activities simultaneously to build a better understanding of the health status of the COVID-19 patient in real-time, as the system will help doctors to detect if the patient has symptoms of the COVID-19. The results obtained experimentally prove how our proposed model can effectively monitor patients and detect emergencies accurately in imbalanced datasets through a case study on a patient with normotensive disorder.

**Keywords** Remote patient monitoring · Internet of Things (IoT) · Ambient Assisted Living (AAL) · Machine learning · Imbalanced datasets

A. Waleed (✉)
Electrical Engineering, The British University in Egypt, Cairo 11837, Egypt
e-mail: ahmed.waleed@bue.edu.eg

S. M. Elghamrawy
Computer Engineering Department, MISR Higher Institute for Engineering and Technology, Mansoura, Egypt
e-mail: sally_elghamrawy@ieee.org



A.-E. Hassanien et al. (eds.), *Advances in Data Science and Intelligent Data Communication Technologies for COVID-19*, Studies in Systems, Decision and Control 378, https://doi.org/10.1007/978-3-030-77302-1_11

## 1 Introduction

As with the emerge of the global pandemic, COVID-19 affected over 210 countries throughout the world. With the spread of the COVID-19 globally, remote patient monitoring is needed throughout the world, as it effected over 74,120,000 people and 1,600,000 deaths over the world. with the high rate of people getting effected with the virus, the ratio of doctors and caretakers against the number of patients is increasing rapidly with the start of the pandemic. The percentage of cases that needs to hospitalized is different from each other, in china, 15–20% of the cases needed to enter a hospital including 5% that needed to enter an Intensive care unit (ICU), however, Italy and Spain had 40–50% of the cases in need to be hospitalized, and an average of 10% needed to enter an ICU [1]. COVID-19 is known to attack the respiratory system, which might cause pneumonia and might also progress into 'acute respiratory distress syndrome' (ARDS). Not only that, but severe cases may have a cardiogenic or a distributive shock. A patient that was diagnosed with the virus has a possibility of entering intensive care unit (ICU). Cases of COVID-19 are generally classified into five categories: asymptomatic, mild, moderate, severe, and critical. With the severity and the fast spread, a huge challenge has impacted health care systems, since they were not ready for this huge number of patients who needed to enter a hospital or be taken care of. Most countries are trying to reduce the number of patients in hospitals by admitting only severe cases and letting mild cases of patients without any underlying chronic conditions to be taken care of at home unless severe symptoms have started to show. In order to take care of the maximum number of patients, while also maintaining the stability of health care systems, the following changes are urged in the systems:

1. Monitoring the health status of patients with mild symptoms while they are home. This allows for early detection of severe symptoms.
2. Monitoring the health of people who have high risk of death from the virus e.g., people over the age of 60 years or with any respiratory conditions.
3. People who were in contact with an infected patient to prevent the outbreak [2].

Remote patient monitoring positively impacted COVID-19 by offering to implement systems on the most used technological device on the planet, smartphones. Since China was the first country to report a COVID-19 case, the Chinese government had to think of an architecture to implement monitoring patients and citizens. At first, it began with a location tracker and heartbeat monitoring through the smartphone camera, then they teamed up with two major companies, Alibaba and Tencent, as they were able to transmit citizens' location to the police and several authorities. This application used to divide the individuals into a range of spectrum depending on the risk of this individual from green (low risk) to red (high risk) [3]. People who were marked as high risk are banned to enter certain locations. Later, after authorities' approval, an app is developed that will transmit four types of data [4]: (1) Personal identification information. (2) Personal health information. (3) Travel history. (4) Health assessment information. Other countries have implemented a system that can track your temperature through a thermal and laser camera. This method was implemented in Singapore with the use of android smartphones [5].

IoT is one the emerging technologies since the late 1990s, IoT has the ability of communication through different protocols and technologies; choosing the desired technology to communicate depends on 3 factors:

1. Range: the distance between the server and the device.
2. Power: total power drained to send the data.
3. Bandwidth: the amount of data being sent [6].

Since the systems are being designed to monitor multiple number patients simultaneously, Big data needs to stated. To be considered as a big data model, the model needs to consist of one of the seven V's components: (value, vagueness, veracity, vocabulary, validity, variability, and venue) [7].

The interest in Ambient Assisted Living (AAL) is increasing lately due to the COVID-19 pandemic. The AAL's abilities make it a perfect solution of the pandemic with real-time ability decision making. To use such systems in the real-world health care, the system should have highly responsive and efficient [8]. An AAL model comprises a patient, sensors, and wireless network to communicate, software and actuators [9]. Ambient sensors to monitor the surroundings of the patient [10]. To meet the patients' standards, context awareness needs to be used to have a personalized model for each patient as every patient can have a different physiological state, which can depend on the patient's location, activity, and age [11]. Using the information above, every patient's data is going to be diverse. The AAL system needs to be specified for each patient, which is hard to implement due to high number of patients. Cloud-enabled platforms are used to overcome this gap which can manage data and easier and quicker due to its high computational power. The proposed platform will have the ability to transfer the raw data into context data and manage and adapt it. A high level of abstraction is used with mature web service protocols.

Section 2 will discuss the background and literature review. Section 3 will present the proposed architecture, Sect. 4 offers the proposed HCM, Sect. 5 will discuss the case study, Sect. 6 will discuss the results, and finally, Sect. 7 will conclude the research.

The main objective of using assisted living systems is to deliver the monitoring services and medical attention and support a patient's needs more efficiently at less costs. Such a system can greatly benefit patients, especially those who cannot frequently go to hospitals, for example, old or handicapped people who have no one to look after them. The main points that motivated us to research this topic are:

- Many healthcare infrastructures depend on assistive technologies that are context-driven. The current architectural solutions of context-aware middleware are only limited to specific services. Additionally, they depend on a local smart agent like a mobile device for context discovery and management [12–15]. Wearable sensors and other mobile devices have limited storage and power, thus limiting only processes that do not need limitless sensor data using decent computational methods. Add to that, the demand for more intelligent assistive services has increased with each new discovery of a new smart sensor or device. Cloud-based computing is capable of handling massive data chunks and deliver versatile services quite efficiently. This has paved the way for us to build a well-collaborative system. The context processing task is moved from the local smart device to a distributed cloud hence reducing the processing time of context generation and conveying complex services.

- Since the traditional systems of context management depend on applications separately on the local server, they cannot simultaneously handle several AAL systems. This particular reason has inspired us to design a cloud-oriented middleware that can handle a sufficiently large number of clients simultaneously. Additionally, each context derived from one AAL system is added to a knowledge bank for other AAL systems inside the same cloud. Therefore, the cloud middleware can be a knowledge source of context and be able to deliver assistive actions more quickly over time.
- Since using an integrated system of cloud computing with AAL can surely increase the variety of services it can deliver; we aim to develop a system that is capable of delivering all services using only one single model. The cloud can be implemented with cloud-based software services, a web service in the form of a set of instructions, or an API that is cloud-based for event alerts. In other words, everything can be processed in the cloud platform, starting from the context generation all the way to the service delivery. However, designing an SOA system that can successfully meet the objectives mentioned above, but has scalable computing facilities is indeed a challenging task.
- An additional challenge is collecting all these real-time data from several body sensors and other various devices at the exact same time. Not only that, but then handling and processing a huge set of data to categorize them properly according to their context is an added challenge.
- The integration of several elements with massive sources of data is a challenge. In AAL systems, generated raw data is possibly either an insignificant stream composed of binary data (a sensor), an analog signal, or even a heavy stream of multimedia content (mainly streaming cameras or microphones). Therefore, received raw data is as diverse as the sensors and devices used, and to accurately process and classify them according to their context quickly is a very challenging task. Additionally, receiving huge data from several systems is undoubtedly more challenging.
- To choose the most suitable needed-service accurately for a given context from a huge set is another challenging task. In addition to informing the user immediately with these actions or services.
- Another challenge is distributing the storage and data migration if the datasets and their services were geographically distributed.

To successfully be able to overcome all these challenges, we are willing to design a framework with cloud computing technology because it enables patients and doctors to collect easily, access, process, visualize, archive, share, and search huge data from different AAL systems and service providers. Additionally, cloud computing makes the processing of highly swift data easier and gives instant responses. The proposed framework works as a backend decision support system. It also offers intelligent services from raw data. With the powerful cloud computational and storage capabilities, the processing, analysis, and storage of huge data of context is possible.

Furthermore, cloud computing allows sharing and computing the same information for different applications, thus minimizing expenses. Additionally, and because most of the processing is cloud-based, sensors and other devices are now focusing on their specialized tasks. Finally, since cloud-based services can be delivered to any system connected to the internet, our framework works on making the processing of each component easier and simpler, thus reducing the computing load on sensors and other devices. Therefore, our solution helps the elderly and disabled people who cannot go to healthcare infrastructures and COVID-19 patients and ultimately reduces the workload on healthcare professionals and caregivers and decreases the hospitals' traffic load.

## 2 Background and Literature Review

RPM models are divided into two types: the first is called store and send, which transmits data stored to whom it may concern. The other is real-time, which involves sending biological data and multimedia if needed [16]. Using a combination of RPM with the IoT can greatly ease the lives of healthcare workers and also patients as it sends data to either data-centres or the cloud to be analysed and compared, see if there is anything else unsuspicious, then finally alarm the patient and doctor if necessary [17]. As the data are being sent, they are divided into several datasets, however, they are imbalanced naturally, therefore, it has been considered to split these data unequally in order to even out the sets. Nevertheless, it should be noted that it is a problem when the sets have varying amount of data in it. For example, a dataset has 80% of all data (here named the majority class) while the other has 20% (here called the minority class); having such an imbalanced dataset negatively affects the system's accuracy because the classifier chooses the majority class more often. Luckily, this problem can be solved by many ways, the most common solution is oversampling: the data in the minority class is sampled until it matches that of the majority class, however, a disadvantage is the repetition of data in the minority class. Another solution is undersampling: removing data from the majority class, a disadvantage is the loss of data.

Leo John Bapist Andrews and et al., have considered using the open source's capabilities of android systems in order to send data to the database. Their developed system can work both online and offline. Their proposed system gathers and sends data through a wrist band that does not bother the patient/user. It also can send six different biological signals including Body Surface Temperature,

Respiration Rate, Pulse Oximetry (SpO2), Blood Pressure (BP), Electrocardiogram (ECG) and Fall Detection. Collected data is transmitted to the database to be analysed and responded to. Their proposed system works on the online mode when connected to the internet (updating the wrist band and server regularly) and can alarm the caretaker if any abnormal signs were sensed. If not connected to the internet, the system operates on the offline mode (the server is not updated) and compares collected data with the stored standard data. Their system consists of four alarm stages: the first is notifying the patient with self-First aid steps and guidance through medication. The next stage involves sending an emergency alert to the nearest hospital using android system's location feature. The message sent contains the patient's conditions along with his/her location. The third stage involves sending the patient's conditions to his/her relatives. The last stage involves advising nearby people to perform first aid following voice instructions [18].

Another group of researchers, Abdelrahman Rashed et al. have proposed an Internet of Medical Things (IoMT) based system with raspberry pi as their gateway. Their system is cloud-based consisting of three layers which are integrated application layer, gateway and network, and Perception. Generally, in IoMT systems, the core layer is the perception layer, in which data are received from the sensors and passed to the next layer: gateway and network. The Perception layer is furtherer separated to two sublayers: Physical interface, the interface at which data is collected from sensors, the second sublayer is the Data Collection layer, where data is collected and sent to the following layer. Two NodeMCUs were the data collectors, each for wearable and non-wearable nodes. The second layer, which is the Gateway and Network Layer, connects the sensors with the database. They used Raspberry Pi 3 as their gateway, which received data wirelessly from the two NodeMCUs through Message Queue Telemetry Transport (MQTT) protocol using python. Because machine to machine and light communication is preferred, they used MQTT protocol. The last layer is the Integrated Application Layer, which is the back-end of their system. They used virtual servers as they are better than traditional ones. Here, the process of saving, backing-up, and analysing data is run to be ready to send alarms (in case of sensing any abnormal signs). An android-based mobile application was developed to visualize all this information. In it, the doctor's interface includes all patients' profiles (assigned to him) with their vital readings. In any emergency, the healthcare providers will receive a notification. On the other hand, the Patient's Interface allows him/her to track their records, history and current visuals [19].

Other researchers, Sindu Divakaranand et al., proposed a system which works the embedded systems (a computerized system used to operate one function based on its program). They integrated embedded systems within a larger system. Their main objective was to provide diagnosis by sending live stream. They used different types of sensors, including: Body Position, Body temperature, Heartbeat Rate, Respiration Rate, Blood Pressure, 3-channel ECG, and Camera. ARM Cortex M4-based microcontroller was used to send data, thus making the system online-based. With this system, doctors or caregivers can know the real time situation of the patient as long as the device has internet connection, and they can access the data (with a username and the password). TCP/IP connection is required until the connection is ended by the patient [20].

Moving to imbalanced datasets, Anantaporn Hanskunatai has proposed a hybrid sampling method which basically combines both the undersampling and oversampling methods. His proposed solution eliminates data that is not important in the majority class, while also oversampling good data in the minority class. His solution consisted of three parts: grouping, sampling, and gathering. Grouping was made with Density-Based Spatial Clustering of Applications with Noise (DBSCAN), which is simply a data-clustering algorithm works on grouping data in space which are closely packed together, thus showing the outlier's regions with low-density. Following grouping, each cluster is sampled separately. In case the cluster consisted of only majority class data, the system removes half of the data nearest to the center. If the cluster had a mix of negative and positive data, the algorithm finds an overlapping area and call it S. It should have instances unfit at k = 5 by kNN, where the opaque instances are defined to be unfit. By then, all the opaque is eliminated. The final step uses SMOTE algorithm on the remaining positive instances and integrates them [21].

Lastly, Sally Elghamrawy et al. considered a cloud-based remote patient monitoring that works both on offline and online modes. Their proposed architecture consists of four layers. The first is AAL with digital communication to remotely monitor the patient. The first layer had three sublayers: My Signals platform (an open-source electronic health platform that has the sensors kits to monitor the patient), ambient sensors (records environment's readings to consider their effects), and finally a sub-layer to collect all these data and send it to the second layer, which is processes data into one contextual state. It also consists of the connectivity checker (CC), which checks whether to run on online or offline modes and the local database that helps the OLMM run in the offline mode. The third layer is the outpatient cloud monitoring module (OCMM), the personal knowledge discovery module and the doctors' encyclopedia to put their rules in. The fourth and last layer classifies data in addition to minimizing errors and false alerts [22].

## 3 The Proposed Intelligent Cloud Computing Context-Aware Architecture for Remote Monitoring COVID-19 Patients

The proposed architecture monitors patients accommodated in their homes, which can communicate using several technologies such as IoT and the ability of Cloud Storage and Computing. The model can handle big data perfectly and begin to process theses data arriving through the sensors as shown in Fig. 1. The proposed model is inspired by the paper of Sally Elghamrawy et al. [23–26]. The proposed architecture is enhancement to the previous IHCAM-PUSH, the previous architecture had an assumption that the caregivers or family members or any of the contact list is available anytime, so the newly proposed architecture is can solve this problem by number of factors. First, each patient will contain prioritized contact list dependent on what is the category the system diagnosed the patient. The more severe the symptoms, the higher the priority the system will contact.

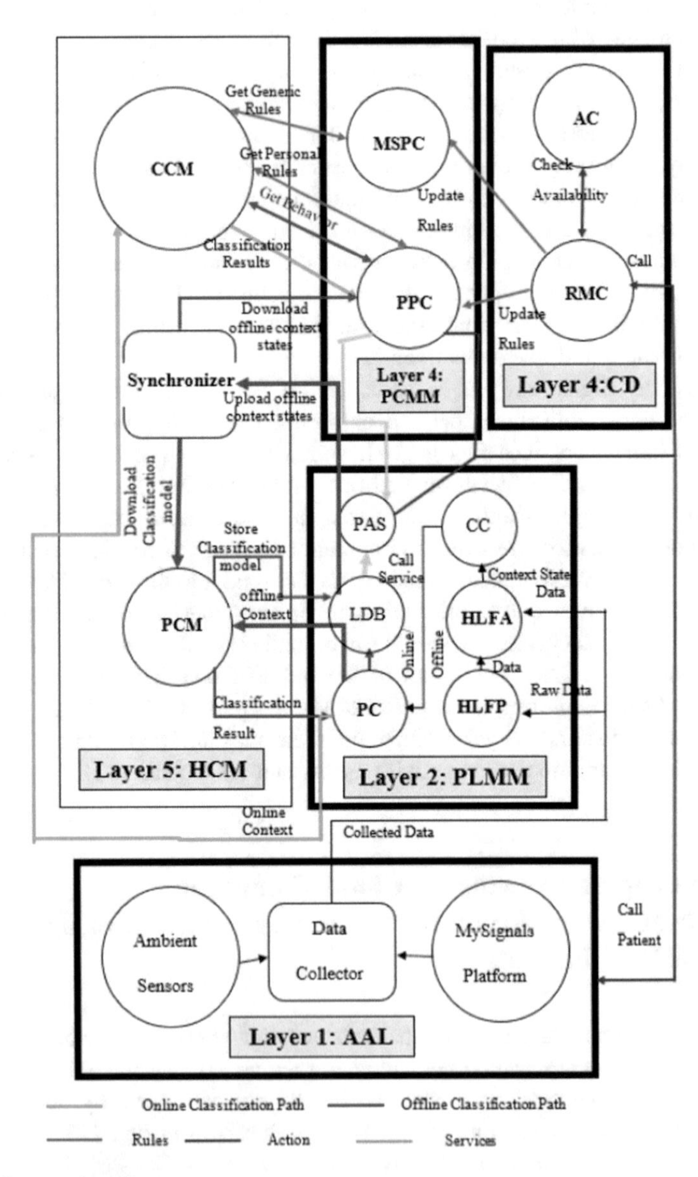


**Fig. 1** Proposed architecture

**Layer 1: AAL**

The first layer will have the ability to monitor the patient vital signs with the assistance of the medical sensors and the ambient signs to the patient. the sensors are chosen according to the type of diseases, due to that change in sensors, the architecture performance is dependent on the disease. The system needs to consist of a patient and a body sensor network (BSN) and surrounding sensors. Each of these sensors and devices is identified using a unique identifier (e.g., number, IP), to maintain the stability of distinguishing the data with the right sensors. This first layers consists of three sub-layers:

1.1 MySignals Platform
To be able to complete the system, MySignals platform is used as it open sourced and helps in the research of the electronic health industry. The patient vital signs can be measured using the sensors that are connected the MySignal platform using WIFI. The kits that they offer can measure various measurements of the patient such as airflow, body temperature, blood pressure, peripheral capillary oxygen saturation (SPO2) and much more. This platform's ability makes it easier to install new sensors on the human body with various options of connection (e.g. WIFI, 4G). These sensors allow the sending it into high level form. These sensors are focused to have a low consuming power and also able to communicate with the MySignals platform. There are no computational calculations are done on the sensors.

1.2 Ambient Devices
Ambient sensors or also known as the atmospheric sensors which include humidity, light detectors, temperature, smoke detectors. Help in diagnosing the patient.

1.3 Data Collector (DC)
As two sets of data types are being sent, the data collector will know how to collect data and send all the at the same time and avoid interference in the data. The MySignal platform is considered to send high-level data and the ambient devices send low-level data. This data is not manipulated in the DC sub-layer, the DC only combines the data and send it the next destination. The DC can be connected either wirelessly or wired to the sensors using several communication technologies.

**Layer 2: Patient Local Monitoring Module (PLMM)**

The PLMM is introduced to handle data in the system. This layer will manipulate and send the combined data from DC and save it afterward in the database. This layer will also check the cloud's connectivity by several tests and monitor the patient until the internet connection is stable.

2.1 High-Level Feature Provider (HLFP)
The HLFP aims to change data from low level to high level, an example of this conversion is converting raw data to a numerical value such as ECG data into

beats per minute (BPM). The output of this layer is then passed to the High-Level Feature Aggregator (HLFA).

2.2 HLFA

The sublayer combines the data from the HLFP and the MySignal platform's data as it is already sent in high-level form into one contextual state. The above contextual data technique is more efficient as it can sense any changes in the signs and ambient data that can perform real-time. Before anything, the HLFA will need to do data processing to check for abnormal clinical values. To have everything represented simultaneously (t) slot, we chose a standard time interval (λ) to sample a context state. This technique will ensure each context state is always sampled at a fixed λ, thus providing our framework with big data's velocity property. It is worth mentioning that the contexts of every domain are summarized in a time slot inside CA.

2.3 Connection Checker (CC)

The CC is responsible for testing the system internet connection with the cloud. This test includes testing internet parameters such as ping, upload, and download speeds. If any of this test fails, CC will automatically switch the system into offline mode to monitor the patient dependent on the information stored offline and store the patient data on Local Database (LDB). Other than that, the CC will state to work on Online mode where the data is stored on in the cloud.

2.4 LDB

As stated above, when the CC states to work in the offline mode, the data will be needed to be stored in an LDB. Also, in the LDB, it stores some of the used classification model as they also trained and tested on the online mode to know which will perform the best on the offline mode.

2.5 Personal Assistive Services (PAS)

This sublayer contains information of the assisting people to each patient, this people will be contacted in case of the alert and emergency cases only. This person could be doctors, family, or caregivers.

**Layer 3: Patient Cloud Monitoring Module (PCMM)**

In this layer, the PCMM will operate if the CC states to work in online mode. The operation if this layer to work as a personalized knowledge discovery module. The layer consists of the two sublayers:

3.1 Patient Personal Cloud (PPC)

Each patient will be accounted for personal storage in the cloud, which is supervised by the hospital. The information stored in this space can vary from personal information such as age, gender, weight, height, etc. It will also contain specific rules for the alert and emergencies taken from the remote monitoring cloud (RMC). It also can contain the patient's behaviors, such as the patient drinks alcohol or smoke and past medical reports. All the data will be sent to the Knowledge Discovery Cloud (KDC).

3.2 Medical Service Provider Cloud (MSPC)
This layer is the knowledge hub for the doctors and medical workers. They will be able to write each disease's symptoms and specify which signs are critical for the corresponding patient, which can be updated easily. Other than that, the MSPC will contain personalised rules for each patient. The knowledge written in this sub-layer will be transferred to the KDC.

**Layer 4: Contact Database (CD)**

The CD will work parallel with the system to contact the people listed in the PAS and check their availability. The following layers will contain:

4.1 RMC
The RMC contains the condition of the which people will be allowed to be called and sorted in a given priority in a specific condition in the alert and emergency situations only. The RMC sends the Availability Checker (AC) to see if the selected contact is available.
4.2 AC
The AC will be able to track if the selected person can be contacted at the moment, the AC can contain the timings or activeness of the contact. In case of emergencies and no contact is available, the AC will respond back to RMC to call an ambulance.

**Layer 5: Hybrid Classification Module (HCM)**

To achieve the highest performance in the system, HCM contains various classification types that are stored in the cloud and in the local storage. This enables the classification process to occur either online or offline. In the online mode, classification process will occur on the cloud, which has large access amount of data but in the offline mode, the local module has the objective to maintain the patient life until connection is back. The process of the HCM is shown in Fig. 2. the following layer consists from the following:

5.1 Knowledge Discovery Cloud (KDC)
This sublayer is the core of the system. The abstracted patient data will be compared health information. This sublayer will use spark to divide the work into different clusters to faster the process.
5.2 Cloud Classification Model (CCM)
The following sublayer will categorize the incoming context data in the online mode. The result will be transferred to PAS to peruse the action.
5.3 Personal Classification Model (PCM)
The following sublayer will categorize the incoming context data in the online mode. The result will be transferred to PAS to peruse the action.

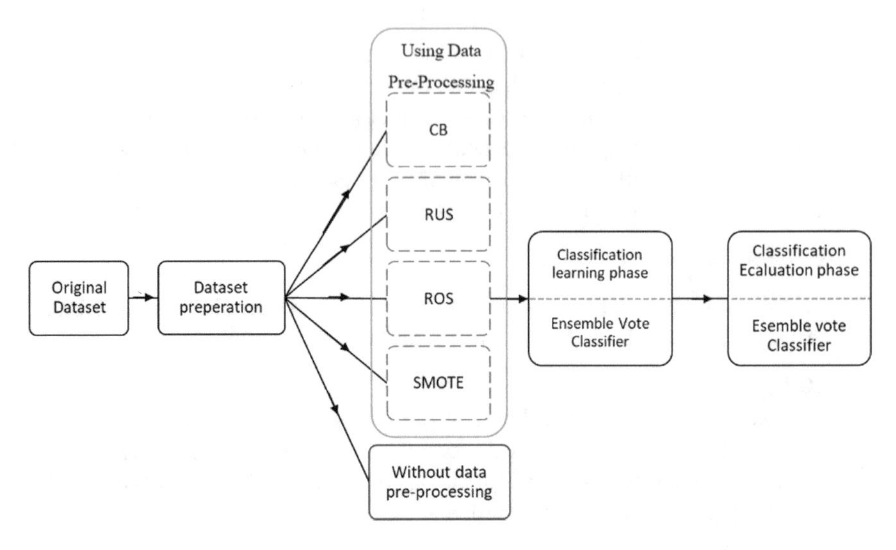


**Fig. 2** HCM process

# 4 Proposed HCM

Two classifiers will be used in the proposed HCM, each mode will assign a classifier. CCM will be used to classify in the online mode and PCM in the offline mode.

## *4.1 The Proposed CCM*

The five stages of the CCM will be implemented using WEKA and Spark Weka; WEKA is a program that can assist with data mining which contain a lot of machine learning algorithms used in the data mining such as classification, regression, clustering, data pre-processing, association rules and visualization as shown in Fig. 3.

**Stage 1: Data splitting**: The data is split into a header file and data file. The data file afterward is divided into parts.

**Stage 2: Dataset Shuffling**: The datasets are shuffled, and the input data is split between managing the minority classes. In the low number of minority classes, samples will be moved to make every part has an appointed sample.

**Stage 3: Data pre-processing**: Sampling techniques are used to achieve equilibrium.

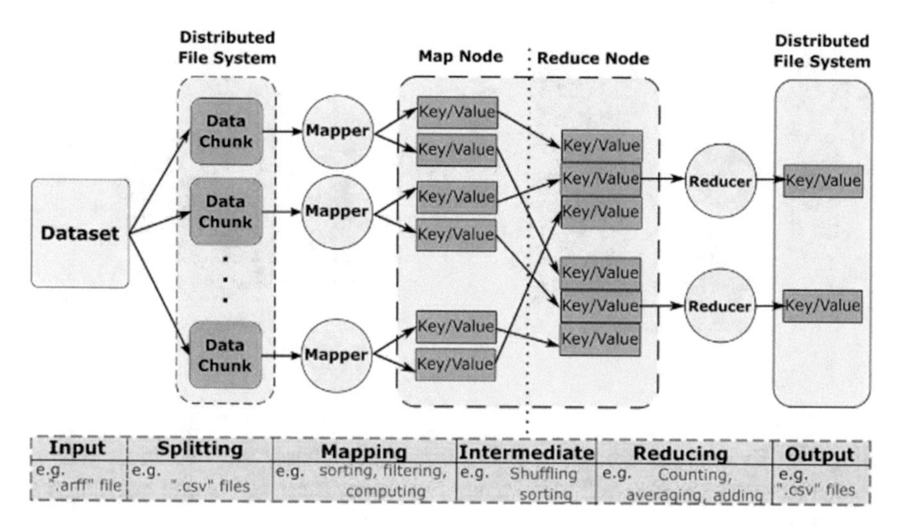


**Fig. 3** Weka flow

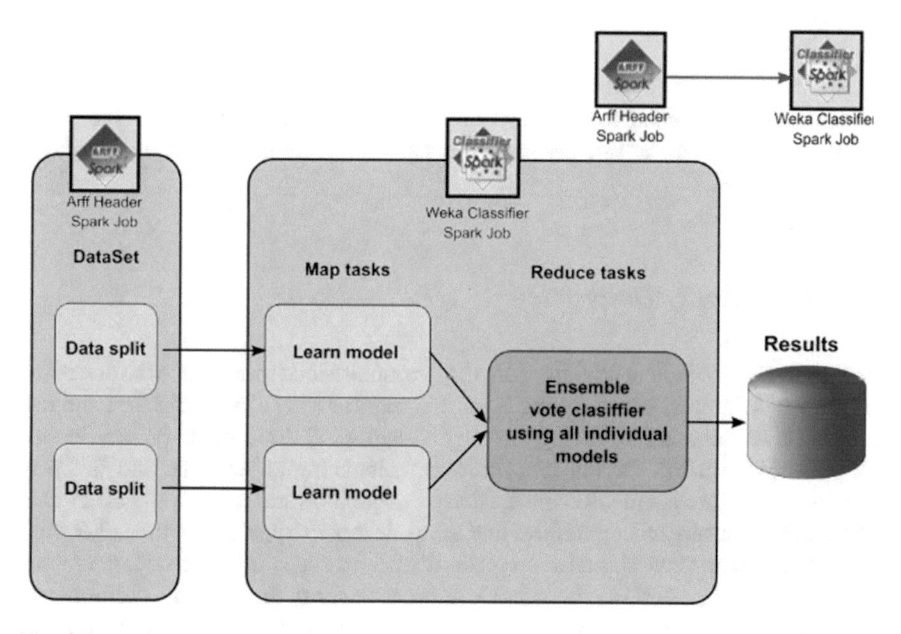


**Fig. 4** Classifier learning phase

**Stage 4: Learning Phase**: This phase will implement various classifiers and sampling techniques. The data will include multiple of classifications according to the amount of data, as shown in Fig. 4.

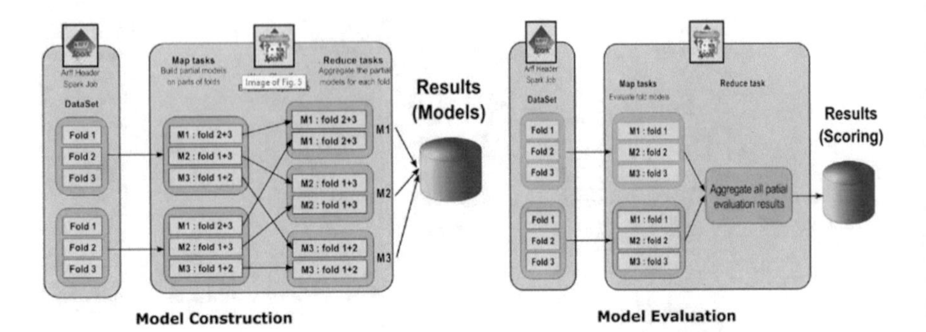


**Fig. 5** Model evaluation

**Stage 5: Evaluation Phase**: Using the learning phase, the classifiers will be evaluated using ten-fold validation using two steps: construction and evaluation as shown in Fig. 5.

## 5 Case Study

A case study is needed to test how our proposed system and its layers are performing. The system in this case study is designed to monitor patients who have chronic diseases.

### *5.1 Case Study Description*

The system monitors a patient who has normal blood pressure (normotensive patient). The interval of measuring is 15 min, run for a year to get big data and use it to build learning models using the plugin distributed cloud mode (Weka-Spark), thus speeding up the classification process maintaining an accurate and fast performance. As the system will convert data to context-aware, it needs to collect data about the environmental conditions and the patient activity, along with his/her vital sign. The system then classifies the patient's health status into one of four classes (normal, warning, alert, or emergency). In order to verify the system's performance in any mode, different classifiers and sampling techniques will be used. In addition to testing the system's ability to handle imbalanced data sets.

## 5.2 Initial Setup

Our proposed system can handle any patient suffering from any disease as long as the proper sensors are used in the AAL layer, however, this case study focuses on BP. Limitations that are taken into consideration are:

1. The most important data collected is the patient's vital signs. Doctors and medical experts who have access to data in the RMC in layer four can be able to indicate the specific vital signs for each patient with a specific disease.
2. In this case study, the main measured body signs are heart rate (HR), systolic blood pressure (SBP), diastolic blood pressure (DBP), mean blood pressure (MBP) respiratory rate (RR), and saturation of peripheral oxygen (SPO2).
3. Since our proposed system is context aware with the vital signs, two other data will be collected which are the room temperature from the ambient sensors in the layer one, and the patient activity, because these two signs can affect the vital signs.
4. In layer one, the AAL setup varies across different diseases, thus monitored patient's vital signs and ambient conditions also vary accordingly. For each case, a suitable IoT sensor must be adopted.
5. General rules of HR and BP ranges will be stored in the MSPC, in addition to other health aspects (patient's age, highest HR expected, the relationship between age and HR, how HR varies with changing levels of activity, etc.).
6. The PPC contains any important information about the patient, such as medical reports, allergies, personalized medical rules, etc.

## 5.3 Data Generation

According to our best knowledge, there is no prolonged patient dataset available on the internet that contains sensor data and focuses on BP disorders which has the style shown in Table 1. In order to train our model, artificial datasets are used that are generated with the help of the same format of MySignals platform in layer one. The Physinet MIMIC-II database has a day-long vital signs data of a real normotensive patient. However, other data like the remaining health signs, environmental conditions, and patient's activities will be in the artificial dataset. The

**Table 1** Sample of the dataset

| Heart rate | *BP_Sys* | BP_Dias | BP_mean | Resp | SPO2 | Temp | Activity | Last activity | Med |
|---|---|---|---|---|---|---|---|---|---|
| 86 | 127 | 87 | 96 | 19 | 95 | 0 | 1 | 1 | 1 |
| 79 | 129 | 67 | 97 | 14 | 98 | 0 | 2 | 1 | 0 |
| 97 | 111 | 96 | 101 | 20 | 100 | 0 | 2 | 2 | 0 |

dataset will be simulated with a prolonged monitoring scheme at a fixed sampling rate (15 min intervals for a year). The artificial dataset considers the following concepts:

1. Time of day is considered, along with the relationship between health signs and activity based on time.
2. The relationship between health signs with environmental conditions and taking medicines.

The normal patient dataset will be generated using MATLAB while considering the real datasets in Physionet MIMIC-II. These artificial datasets have proved reliable to generate big data similar to monitoring over prolonged periods. Previously analyzed biomedical data also proved reliable in previous studies.

Normotensive patients' symptoms are listed in Table 2. Table 3 shows all the used attributes and their types in this case study. The personalized rules on classifying patient's health-class are shown in Table 4. Finally, Table 5 lists the actions of each class.

**Table 2** Patient symptoms

| Type | Symptoms | Value (binary) |
|---|---|---|
| Normotensive | Uncomfortable; anxiety; headache; fatigue; severe headache; dizziness | 6 bit binary (value: 0–63) |

**Table 3** Patient signs

| Name | Features | Type | Range value |
|---|---|---|---|
| Vital signs | HR | Numerical | [30–200] |
| | SEP | Numerical | [50–230] |
| | DBP | Numerical | [30–140] |
| | RR | Numerical | [5–30] |
| | $SPO_2$ | Numerical | [40–100] |
| Activity | Current activity and last activity | At rest | 1 |
| | | Sleeping | 2 |
| | | Walking | 3 |
| | | Eating | 4 |
| | | Exercising | 5 |
| | | Household | 6 |
| Ambient conditions | Room temperature | Normal | 0 |
| | | Hot | 1 |
| | | Cold | 2 |
| Medication | Taken/not taken | Boolean | 0 or 1 |
| Symptoms | Symptoms | Boolean | [0–63] |

**Table 4** Patient classification

| Class | Classification |
|---|---|
| Normal | The values of HR, SBP, DBP, RR and $SPO_2$ all fall in the expected range according to the current activity and symptoms = 0 |
| Warning | Either:<br>• Any of the HR, SBP, DBP, RR or $SPO_2$ values fall in the warning range<br>• Medication not taken<br>• Symptoms >0 |
| Alert | Either:<br>• Any of the HR, SBP, DBP, RR or $SPO_2$ values fall in the alert range<br>• Two or more vital signs fall in the warning range<br>In addition to:<br>• Medication not taken<br>• Or symptoms >0 |
| Emergency | Either:<br>• Any of the HR, SBP, DBP, RR or $SPO_2$ values fall in the emergency range<br>• Two or more vital signs fall in the alert range<br>In addition to:<br>• Medication not taken<br>• Or symptoms >0 |

**Table 5** Patient classification actions

| Case | Action |
|---|---|
| Class = Normal | No action |
| Class = Warning | Warn the patient via mobile phone, monitor, SMS, etc |
| Class = Alert | SMS or phone call the doctor in charge to review the case |
| Class = Emergency | Call an ambulance either directly or after the confirmation of the doctor in charge |
| Medication = 0 | Alert the patient or caregiver via mobile phone, monitor, SMS, etc |

**Table 6** Comparison

| Patient | No. of contexts | Generic Rules | | IHCAM-PUSH | | | |
|---|---|---|---|---|---|---|---|
| | | Normal | Abnormal | Normal | Warning | Alert | Emergency |
| Normotensive | 35,232 | 1 | 35,231 | 12,517 | 21,421 | 1186 | 108 |

## 5.4 *Dataset Exploration*

Table 6 shows that using general medical rules alone in classifying context states is not reliable. It wrongly classified many normal cases as abnormal, thus risking the patient's health by generating too many false negatives. However, implementing IHCAM-PUSH improved the system's efficiency, as it could classify context states successfully into different categories just by adding context awareness that helped our system differentiate between real risks and normal daily activities more

**Table 7** Importance Scorer

| Name | Detailed setting |
|---|---|
| Hardware | |
| CPU | Intel ® Core™ I3 6100 |
| Frequency | 3.7 GHz |
| RAM | 8 GB |
| Hard drive | 1 TB |
| Software | |
| Operating System | Windows 10 64 bit |
| Software | MATLAB 2019a 64bit<br>Weka 3.8.3<br>Plugins:<br>DistributedWekaBase version (1.0.17)<br>DistributedWekaSpark version (1.0.9)<br>SMOTE version (1.0.3) |

accurately. However, imbalanced datasets need to be processed with different sampling methods. After completely generating the dataset, we applied different Weka and Spark data-processing and data-mining techniques.

### *5.5 Tools*

Table 7 lists the specifications of the PC that performed all experiments. It is recommended to install the latest Weka version before installing the Distributed Weka Base package and Distributed Weka Spark package.

## 6 Results

Several experiments are made to examine how architecture is performing well using different classifiers and sampling techniques. The four parameters used to measure that are:

1. Accuracy
2. Weighted F-measure
3. Emergency F-measure
4. Time

Those parameters are used to identify the best classifier and sampling technique for the proposed architecture. The importance of the parameters is shown in Table 8.

**Table 8** Importance Scorer

| Importance | Parameter |
|---|---|
| 1 | F-measure for Emergencies |
| 2 | F-measure |
| 3 | Accuracy |
| 4 | Time |

$$Accuracy = \frac{TP + TN}{TP + TN + FP + FN} \quad (1)$$

$$\text{Recall} = \frac{TP}{TP + FN} \quad (2)$$

$$Precision = \frac{TP}{TP + FP} \quad (3)$$

$$F - Mesaure = \frac{2 \times Precision \times Recall}{Precision + Recall} \quad (4)$$

where

- TP = True Positive
- TN = True Negative
- FP = False Positive
- FN = False Negative

Table 9 shows the performance of the classifiers on a normal patient (normotensive patient). The bolded rows will show the best algorithm all together in every classifier, and the italic row shows the best algorithm overall. According to the results, the best in F-measure and Accuracy is the Random Forest (RF) but the best in time is the J48 classifier. Naïve Bayes (NB), Ripper Classifier (JRIP) and RF has a high success classification rate. The worst in term of times is the Support Vector Machine (SVM) and the worst in terms of success rates are SVM and neighbours' classifiers (IBK) as shown in the graph in Fig. 6

The rules based and decision trees classifiers are easy to read by medical workers as they can visualised in terms of if–then cases. The best sampling techniques are SMOTE and Class Balancer (CB).

## 7 Conclusion

The proposed architecture can monitor and diagnose the severe symptoms of COVID-19 patients. The architecture has a higher hit ratio of the medical staff's response. The architecture was able to analyze a huge set of data in real time and it can be adopted easily in modern health care system. The proposed system can be

**Table 9** Results

| | Accuracy (%) | F-measure | F-measure (E) | Time(s) |
|---|---|---|---|---|
| *JRIP* | *99.9* | *0.999* | *0.972* | *12* |
| JRIP + CB | 99.65 | 0.997 | 0.846 | 9 |
| JRIP + RUS | 94.7945 | 0.949 | 0.577 | 2 |
| JRIP + ROS | 99.5404 | 0.995 | 0.85 | 7 |
| JRIP + SMOTE | 99.889 | 0.999 | 0.932 | 14 |
| *NB* | *96.3272* | *0.96* | *0.749* | *7* |
| NB + CB | 89.4641 | 0.913 | 0.17 | 7 |
| NB + RUS | 88.8596 | 0.909 | 0.296 | 7 |
| NB + ROS | 90.7452 | 0.91 | 0.28 | 8 |
| NB + SMOTE | 95.9752 | 0.959 | 0.692 | 3 |
| SVM | 90.7726 | 0.907 | 0.633 | 49 |
| SVM + CB | 84.234 | 0.854 | 0.442 | 34 |
| SVM + RUS | 81.4248 | 0.821 | 0.446 | 12 |
| SVM + ROS | 84.6214 | 0.862 | 0.483 | 150 |
| *SVM + SMOTE* | *91.1559* | *0.9102* | *0.724* | *45* |
| J48 | 99.9035 | 0.99 | 0.896 | 7 |
| J48 + CB | 99.3387 | 0.994 | 0.62 | 5 |
| J48 + RUS | 95.9412 | 0.961 | 0.643 | 5 |
| J48 + ROS | 99.9789 | 0.999 | 0.951 | 10 |
| *J48 + SMOTE* | *99.9659* | *1* | *0.972* | *17* |
| RF | 99.8325 | 0.998 | 0.869 | 17 |
| *RF + CB* | *99.983* | *1* | *0.991* | *12* |
| RF + RUS | 96.6394 | 0.968 | 0.605 | 9 |
| RF + ROS | 99.7425 | 0.999 | 0.984 | 10 |
| RF + SMOTE | 99.9972 | 1 | 0.995 | 12 |
| IBK | 91.4737 | 0.913 | 0.579 | 47 |
| IBK + CB | 90.7369 | 0.987 | 0.967 | 61 |
| IBK + RUS | 60.9985 | 0.616 | 0.208 | 6 |
| IBK + ROS | 89.0156 | 0.984 | 0.751 | 90 |
| *IBK + SMOTE* | *98.7341* | *0.987* | *0.962* | *66* |

adjusted easily to implement a dataset into detecting from the patient's monitored data.

The LDB is used to keep the system running even if no internet connection can be established, so the system can keep running and monitoring in the offline mode. The proposed architecture is highly efficient, fast and can carry working even if any components failed to operate. The proposed architecture can help the doctors monitor patients that might encountered or have been in touch with a person diagnosed with COVID-19. The architecture will help doctors to also classify the cases into the five categories that was mentioned above. This will help doctors and

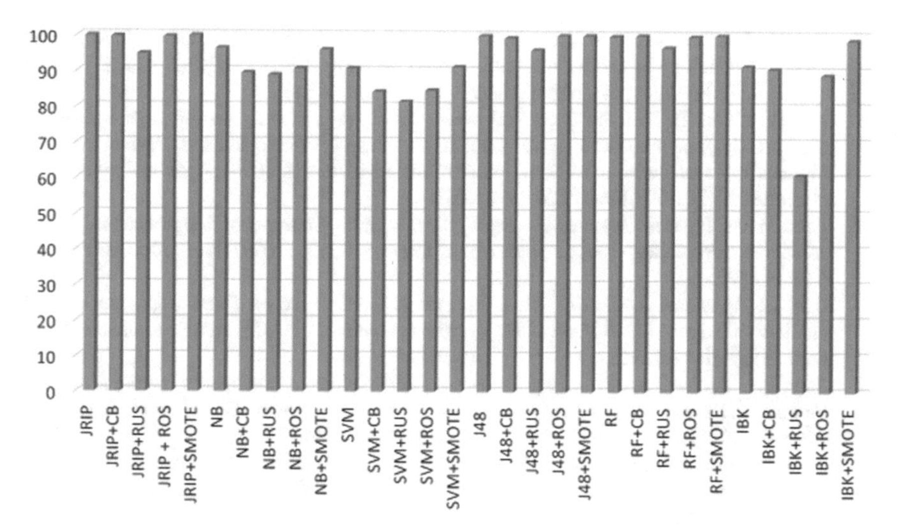


**Fig. 6** Graph with accuracy along with classifiers and sampling techniques

healthcare systems avoid virus transmission between patients as people won't be needed to enter the hospital unless in severe cases.

In severe cases, the architecture will call the given caretaker and call an ambulance to transfer the patient with rush into the hospital as this architecture can monitor the patients in real-time with having into their consideration their activities. Lastly, the architecture will help patients remind them of their medicines and tell them when they have medicine to take.

## References

1. Lazzerini, M., Putoto, G.: COVID-19 in Italy: momentous decisions and many uncertainties. The Lancet Global Health, vol. 8, no. 5. Elsevier Ltd, pp. e641–e642 (2020). https://doi.org/10.1016/S2214-109X(20)30110-8
2. Ding, X.R., et al.: Wearable sensing and telehealth technology with potential applications in the coronavirus pandemic. IEEE Rev. Biomed. Eng. (2020). https://doi.org/10.1109/RBME.2020.2992838
3. Peng, F., et al.: Management and treatment of COVID-19: the Chinese experience. Can. J. Cardiol. **36**, 915–930 (2020)
4. Behar, J.A., et al.: Remote health diagnosis and monitoring in the time of COVID-19. Physiol. Meas. **41**(10) (2020). https://doi.org/10.1088/1361-6579/abba0a
5. Cho, H., Ippolito, D., Yu, Y.W.: Contact tracing mobile apps for COVID-19: privacy considerations and related trade-offs. arXiv (2020)
6. Motlagh, N.H., Mohammadrezaei, M., Hunt, J., Zakeri, B.: Internet of things (IoT) and the energy sector. Energies **13**(2), 1–27 (2020). https://doi.org/10.3390/en13020494
7. Khan, M.A.U.D., Uddin, M.F., Gupta, N.: Seven V's of big data understanding big data to extract value. In: Proceedings of the 2014 Zone 1 Conference of the American Society for Engineering Education—"Engineering Educ. Ind. Involv. Interdiscip. Trends", ASEE Zo. 1 2014 (2014). https://doi.org/10.1109/ASEEZone1.2014.6820689

8. Johnson, K.L., Bamer, A.M., Yorkston, K.M., Amtmann, D.: Use of cognitive aids and other assistive technology by individuals with multiple sclerosis. Disabil. Rehabil. Assist. Technol. **4**(1), 1–8 (2009). https://doi.org/10.1080/17483100802239648
9. Sufi, F., Khalil, I., Tari, Z.: A cardiod based technique to identify Cardiovascular Diseases using mobile phones and body sensors. In: 2010 Annual International Conference of the IEEE Engineering in Medicine and Biology Society, EMBC'10, vol. 2010, pp. 5500–5503 (2010). https://doi.org/10.1109/IEMBS.2010.5626578
10. Maskeliunas, R., Damaševicius, R., Segal, S.: A review of internet of things technologies for ambient assisted living environments. Futur. Internet **11**(12) (2019). https://doi.org/10.3390/FI11120259
11. Riva, G., Vatalaro, F., Davide, F.: 9 Interactive Context-Aware Systems Interacting with Ambient Intelligence. IOS Press (2005). Accessed: Dec. 17, 2020 [Online]. Available: http://www.ambientintelligence.org
12. Zhou, F., Jiao, J., Chen, S., Zhang, D.: A case-driven ambient intelligence system for elderly in-home assistance applications. IEEE Trans. Syst. Man Cybern. Part C Appl. Rev. **41**(2), 179–189 (2011). https://doi.org/10.1109/TSMCC.2010.2052456
13. Taleb, T., Bottazzi, D., Guizani, M., Nait-Charif, H.: Angelah: a framework for assisting elders at home. IEEE J. Sel. Areas Commun. **27**(4), 480–494 (2009). https://doi.org/10.1109/JSAC.2009.090511
14. Paganelli, F., Spinicci, E., Giuli, D.: ERMHAN: a context-aware service platform to support continuous care networks for home-based assistance. Int. J. Telemed. Appl. **2008** (2008). https://doi.org/10.1155/2008/867639
15. Gu, T., Pung, H.K., Zhang, D.Q.: Toward an OSGi-based infrastructure for context-aware applications. IEEE Pervasive Comput. **3**(4), 66–74 (2004). https://doi.org/10.1109/MPRV.2004.19
16. Kamel, M.B.M., George, L.E.: Remote patient tracking and monitoring system. Int. J. Comput. Sci. Mob. Comput. **2**(May), 88–94 (2013) [Online]. Available: www.ijcsmc.com
17. Archip, A., Botezatu, N., Şerban, E., Herghelegiu, P.C., ZalĂ, A.: An IoT based system for remote patient monitoring. In: Proceedings of 2016 17th International Carpathian Control Conference ICCC 2016, no. September 2020, pp. 1–6 (2016). https://doi.org/10.1109/CarpathianCC.2016.7501056
18. Andrews, L.J.B., Raja, L.: Remote based patient monitoring system using wearable sensors through online and offline mode for android based mobile platforms. In: 2017 International Conference on Infocom Technologies and Unmanned Systems Trends and Future Directions ICTUS 2017, vol. 2018-Janua, pp. 602–606 (2018). https://doi.org/10.1109/ICTUS.2017.8286080
19. Abdelrahman Rashed, A.K., Ibrahim, A., Adel, A., Mourad, B., Hatem, A., Magdy, M., Elgaml, N.: Integrated IoT medical platform for remote healthcare and assisted living. In: Japan-Africa Conference on Electronics, Communications and Computers, pp. 160–163 (2017)
20. DIvakaran, S., Manukonda, L., Sravya, N., Morais, M.M., Janani, P.: IOT clinic-Internet based patient monitoring and diagnosis system. In: IEEE International Conference on Power, Control, Signals and Instrumentation Engineering ICPCSI 2017, pp. 2858–2862 (2018). https://doi.org/10.1109/ICPCSI.2017.8392243
21. Hanskunatai, A.: A new hybrid sampling approach for classification of imbalanced datasets. In: 2018 3rd Internationa Conference on Computer and Communication System ICCCS 2018, pp. 278–281 (2018). https://doi.org/10.1109/CCOMS.2018.8463228
22. Hassan, M.K., El Desouky, A.I., Elghamrawy, S.M., Sarhan, A.M.: Intelligent hybrid remote patient-monitoring model with cloud-based framework for knowledge discovery. Comput. Electr. Eng. **70**, 1034–1048 (2018). https://doi.org/10.1016/j.compeleceng.2018.02.032
23. Elghamrawy, S.: Healthcare informatics challenges: a medical diagnosis using multi agent coordination-based model for managing the conflicts in decisions. In: Advances in Intelligent Systems and Computing, vol. 1261 AISC, pp. 347–357 (2021). https://doi.org/10.1007/978-3-030-58669-0_32

24. Abdel-Hamid, N.B., ElGhamrawy, S., El Desouky, A., Arafat, H.: A dynamic spark-based classification framework for imbalanced big data. J. Grid Comput. **16**(4), 607–626 (2018). https://doi.org/10.1007/s10723-018-9465-z
25. El-Ghamrawy, S.M.: A knowledge management framework for imbalanced data using frequent pattern mining based on bloom filter. In: Proceedings of 2016 11th International Conference on Computer Engineering and Systems, ICCES 2016, pp. 226–231 (2017). https://doi.org/10.1109/ICCES.2016.7822004
26. ELGhamrawy, S., Hassanien, A.E.: Diagnosis and Prediction Model for COVID19 Patients Response to Treatment based on Convolutional Neural Networks and Whale Optimization Algorithm Using CT Images (2020). https://doi.org/10.1101/2020.04.16.20063990

# The Relationship Between the Government's Official Facebook Pages and Healthcare Awareness During Covid-19 in Jordan

**Zeyad Mahmoud Al-Shakhanbeh and Mohammed Habes**

**Abstract** Today new trends in new media and advertising have revolutionized the whole world. Social media have enabled health sectors to access the potential audience and persuade them effectively. In this regard, coronavirus is currently a prevalent disease as it is disseminated with great speed and a distinct composition in 2020. In addition, many people are still unaware of its severity and lack any relevant information and this is burdening the healthcare system, especially in Jordan. This chapter aims to explore the role of Facebook in health awareness concerning Covid-19 in Jordan. It has adopted the descriptive approach, using a content-analysis form that consists of main categories and sub-topics carefully selected to analyze three Facebook pages for the official Jordanian institutions. These Facebook pages are directly linked to Covid-19, namely: The Prime Ministry, the Ministry of Health, and the Crisis Management Center (CMC). The study analyzed their posts regarding Covid-19 awareness during the period from April 1, 2020 to April 30, 2020. Through the analysis process, six topics were revealed in terms of health awareness: its attractive approach, source of information, factors, size of interaction, and method of presentation and language. The success of using Facebook for Covid-19 awareness requires that other official institutions raise their level of Facebook use in solving different issues, guiding citizens as to how the work is done through it and strengthening the partnership between official institution and citizens through this network. This opens the door to further research that independently addresses awareness topics and analyses of other Facebook pages to reveal more about Covid-19 and to more studies addressing the use of official institutions' pages by social media to spread awareness of the current health crisis worldwide.

**Keywords** Facebook · Social media · Healthcare awareness · Covid-19 · Coronavirus · Pandemic · Jordan

Z. M. Al-Shakhanbeh
University of Petra, Amman, Jordan

M. Habes (✉)
Department of Radio and Television, Mass Communication College, Yarmouk University, Irbid, Jordan



A.-E. Hassanien et al. (eds.), *Advances in Data Science and Intelligent Data Communication Technologies for COVID-19*, Studies in Systems, Decision and Control 378, https://doi.org/10.1007/978-3-030-77302-1_12

# 1 Introduction

Today, people find themselves forced to use, and adapt to, new technology and take advantage of it after it has penetrated all fields of life [1–5]. In this regard, the use of social networks facilitates the spread of ideas and the exchange of information, especially during times of crisis or during issues of concern to society [6–8], as is the case with the Covid-19 pandemic that has stopped the movement of the world. The social networks are crowded with health information about Covid-19, which is mostly directed by government officials to reduce the spread of the epidemic [9]. To raise awareness further, the World Health Organization is also collaborating with local media platforms. According to [10], to counteract an increased rate of coronavirus, social awareness is important. For this purpose, the global community has introduced several instructions, especially concerning mobility and traveling. In addition, many healthcare advertising campaigns have also been initiated to educate people about the transmission and the effects of this pandemic [11]. As coronavirus is increasing, we are learning new things about it every day [12, 13]. For this reason, informing the public about the disease's origin, causes, and precautionary measures is the need of the day [14]. Moreover, media sources can play a prominent role in raising public awareness about coronavirus as a fatal respiratory disease [15]. Wogu also highlighted the importance of media-based healthcare campaigns and considered the campaigns a source of effective information. According to the figures, the number of Facebook users in Jordan has reached 67% and the overall number of internet users is approximately 8.7 million. So, in Jordan, Facebook is the main digital platform for monitoring and transmitting news of Covid-19 and for following the procedures imposed by the government in response to the disease outbreak [16, 17]. Thus, it is important to examine prevailing pandemic awareness through social media, which is today considered comparatively more influential than traditional media. Additionally, it is also necessary to figure out the role of official institutions' pages via Facebook in terms of pandemic-related information in Jordan where the number of users of Facebook, compared with the estimated population of more than 10 million, is about 57%. It is also important to examine the use of Facebook since it is the most widely used resource for health awareness, news, and public mobilization [18]. There is also a need to figure out the content, aimed at raising awareness of Covid-19, published by the official Jordanian institutions concerned with the protection of the people, and the extent to which the public interacts with these institutions' posts. Health awareness is defined as the sum of the health communication and information activities directed at people concerning the risks of diseases and aimed at educating society on better health values [19]. This awareness is divided into health education, health guidance, health prevention, health statistics, health obligations with instructions, and health warnings [20, 21]. Previous studies have emphasized Facebook's role in spreading healthcare awareness during previous outbreaks. This has motivated us to do further investigation into the same topic in Jordan during the Covid-19 pandemic. On one hand, the first contribution of this work is to confirm Facebook's role in such matters and to highlight its importance in spreading healthcare awareness. On the other hand, another implication of this contribution is that Jordanian Facebook pages are largely concerned with the healthcare system, including the number of cases, days of quarantine, etc., as these issues are not yet covered by previous studies concerning healthcare awareness. This work also contributes to reporting on the most active bodies regarding digital media-based healthcare activism, which has also not been addressed in previous work.

## 2 Literature Review

According to the World Health Organization, the term "coronavirus" encompasses a wide range of viruses that cause severe respiratory ailments in humans and animals [22]. The recently discovered Covid-19 virus is an infectious disease caused by the most recent coronavirus strain [23]. The spread of coronavirus from China to other countries was very brisk and many studies on this disease also found a basic symbol (R0) for its reproduction, which is understood by the science of epidemiology and infectious diseases [24]. Covid-19 caused major concerns for health experts, creating certain challenges for doctors and researchers as their estimates were consistent with the statement issues but the World Health Organization identified that two out of every people were victims of coronavirus. It is therefore obvious that, unlike other diseases, coronavirus spreads through direct human contact [25]. Researchers in media and marketing are also obliged to play their part in spreading awareness regarding this major health concern by using different strategies. More particularly, using media advertisements as public-service messages can help the fight against coronavirus [26]. We can employ these marketing messages to educate the public about how to protect themselves and about measures to reduce its spread as many individuals are still ignoring the seriousness of this disease [27–29]. This disease includes seasonal influenza, in addition to bacterial pulmonary diseases, derived from the SARS virus as some studies have shown that this virus is transmitted from infected animals to humans and is present in the marine food market in Wuhan, China [30]. The process of ascertaining the pathological injury takes place through verification at the beginning: does the person communicate with anyone who was outside the country, or who is carrying the disease? It is then necessary to perform lung imaging and a clinical examination. This most important examination delivers the final result, showing whether the person is infected or not through RT-PCR. If the result is positive, the individual is a carrier of the virus, but if it is negative, the individual is not infected [24]. However, this disease is much more lethal than previous epidemics such as SARS and MERS [24].

### 2.1 Facebook and Healthcare Awareness

Modern healthcare-awareness policies are dependent on integrating modern media across networks. Social media make it easier for health organizations to manage their content as well as to interact with social media users and help to find required information [31, 32]. According to Al-Dweeri et al. [33], social networks provide a great opportunity to facilitate the flow of health information through dynamic and evolving social-networking platforms. These further contribute to increasing public awareness by accelerating the process of information exchange. Also, Facebook is a more attractive site for the health sector to use to raise awareness and adapt health-literacy strategies. According to Salloum et al. [2], social media have introduced new communication technologies to expand the scope of educational opportunities so as to increase healthcare knowledge and develop initiatives to search for appropriate information and bridge the health gap. One way of further validating this phenomenon [34] is to scrutinize how different profit and non-profit medical associations use Facebook advertising to raise healthcare awareness. The researchers conducted a content analysis of $n = 1760$ comments and posts by different healthcare organizations. Findings revealed that non-profit organizations were comparatively more active on Facebook platforms than were other organizations. Also, these organizations were using more interactive features to educate the public about healthcare programs and their importance. Overall, the organizations were tactfully using Facebook to advertise their awareness messages. However, the researchers suggested more strategic and organized usage to produce the desired outcomes [35]. As social media offers a highly cost-effective means of spreading healthcare awareness, and because growing competition and expensive marketing are features of traditional media forms, social media has become comparatively preferable and more accessible to a worldwide audience. For this reason, healthcare professionals and government resources have interlinked public healthcare advertising with digital media to gain even more positive outcomes [35].

### 2.2 The Facebook Pages of Official Institutions

According to Triñonal [36], the success and acknowledgment of healthcare awareness are highly reliant upon the messages' ability to attract the potential audience. These messages are also capable of sharing a concise, clear, and easily understandable message with the audience. The researcher also analyzed the role of Facebook as one of the most preferred social networking sites for spreading tuberculosis awareness in the Philippines [36]. Data gathered by using online surveys from $n = 50$ respondents showed that they often read the information available on Facebook regarding tuberculosis. However, this data also revealed that they kept information concerning HIV and other prevailing health ailments. Therefore, the researcher concluded that healthcare information was informative, addressing different aspects of the disease and other possible consequences. According to Anand et al. [37] the use of social media for healthcare awareness is a highly acknowledged phenomenon. Although healthcare organizations mainly prefer traditional media for spreading public-service messages, the role of social media is also under great consideration. Here people are empowered to actively select the relevant information, which further guarantees positive outcomes. Qasimi [38] revealed that using Facebook increased the participation and interaction rate concerning Ebola and that the most common motivation for such use was health awareness. Al-Afif et al. [39] showed that 37% of the medical cities and government hospitals in Riyadh, Saudi Arabia, did not have social-media platforms to raise awareness of Covid-19 (13). Xu et al. [40] also found that the health media on the Facebook page of the Jordanian Ministry of Health were characterized by variety and that the health awareness percent increased after there was a focus on posting through Facebook [41]. A recent study conducted by Nooh et al. [15] revealed that the most popular tools for health education were text + link + image. They further focused on health prevention in the awareness process [21]. Meanwhile, Melissa showed that Facebook was the most widely used social media by Dutch health institutions in health affairs [19, 42]. Furthermore, Jain et al. [16] asserted that Facebook was most influential in raising health and behavior issues and using emotive types of content [43].

## 2.3 Facebook and Health Awareness Regarding Coronavirus in Jordan

Facebook is one of the most prominent means of transferring information on international issues and events [44, 45]. It is a reliable source of awareness, information, education, and correction. Facebook has proved its central role globally in terms of information related to Covid-19 and it is a key means for government agencies and international organizations to spread health instructions and information. Given the role played by Facebook and its impact, the Jordanian Ministry of Health signed an agreement with the Facebook company on March 6, 2020 to implement an awareness campaign [48] in which Facebook published the preventive awareness content issued by the Health Ministry so that it would appear in the form of alerts to users opening the Facebook platform from Jordan. These alerts then directed them to the official website of the Jordanian Ministry of Health. On March 17, 2020, Facebook published a statement saying that the company was working closely to promote efforts to combat misinformation regarding Covid-19 [6, 13]. Facebook is constantly working to show "trusted" content on combating Covid-19 on users' pages in the form of a "center for information" about the Covid-19 and it shows videos from the WHO and well-known experts to encourage people to apply social distancing [49]. For this purpose, Banerjee and Dash [49] examined the opinion of different healthcare professionals about disease-awareness advertisements in Jordan. The basic aim was to document the opinion of medical professionals concerning the quality of these online communication activities. Results indicated that the majority of healthcare professionals considered these online healthcare-awareness campaigns as a positive constructive step in reinforcing a healthy lifestyle. However, they also suggested that these awareness campaigns should be co-sponsored by different non-profit organizations to increase their effectiveness aside from commercial gain [50].

## 3 The Systematic Framework of the Chapter

"Analysis" is a research method in which features of textual, visual, or aural material are systematically categorized and recorded so that they can be analyzed [44]. The study sample [45] consists of three Facebook pages of official Jordanian institutions directly involved with Covid-19:

- The Jordanian Prime Ministry's Facebook page (https://web.facebook.com/PMOJO) which has more than half-a-million followers (571,174).
- The Jordanian Ministry of Health's Facebook page (https://web.facebook.com/mohgovjordan), with a quarter-of-a-million followers at the time of writing (254,994).
- The National Center for Security and Crisis Management (https://web.facebook.com/ncscmjordan), followed by 82,651 people. This is directly concerned with educating citizens about the Covid-19 epidemic.

As for the posts under analysis, these are the health-awareness posts concerning Covid-19 that were published on the three pages from April 1, 2020 to April 30, 2020 [53]. During this period, the local government imposed Jordan's defense law to confront Covid-19, signaling a potential crisis during which daily routine life in Jordan and most of the world stopped. The content-analysis form was divided into several categories with each category containing several topics (see Appendix 1) as follows:

(A) **Health guidance**: i.e. informing the public of the proper ways to deal with this virus.
(B) **Health education**: i.e. increasing information about this virus. **Health prevention**: informing the public about the necessary measures to prevent virus transmission.
(C) **Health statistics**: the number of confirmed cases, active cases, and deaths. This statistical record is daily and increases awareness by determining the magnitude of the prevalence.
(D) **Health obligation**: i.e. laws and regulations imposed by the country on citizens in order to fight this epidemic.
(E) **Health warning**: Any warning against behaviors that might cause injury, or warning against violating official instructions.
(F) **The type of attraction**: the approach used in the awareness message, including mental, emotional, and double attractions.

(G) **Source category**: this includes the presence or absence of information sources in the education posts.

(H) **Category of factors**: the agency posting the topic of health education, including the Prime Minister, Ministry of Health, Crisis Management Center, Minister of Information and medical staff.

(I) **Interaction category**: the extent to which the public interacts with posts, including the number of comments on a post, the number of shares of the post, and the number of likes.

(J) **Post's display category**: the form in which the information of the post is presented: text, image, video, image + text, image + text + video, text + video, infographic design, text + infographic design, text + link, mixed.

(K) **Post's language category**: including classical Arabic, colloquial Arabic, and English. The previously proposed framework was taken and altered in the context of digital media to develop relations. The evaluation of the measurement model is conducted by using the Statistical Package for the Social Sciences (SPSS).

## 4 Reliability of the Measuring Instrument

The researchers conducted a post-test using Holsti's equation, which states:

$$\text{Reliability} = \frac{2m}{2!N1 + N2} = 41 \times 2 = 82 = 91.11\%\ 45 + 45 = 90$$

*M* in the equation symbolizes the number of coding agreements on which coders agree. N1 + N2 represents the total number of coding agreements by the coders. For this purpose, many coders (analysts) were trained journalists, such as Mohammed Al-Fuqahaa/Jordan News Agency (Petra) and (Dr Marcel Jwenat/ University of Al-Khwarizmi). They analyzed individually the content of 10% of the posts in the analysis and it turned out that the number of agreements was 41 out of 45. In other words, the reliability degree was 91.11%. This confirms that the tool is highly applicable.

## 5 Findings and Discussion

### *5.1 Most Popular Health-Awareness Content Regarding Covid-19 on the Facebook Pages of Jordanian Institutions Concerned with Fighting Covid-19*

Table 1 shows that the pages of the Jordanian Ministry of Health (44%) and the Jordanian Prime Ministry (47%) focused on raising awareness of "health statistics", i.e. the pages demonstrate the ministries' interest in publishing the numbers of confirmed cases, the number of patients that have recovered, the locations of the cases and the number of daily intended and random tests. This is inconsistent with Abdul Nour, which focused on the health-prevention aspect of the awareness process through Facebook. It comes in line with the page of the Jordanian Crisis Management Center, interested in publishing the two aspects: health prevention and health statistics with the same percentage of 24%. The crisis management center's interest was also similar for health prevention since the prevention process falls within its tasks in dealing with the crisis with all its types.

### *5.2 Interaction with Posts for Healthcare Awareness Through the Facebook Pages of Jordanian Institutions Involved in Fighting Covid-19*

Table 2 shows that the Jordanian Ministry of Health's page ranked first in the rate of interaction regarding posts related to Covid-19, with a total of 1,671,280. This confirms the interest of the Jordanian public in interacting with the page of the Ministry of Health rather than with the page of the Prime Ministry, which had a total interaction figure of 259,618. Third place went to the Crisis Management Center's page which had an interaction rate of 4920. Therefore, the previous results are consistent with Lubna Kassimi (2015) which showed a high participation and interaction rate in regard to Ebola because of the use of Facebook in the awareness process.

The Ministry of Health's Facebook page contained the highest number of posts on health-awareness topics with $n$ = 209 posts, 47% of the total number of posts for the three pages. This is to be expected as it is directly concerned with Covid-19, whereas the Prime Ministry and the Crisis Management Center are concerned only with planning, monitoring and guidance, which do not require a high number of posts. It appears that the Jordanian institutions tend to use digital means of communication for raising awareness of Covid-19. This can be compared with the Saudi health institutions as Khaled al-Ferm has shown that 37% of the medical cities and government hospitals in Riyadh do not have social-communication platforms through which they educate people, as mentioned in Table 3.

More than half-a-million Facebook users are following the official Facebook page of the Jordanian Ministry. However, the total interaction with its health education posts is less than half the number of followers, as shown in Table 4. On the other hand, the official Facebook page of the Jordanian Ministry of Health has 254,994 followers up to the time of writing this chapter. Yet the total interaction with its posts on health education is more than 1,600,000, which indicates a large following by the public who confidently rely on this page's credibility during the pandemic. This is consistent with the study conducted by Mohammed Fadhil Ali, which showed that there was a variety of healthcare topics on the official Facebook

**Table 1** The popular health-awareness regarding Covid-19 at Facebook page

| | Health-awareness topics | Health guidance | Health education | Health prevention | Health statistics | Health obligations with instructions | Health warnings | Total |
|---|---|---|---|---|---|---|---|---|
| The Jordanian Ministry of Health Facebook page | Topic frequency | 29 | 38 | 18 | 93 | 18 | 13 | 209 |
| | Percent (%) | 14 | 18 | 9 | 44 | 18 | 6 | 100 |
| The Jordanian Prime Ministry Facebook page | Frequency | 11 | 13 | 12 | 59 | 14 | 17 | 126 |
| | Percent (%) | 9 | 10 | 10 | 47 | 11 | 13 | 100 |
| The National Center of Security and Crisis Management (NCSCM) Facebook page | Topic frequency | 16 | 19 | 27 | 27 | 14 | 8 | 111 |
| | Percent (%) | 14 | 17 | 24 | 24 | 13 | 7 | 100 |

**Table 2** Jordanian Ministry of Health's page ranked

| Pages | Likes | Comments | Posts | Total interaction |
|---|---|---|---|---|
| Ministry of Health's page | 1.631.890 | 28.375 | 10.875 | 1.671.280 |
| Prime Ministry's Page | 225.200 | 26.405 | 8.013 | 259.618 |
| Crisis Management Center's page | 36.264 | 4.691 | 3.965 | 44.920 |

**Table 3** Number of health education posts on the pages

| Pages | Number of posts | Percentage of the total number of posts (%) |
|---|---|---|
| Ministry of Health's page | 209 | 47 |
| Prime Ministry's page | 126 | 28 |
| Crisis Management Center's page | 111 | 25 |
| Total | 446 | 100 |

**Table 4** Number of awareness posts versus page followers

| Pages | Total page followers | Total interaction with health-awareness posts |
|---|---|---|
| Ministry of Health's page | 254.994 | 1.671.280 |
| Prime Ministry's page | 571.174 | 259.618 |
| Crisis Management Center's page | 82.651 | 44.920 |

page of the local Ministry of Health, resulting in increased healthcare awareness among Jordanian citizens.

As for the National Center for Security and Crisis Management's page, total interaction amounted to nearly half of the page's followers. This requires developing the page and increasing the volume of followers and interactions.

## 5.3 *Persuasive Appeals Used in Awareness Posts About COVID-9 on the Facebook Pages of Jordanian Institutions Involved in Fighting Covid-19*

The results show that the most common appeal used in health education is the psychological/emotional appeal, totaling 68% on the page of the Jordanian Ministry of Health, followed by 60% for the Jordanian Prime Ministry's page and 75% for the Jordanian Crisis Management Center's page. This is inconsistent with (Melissa 2014) which revealed the interest of the Dutch health institutions in mental-health education through focusing on the right methods of treatment and on educating

**Table 5** Distribution of the attractions on the three pages

| Pages | Mental attractions | Emotional attractions | Double attractions | Total |
|---|---|---|---|---|
| Ministry of Health's page | 142 | 39 | 28 | 209 |
| | 68% | 19% | 13% | 100% |
| Prime Ministry's Page | 76 | 26 | 24 | 126 |
| | 60% | 21% | 19% | 100% |
| Crisis Management Center's page | 83 | 15 | 13 | 111 |
| | 75% | 14% | 12% | 100% |

people about the diseases and their causes, as mentioned in Table 5. The interest of the Jordanian institutions in using emotional appeals indicates an interest in psychological discourse directed at people, asking them to abide by the instructions and procedures issued by the local government. This indicates a great percentage in terms of the awareness, education and commitment of the Jordanian people, especially during crises. This appeal is followed mainly by emotional attraction, using texts, images and videos that motivate people to commit to love and fear regarding themselves and their humanity. In terms of frequencies and percentages, emotional attractions rank second since their focus is mainly on mental and emotional attractions, rather than on double attractions.

## 5.4 *Factors Containing Health-Awareness Posts on the Pages of Jordanian Institutions Concerned with Fighting Covid-19*

The results in Table 6 show that, on the official page of the Ministry of Health, Dr Saad Jaber was the most active user with a posts total of 48%. However, this was to be expected as Dr Saad is the most responsible official concerned with this crisis. Again, it is unsurprising that posts from "Ministry of Health" ranked second with a total of 28%, including posts concerned with procedures, issues and directives about Covid-19. On the Official Page of the Prime Ministry, Dr Omar Al-Razaz was the most active user of the Official Account owned by the Prime Ministry with a total percentage of 27%. Usually, this page focuses on the Prime Minister who is also the Minister of Defense, deciding to work in defense in order to deal with this pandemic. Similarly, the "Prime Ministry" involved 25% of Facebook posts, mainly concerned with issuing daily instructions and directives. On the Official Facebook Account owned by the Crisis Management Center, the Crisis Management Center itself was the most active in terms of health-education posts with 43% of posts. As the crisis-management cell is located in the center, and this is where ministers and all experts and concerned bodies conduct emergency and regular meetings, it is to be expected that the respective page focuses more on Covid-19. Likewise, the

**Table 6** Posts on the pages of the Jordanian institutions

| Factors pages | Prime Ministry | Ministry of Health | Crisis Management Center | Prime Minister | Minister of Health | Minister of Information | Medical Staff | Total |
|---|---|---|---|---|---|---|---|---|
| Ministry of Health | 4 | 59 | 13 | 5 | 101 | 14 | 13 | 209 |
| | 2% | 28% | 6% | 2% | 48% | 7% | 6% | 100% |
| Prime Ministry | 31 | 10 | 4 | 34 | 21 | 23 | 3 | 126 |
| | 25% | 8% | 3% | 27% | 17% | 18% | 2% | 100% |
| Crisis Management Center | 7 | 3 | 48 | 9 | 19 | 19 | 6 | 111 |
| | 6% | 3% | 43% | 8% | 17% | 17% | 5% | 100% |
| Total | 42 | 72 | 65 | 48 | 141 | 56 | 22 | 446 |

Minister of Health and the Minister of Information, Amjad Al-Adayla, posted with the same ratio (17%). However, previous studies have not addressed the most active bodies regarding digital media-based healthcare activism; therefore, the result will also open the door for future studies to concentrate on this crucial sector in analyzing social networks.

### *5.5 Forms, Language and Sources (Health-Awareness Posts) on Official Facebook Pages Concerned with Confronting Covid-19*

Table 7 shows that the most preferred form of presentation used by the three pages in health education is "textual videos" with a frequency of $n = 133$. Also, the integration of the text with the video content is highlighted and acknowledged by people. The second form is "images accompanied by the text", with a frequency of $n = 132$, and this indicates an explicit interest in the integration of text with images and video content. Moreover, the third form is "info-graphic design + text" which is a contemporary way of conveying persuasive information to the public. In this regard, again, previous studies are not consistent with the result of Nooh et al.'s [15] study, which stated that the most popular method of posting on health education on Facebook pages in Algeria was a mixture of text + link + image.

Regarding the type of language used in healthcare-awareness posts on the official pages regulated by Jordanian institutions, Table 8 shows that the three pages primarily use "standard Arabic" in the awareness posts, as compared with "colloquial Arabic" which appeared in some posts aimed at groups that were using colloquial Arabic for the sake of awareness. Finally, despite there being a significant number of members of the international community in Jordan, we found no post using the English language to spread awareness. Thus, we can assume that healthcare awareness information is hard to understand for foreigners living in Jordan. Once again, this result is inconsistent with previous studies that have not addressed the subject of the language used in awareness posts. Therefore, this is an important result that should be highlighted, especially regarding healthcare awareness.

Regarding the types of sources of healthcare awareness, it appears from the previous table that the Ministry of Health is the most prominent page with the source of most posts being the Ministry itself (90%). This confirms the activity of the Ministry and its contribution to fighting Covid-19. The Prime Ministry appeared as the second source, publishing 81% of the total posts. The Crisis Management Center, being an umbrella for all institutions and not so much an executive body, but rather a body for planning, directing and coordinating with various institutions, came in third place in terms of posting relevant content. This result is also inconsistent with previous studies that did not address the sources of health awareness. This is an important result which reveals the role and frequency of awareness, indicating an important concern to be studied (Table 9).

**Table 7** Forms of health-awareness posts on the Facebook pages

| Forms of presentation | Ministry of Health | Prime Ministry | Crisis Management Center | Total |
|---|---|---|---|---|
| Text | 12 | 22 | 19 | 53 |
| Image | 12 | 1 | 3 | 16 |
| Video | 13 | 0 | 0 | 13 |
| Image + Text | 105 | 9 | 18 | 132 |
| Image + Text + Video | 0 | 0 | 0 | 0 |
| Text + Video | 15 | 63 | 55 | 133 |
| Text + Infographic Design | 41 | 31 | 15 | 87 |
| Text + Link | 6 | 0 | 1 | 7 |
| Mixed | 5 | 0 | 0 | 5 |

**Table 8** The languages used in health-awareness posts

| Pages | Standard Arabic | Colloquial Arabic | English | Total |
|---|---|---|---|---|
| Ministry of Health's page | 197 | 12 | 0 | 209 |
| | 94% | 6% | 00% | 100% |
| Prime Ministry's page | 117 | 9 | 0 | 126 |
| | 93% | 7% | 00% | 100% |
| Crisis Management Center's page | 104 | 7 | 0 | 111 |
| | 94% | 6% | 0 | 100% |
| Total and ratio out of (446) | 418 | 28 | 0 | 446 |
| | 94% | 6% | 00% | 100% |

**Table 9** Sources used in health-awareness posts

| Pages | Same institution | Another source |
|---|---|---|
| Ministry of Health's page | 188 | 21 |
| | 90% | 10% |
| Prime Ministry's page | 102 | 24 |
| | 81% | 19% |
| Crisis Management Center's page | 71 | 40 |
| | 64% | 36% |
| Total and ratio out of (446) | 361 | 85 |
| | 81% | 19% |

## 6 Conclusion and Future Research

The purpose of this chapter was to study the role of Facebook in health awareness in Jordan. Our results are confirmed by previous studies regarding the global role played by Facebook in previous outbreaks and the importance of relying on the site for rising healthcare awareness. It appears that Jordanian Facebook pages are largely concerned with the healthcare system, particularly numbers of cases, days of quarantine, mortality rates, both locally and globally, and the amount of equipment available. This further opens the doors for future research as it is yet not covered by previous studies concerning healthcare awareness. It was also evident that Facebook users in Jordan interacted with government institutions as they chose to pay special attention to accountable stakeholders during the Covid-19 pandemic. We found that their exposure to posts from the Jordanian Ministry of Health was greater than their interaction with other pages, such as the Prime Ministry and the Center for Crisis Management, which were more concerned with dealing with strategic management and planning. The Jordanian institutions were interested in using highly persuasive discourse in addressing Jordanian society about health awareness, and in so doing they chose to use emotional persuasion for a society with a high percentage of learners. These results were in agreement with the recent studies in the domain [6, 14, 54–57]. Finally, the government institutions should develop their Facebook pages, raise trust and interact more with their citizens. Similarly, they should consider their opinions and promote more publishing about healthcare measures. Furthermore, the three pages in question should also continue to publish on other matters concerning healthcare awareness that need strong consideration. The agricultural, cultural and educational institutions should also focus on posting on their issues through Facebook, at the same pace as dealing with Covid-19, to guide the community towards adopting constructive behavior. This work is limited to these three Jordanian official institutions: The Prime Ministry, the Ministry of Health, and the Crisis Management Center (CMC). In future work, we will expand the range of other Facebook pages and observe the Covid-19 awareness topics on such pages. In addition, we will conduct more research aimed at detecting the use of Facebook pages in other crises beyond coronavirus. Last but not least, we will increase the geographical area and include countries other than Jordan. Other social media, such as Twitter, YouTube, etc. can also be included.

## References

1. Agnoletto, R., Queiroz, V.C.: COVID-19 and the challenges in Education. CEST Bullit. **5**(2), 1–2 (2020)
2. Salloum, S.A., Al-Emran, M., Khalaf, R., Habes, M., Shaalan, K.: An innovative study of e-payment systems adoption in higher education: theoretical constructs and empirical analysis. Int. J. Interact. Mob. Technol. **13**(6) (2019)

3. Salloum, S.A., Al-Emran, M., Abdallah, S., Shaalan, K.: Analyzing the Arab Gulf Newspapers Using Text Mining Techniques, vol. 639 (2018)
4. Salloum S.A., Shaalan, K.: Adoption of e-book for university students. In: International Conference on Advanced Intelligent Systems and Informatics, pp. 481–494 (2018)
5. Salloum, S.A., Mhamdi, C., Al Kurdi, B., Shaalan, K.: Factors affecting the adoption and meaningful use of social media: a structural equation modeling approach. Int. J. Inf. Technol. Lang. Stud. **2**(3) (2018)
6. Habes, M., Alghizzawi, M., Ali, S., SalihAlnaser, A., Salloum, S.A.: The relation among marketing ads, via digital media and mitigate (COVID-19) pandemic in Jordan. Int. J. Adv. Sci. **29**(7), 2326–12348 (2020)
7. Al Mansoori, S., Almansoori, A., Alshamsi, M., Salloum, S.A., Shaalan, K.: Suspicious Activity Detection of Twitter and Facebook using Sentimental Analysis
8. Al-Maroof, R.S., Salloum, S.A., Hassanien, A.E., Shaalan, K.: Fear from COVID-19 and technology adoption: the impact of Google Meet during Coronavirus pandemic. Interact. Learn. Environ. (2020). https://doi.org/10.1080/10494820.2020.1830121
9. Brennen, A.J.S., Simon, F.M., Howard, P.N., Nielsen, R.K.: Types, Sources, and Claims of COVID-19 Misinformation. Oxford University Press, no. April, pp. 1–13 (2020)
10. World Health Organization: Coronavirus Disease 2019, vol. 2019, no. March, p. 2633 (2020). https://doi.org/10.1001/jama.2020.2633
11. Huang, R.H., Liu, D.J., Tlili, A., Yang, J.F., Wang, H.H.: Handbook on facilitating flexible learning during educational disruption: the Chinese experience in maintaining undisrupted learning in COVID-19 Outbreak. Smart Learn. Inst. Beijing Norm. Univ. UNESCO, 1–54 (2020)
12. Alhawamdeh, A.K., Alghizzawi, M., Habes, M.: The Relationship Between Media Marketing Advertising and Encouraging Jordanian Women to Conduct Early Detection of Breast Cancer The Relationship Between Media Marketing Advertising and Encouraging Jordanian Women to Conduct Early Detection of Breast Canc, no. May (2020). https://doi.org/10.7176/EJBM/12-12-11
13. Habes, M., Alghizzawi, M., Salloum, S.A., Mhamdi, C.: Effects of facebook personal news sharing on building social capital in Jordanian Universities. In: Recent Advances in Intelligent Systems and Smart Applications. Springer, 2020, pp. 653–670.
14. Yang, P., Wang, X.: COVID-19: a new challenge for human beings. Cell. Mol. Immunol. **17**(5), 555–557 (2020). https://doi.org/10.1038/s41423-020-0407-x
15. Nooh, H.Z., et al.: Public awareness of coronavirus in Al-Jouf region, Saudi Arabia. J. Public Heal. (2020). https://doi.org/10.1007/s10389-020-01209-y
16. Jain, M.R., Gupta, P., Anand, N.: Impact of social networking sites in the changing mindset of youth on social issues-a study of Delhi-NCR youth. Res. World, **3**(2 Part 2), 36 (2012)
17. Almuhaisen, O., Habes, M., Alghizzawi, M.: An empirical investigation the use of information, communication technologies to english language acquisition: a case study from the Jordan. Development, **7**(5) (2020)
18. Ali, S.: Social media usage among teenage girls in Rawalpindi and Islamabad. Glob. Media J. **16**(31), 1 (2018)
19. Verhaag, M.L.: Social media and healthcare–hype or future?: status update of the social media use in the healthcare industry. University of Twente (2014)
20. Al-Khazaleh, M.S.F.: The impactof social networking websites on the system of university values among students of Al Ain University of Science and Technology in the UAE. Dirasat Hum. Soc. Sci. **47**(1) (2020)
21. Sharaydih, R., Abuloha, S., Wazaify, M.: Promotion of appropriate knowledge and attitude towards medicines among schoolchildren in Jordan: the role of teachers. Int. J. Pharm. Pract. **28**(1), 84–91 (2020)
22. WHO: Coronavirus disease 2019 (67). World Heal. Organ. **2019**(March), 2633 (2020). https://doi.org/10.1001/jama.2020.2633
23. W. H. Organization: Rational use of personal protective equipment for coronavirus disease (COVID-19): interim guidance, 27 February 2020. World Health Organization (2020)

24. Peeri, N.C., et al.: The SARS, MERS and novel coronavirus (COVID-19) epidemics, the newest and biggest global health threats: what lessons have we learned? Int. J. Epidemiol. (2020)
25. Liu, Y., Gayle, A.A., Wilder-Smith, A., Rocklöv, J.: The reproductive number of COVID-19 is higher compared to SARS coronavirus. J. Travel Med. (2020)
26. Zu, Z.Y., et al.: Coronavirus disease 2019 (COVID-19): a perspective from China. Radiology, 200490 (2020)
27. Debatin, B., Lovejoy, J.P., Horn, A.-K., Hughes, B.N.: Facebook and online privacy: attitudes, behaviors, and unintended consequences. J. Comput. Commun. **15**(1), 83–108 (2009)
28. Huh, S.: How to train the health personnel for protecting themselves from novel coronavirus (COVID-19) infection during their patient or suspected case care. J. Educ. Eval. Health Prof. **17**, 10 (2020)
29. Schiffman, L.G., Wisenblit, J.L.: Consumer behavior (Vol. 11). English Pearson Education Ltd. (2015)
30. Lai, C.-C., Shih, T.-P., Ko, W.-C., Tang, H.-J., Hsueh, P.-R.: Severe acute respiratory syndrome coronavirus 2 (SARS-CoV-2) and corona virus disease-2019 (COVID-19): the epidemic and the challenges. Int. J. Antimicrob. Agents, 105924 (2020)
31. Habes, M., Alghizzawi, M., Khalaf, R., Salloum, S.A., Ghani, M.A.: The relationship between social media and academic performance: facebook perspective. Int. J. Inf. Technol. Lang. Stud. **2**(1) (2018)
32. Al-Emran, M., Salloum, S.A.: Students' attitudes towards the use of mobile technologies in e-evaluation. Int. J. Interact. Mob. Technol. **11**(5), 195–202 (2017). https://doi.org/10.3991/ijim.v11i5.6879
33. Al-dweeri, R.M., Obeidat, Z.M., Al-dwiry, M.A., Alshurideh, M.T., Alhorani, A.M.: The impact of e-service quality and e-loyalty on online shopping: moderating effect of e-satisfaction and e-trust. Int. J. Mark. Stud. **9**(2), 92 (2017)
34. Park, H., Rodgers, S., Stemmle, J.: Health organizations' use of facebook for health advertising and promotion. J. Interact. Advert. **12**(1), 62–77 (2011). https://doi.org/10.1080/15252019.2011.10722191
35. Pillai, P.: Social Media in Healthcare Making the Case (2012)
36. Triñonal, J.P.: Social Networks for Health Information: Increased Awareness on Tuberculosis Through Viral Marketing Schemes in Facebook Jerome P. Triñona 1, pp. 1–30 (2012)
37. Anand, S., Gupta, M., Kwatra, S.: Social media and effective health communication. Int. J. Soc. Sci. **2**(8), 39–46 (2013)
38. Qasimi, L.: The Role of Social Networks in Health Awareness on Ebola: Facebook Pages as Model, al-Arabi bin Mehadi um Bouaqi university, Algeria (2015)
39. Al-Afif, K.A.M., et al.: Understanding the burden of atopic dermatitis in Africa and the Middle East. Dermatol. Ther. (Heidelb) **9**(2), 223–241 (2019)
40. Xu, Z., et al.: Pathological findings of COVID-19 associated with acute respiratory distress syndrome. Lancet Respir. Med. (2020)
41. J. Ministry of Health: COVID-19 Updates in Jordan (2020)
42. Habes, M.: The influence of personal motivation on using social TV: a uses and gratifications approach. Int. J. Inf. Technol. Lang. Stud. **3**(1) (2019)
43. Salloum, S.A., Shaalan, K.: Investigating students' acceptance of e-learning system in higher educational environments in the UAE: applying the extended technology acceptance model (TAM). The British University in Dubai (2018)
44. Alghizzawi, M., Salloum, S.A., Habes, M.: The role of social media in tourism marketing in Jordan. Int. J. Inf. Technol. Lang. Stud. **2**(3) (2018)
45. Al Muhaisen, O., Habes, M., Alghizzawi, M.: An empirical investigation the use of information, communication technologies to english language acquisition : a case study from the Jordan. Int. J. Innov. Eng. Sci. **7**(5), 261–269 (2020)

46. Alghizzawi, M., Habes, M., Salloum, S.A.: The relationship between digital media and marketing medical tourism destinations in Jordan: facebook perspective. In: International Conference on Advanced Intelligent Systems and Informatics, pp. 438–448 (2019)
47. Trivedi, N., Krakow, M., Hyatt Hawkins, K., Peterson, E.B., Chou, W.-Y.S.: 'Well, the message is from the institute of something': exploring source trust of cancer-related messages on simulated facebook posts. Front. Commun. **5**, 12 (2020)
48. Holthus, B., Gagné, I., Manzenreiter, W., Waldenberger, F.: Japan Through the Lens of the Tokyo Olympics. Taylor & Francis (2020)
49. Banerjee, S., Dash, S.K.: Effectiveness of disease awareness advertising in emerging economy: views of health care professionals of India. J. Med. Mark. Device, Diagnostic Pharm. Mark. **13**(4), 231–241 (2013). https://doi.org/10.1177/1745790413516479
50. Coe, K., Scacco, J.M.: Quantitative Content Analysis (2017)
51. Habes, M., Salloum, S.A., Alghizzawi, M., Alshibly, M.S.: The role of modern media technology in improving collaborative learning of students in Jordanian universities. Int. J. Inf. Technol. Lang. Stud. **2**(3) (2018)
52. World Bank: Tertiary Education and COVID-19: Impact and Mitigation Strategies in Europe and Central Asia (2020). https://doi.org/10.1017/CBO9781107415324.004
53. Alghizzawi, M., et al.: The impact of smartphone adoption on marketing therapeutic tourist sites in Jordan. Int. J. Eng. Technol. **7**(4.34), 91–96 (2018)
54. Habes, M., Alghizzawi, M., Salloum, S.A., Ahmad, M.F.: The use of mobile technology in the marketing of therapeutic tourist sites: a critical analysis. Int. J. Inf. Technol. Lang. Stud. **2** (2) (2018)
55. Salloum, S.A., Al-Emran, M., Habes, M., Alghizzawi, M., Ghani, M.A., Shaalan, K.: Understanding the impact of social media practices on e-learning systems acceptance. In: International Conference on Advanced Intelligent Systems and Informatics, pp. 360–369 (2019)
56. Haleem, A., Javaid, M., Vaishya, R.: Effects of COVID-19 pandemic in daily life. Curr. Med. Res. Pract. (April), 10–12 (2020). https://doi.org/10.1016/j.cmrp.2020.03.011
57. Alhumaid, K., Ali, S., Waheed, A., Zahid, E., Habes, M.: COVID-19 & Elearning : Perceptions & Attitudes Of Teachers Towards E- Learning Acceptancein The Developing Countries, vol. 6, no. 2, pp. 100–115 (2020). https://doi.org/10.5281/zenodo.4060121

# The Influence of YouTube Videos on the Learning Experience of Disabled People During the COVID-19 Outbreak

**Khalaf Mohammed Tahat, Walaa Al-Sarayrah, Said A. Salloum, Mohammed Habes, and Sana Ali**

**Abstract** A record number of individuals are not attending educational institutions during the Covid-19 outbreak. This situation is highly thought-provoking, yet; the use of new media technology can overcome this challenge. In this regard, this chapter investigated the influence of YouTube usage as an educational tool on the learning process of disabled people during the COVID-19 pandemic. The researchers selected n = 60 individuals working as disability specialists and used structural equation modeling to examine the proposed study model. The results revealed that there is a positive relationship between YouTube videos and e-Learning among disable individuals. Moreover, the quality, ease of use, and texts in the video also contribute to improving the disabled people's learning experiences. Thus, the results highly supported technology acceptance and usage during the global healthcare crisis. Moreover, the researchers also recommended evaluating other aspects of YouTube Videos that can influence people with disabilities about their social media usage, especially during an emergency like Covid-19.

**Keywords** ICTs learning · YouTube videos · Social Networking Sites (SNSs) · Disabilities · Technology acceptance · Covid-19

---

K. M. Tahat
Journalism Department, Mass Communication College, Yarmouk University, Irbid, Jordan

W. Al-Sarayrah
Faculty of Computer Science, Jordan University of Science and Technology, Irbid, Jordan

S. A. Salloum (✉)
Machine Learning and NLP Research Group, Department of Computer Science, University of Sharjah, Sharjah, UAE
e-mail: ssalloum@sharjah.ac.ae

S. A. Salloum
School of Science, Engineering, and Environment, University of Salford, Salford, UK

M. Habes
Department of Radio and Television, Mass Communication College, Yarmouk University, Irbid, Jordan

S. Ali
Department of Mass Communication, Allama Iqbal Open University, Islamabad, Pakistan



A.-E. Hassanien et al. (eds.), *Advances in Data Science and Intelligent Data Communication Technologies for COVID-19*, Studies in Systems, Decision and Control 378, https://doi.org/10.1007/978-3-030-77302-1_13

## 1 Introduction

During the current pandemic, healthcare workers are always making efforts to overcome the outbreak [1, 2]. Despite many struggles, the situation is still out of control and raising many challenges. According to Alhawamdeh et al. [3], social distancing and quarantine measures are raising several concerns for individuals all over the world. The current outbreak has adversely affected healthcare, economy, and labor, leading to gigantic social and psychological issues. The World Health Organization declared Covid-19 as a healthcare emergency that affects the healthcare system and challenging the other fields of life [4]. In this regard, education is an important sector, facing specific challenges during the pandemic as there is an ongoing struggle to sustain educational matters through a strategic management process [5, 6]. Despite institutional closure is a positive step towards social distancing for hindering the growth of infection spread, both students and instructors are facing several challenges to resume the learning process [6]. To maintain the flow of communication, many institutions resort to online learning education systems. They have online systems, which facilitate the learning procedures by sustaining social distancing. Countries who already had a digital learning system are not facing many problems as compared to those who have weak web services, relying mainly on the formal educational methods [7]. As the situation is still not controlled, the future of education is uncertain concerning the formal educational process. The multi-layered and multidimensional crisis of Covid-19 is leading to the adoption of equally helpful substitutes.

To mitigate the educational crisis, the use of Social Networking Sites is widely facilitating both teachers and students to learn efficiently without halting their educational activities [8]. Social Media platforms such as YouTube, Facebook, Zoom, Skype, WhatsApp, and others provide interactive platforms to learn at the same pace as the students are equally using these sites to continue their learning process. The rise of second-generation web-based technologies also triggered technology acceptance and usage among the students. Web-based learning is not easily available, but also, people can save it for future correspondence. These eLearning services provide flexible learning methods, communicating, and performing the required educational tasks [9]. Thus, despite Covid-19 is a quasi-adaptive challenge, continuing the educational journey is inevitable. For this purpose, the swift transformation and adoption of digital media resources is a significant step towards mitigating the current pandemic [8]. As noted by Alhumaid et al. [6], although the closure of educational institutions hampered the formal education system worldwide, virtual learning and accessibility to web-based education resumed the educational activities. Nonetheless, during the current pandemic, accepting e-learning was the only option to facilitate the students. Now, both instructors learn to acknowledge e-Learning as a substitute for formal learning during the current healthcare crisis. Thus, the current global healthcare crisis also raised several challenges for the education system in Jordan. The local Ministry of Education is searching for effective alternatives to mitigate the impacts of Covid-19 on the education and learning process [10]. In this context, the current chapter highlights the importance of Youtube-based learning as a substitute for the formal education environment, especially for the Jordanian disable students [11]. In the first phase, the researchers addressed the educational challenges during Covid-19 and cited the relevant literature supporting e-learning adoption. In the second and third phases, the researchers proposed the conceptual framework and highlighted the suitable study methods, respectively. Finally, in the fourth phase, the researchers conducted statistical analysis, obtained the results, and made the conclusions accordingly [10, 11].

## 1.1 Videos Impact on the Education Process

Nowadays, Information and computer technologies (ICTs) are used extensively in almost all aspects of life. ICTs provide new ways to spread information and knowledge, facilitate communication, and present motivational learning tools. Currently, ICTs such as smartphones, tablets, social networking sites (SNSs), etc., have become a companion of most areas of our life. They are of greater significance for the educational institutions to cope with the wider technological spread. An example of technological advancement in education is deploying videos to lectures and tutorials. Many institutions use computer-assisted learning techniques to improve the students' learning experience and give them a rich learning environment [12]. Allowing the students to review lectures repeatedly [11] helps in distance learning, allowing them to listen and view lectures similar to traditional classroom lectures [13]. TED, Coursera, and Udemy are significant examples of open learning systems that use video-based courses. Besides, video-based lectures are better than text in motivating students during class [14, 15]. The human mind remembers and understands better the information shown and heard during the eLearning process. Also, studies conducted by [16, 17], and [13] affirmed a better understanding of the audio and video lectures among eLearners. Similarly, YouTube offers intelligent media services of different kinds. YouTube has a pool of nifty features that involve every topic, including music, film clippings, news programs, and instructive videos [18]. YouTube is now preferred for its visual media technology and is one of the most popular social media platforms [19]. Thus, deploying different content into videos holds great promises for the audience, especially students and instructors [20].

## 1.2 Using YouTube in the Education Process

Multiple online video platforms are used in educational institutions, i.e., Teacher Tube, Vimeo, and YouTube [21]. These platforms allow students to view, share, or even download the video content. In this research, YouTube—a Web 2.0 tool—is targeted as an educational tool since it is the third most preferred website right after Google and Facebook [22]. YouTube is used by people of all ages, especially by the youth [23]. Also, it can be used as a source of entertainment by users. Several studies witnessed that the integration of YouTube into the educational sector has positive advantages [12, 24]. Any academic topic can be implemented and assisted through YouTube, such as medicine [25], nursing [24], teaching English [26], and business [27]. One of the main advantages is that searching and sharing educational topics became convenient. Besides, it assists teamwork and student–teacher collaboration and enhances the presented topic's quality by facilitating gathering feedback [28]. YouTube makes the learning experience more attractive, excited, creative and enhances the students' engagement. One of the reasons behind considering YouTube a valuable educational source is that students are familiar with this technology. They use it for educational purposes as it grabs their attention faster than the traditional ways. YouTube videos are conveniently accessible, and are free of cost. In other words, YouTube helps students to join educational channels and bring them a step closer to self-education and eLearning.

## 1.3 Using YouTube in the Education Process of People with Disabilities

According to several studies [29, 30], the extraordinary features of ICTs are why they are preferred as an educational aid to help people with disabilities. The main difficulties they face when using ICTs are accessibility and usability, which causes exclusion from the huge technological spread. As a result, it leads them to acquire poor reading, understanding, and lack of coordination between their eyes and hands [31]. In 2010, YouTube hosted more than two billion videos every day [32]. It helped the educational videos to reach a larger number of people with disabilities. YouTube provides an opportunity for people to connect, socialize, communicate, and share knowledge across the globe. YouTube also contains a big library of self-education videos available free of charge, facilitating people with disabilities to learn new things [33]. Also, these people can share their experiences with others as well. Furthermore, a study conducted by Alghizzawi et al. [34] found that YouTube was the most used online platform by people with disabilities in 2012. Another benefit of integrating YouTube into people with disabilities lives is to improve society's perception of disability and reduce the stigma associated with it.

## 1.4 Guidelines for Effective Educational YouTube Videos

To gain the maximum benefit from the educational videos on YouTube, the following guidelines and characteristics are proposed [35, 36]:

- Make the video focused on the learning goals and keep it brief.
- Enhance students' engagement by using styles that are enthusiastic and conversational.
- Add complementary and supportive elements to the videos, such as audio and visual elements.
- Maintain a sufficient production quality i.e., use HD videos, less background noise.
- Make the video timing suitable for the students to follow the presented content. The video should not be too fast or not too slow with natural breaks.
- Decrease the use of on-screen text by using narration and optional captions.
- Support the students while learning through on-screen labels and cues to guide their attention, spoken prompts, and questions.
- Present the content in an organized structure.

This study tends to investigate the influence of using YouTube as an educational tool on the learning process of people with disabilities. For this purpose, the technology acceptance model (TAM) is applied to the designated determinants.

## 2 Research Model and Study Hypotheses

Figure 1 presents the theoretical research model of this study. This model examines the relationships between the learning experience of people with disabilities and the determents, including video, text, quality, usability, and perceived usefulness. The following subsections contain a brief description of each factor and the developed hypotheses of this research.

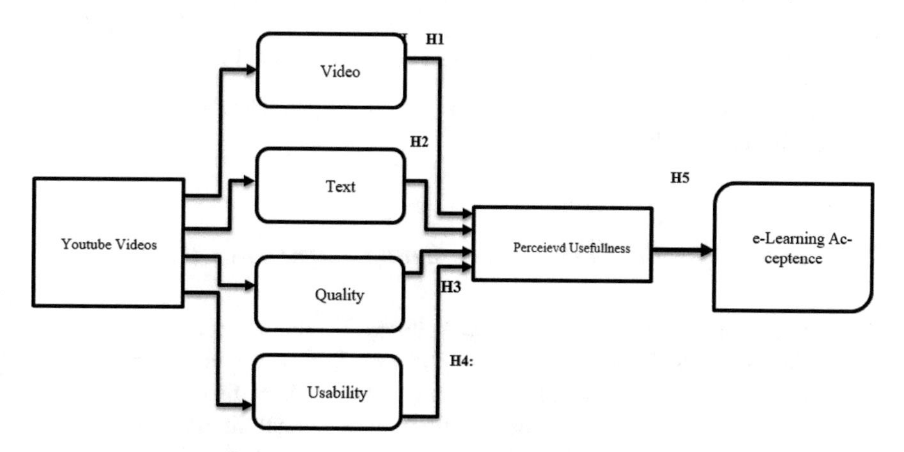


**Fig. 1** Research model

## 2.1 Video Type (VT) and Learning Experience (LE)

Generally, there are many valuable styles of eLearning videos online. They can differ in techniques but having the same purposes. These may include Screencasts, Talkinghead, Presentations, Lecture Recordings, Classroom Recording, Simulations, Animations, and others. These visual aids are of greater pertinence as they serve to make communication quick and easier and use visual acts as simulations for the emotions. Additionally, they help the learners to save information for longer intervals. Resulting in better comprehension and motivating the learners [37]. These videos can also enhance learning abilities at lower costs with ease of access to spacious study material [38]. Therefore, by keeping in view the types and importance of videos in eLearning, the following hypothesis is proposed:

**H1**: Video Type (VT) positively influences YouTube's perceived usefulness (PU) of people with disabilities.

## 2.2 YouTube Text (YT) and Learning Experience (LE)

YouTube Text (YT) is metadata that helps to understand the video content. YouTube text is of greater pre-eminence to increase the understanding and context of a video. A description, particularly in eLearning videos, can provide information about the learners' essential key points [39]. Adding text and description can help to increase the learning and accessibility to YouTube eLearning videos. Most importantly, with close captions on YouTube, a video can read more than 360 million people worldwide who have hearing issues. These captions and subtitles are created before releasing the video and it is ensured that they are synchronized with the video [40]. More than 100 studies affirmed the importance of YouTube text as equally beneficial for learners of all ages. They can help people who do not understand the video language or have hearing problems [41]. Thus, the argumentation validating the importance of YouTube text-enabled us to assume that:

**H2**: Video Text (VT) positively influences YouTube's perceived usefulness (PU) of people with disabilities.

## 2.3 Videos Quality (VQ) and Learning Experience (LE)

The use of better quality video content to facilitate eLearning is essential. A study in 2015 affirmed that 93% of instructors believe that using video material for eLearning improves the learning experience. These videos also provide a wider understanding of the phenomenon. They provide digital literacy and ease of communication [42]: a high-quality video guarantees better interaction and improved performance of a phenomenon. Significantly, people who are visual learners get highly benefitted from visualized eLearning lessons. Even by using animations and graphical representations, these videos engage the eLearners [43]. Only a high-quality video can ensure better engagement and motivation for the eLearners. The creative digital and logistics can arise, but the eLearners can get greatly benefitted [44]. Accordingly, the researchers proposed the following hypothesis:

**H3**: Video Quality (VQ) positively influences YouTube's perceived usefulness (PU) of people with disabilities.

## 2.4 YouTube Usability (YU) and Learning Experience (LE)

YouTube usability makes it one of the most successful and preferred social networking sites [45]. Youtube contains easily accessible online learning material as additional study material for the eLearners. The purpose of usable and easy to access study material is to increase learning opportunities. Youtube witness positive learning outcomes for the students [39]. These Youtube videos are as significant as classroom lectures rigour and easy to understand. They have greater usability and useful outcomes for the eLearners. Students can stop and replay the segments they want to see again. It is essential for an eLearning study material to be easily useable and accessible for the learners [43]. To further validate this argumentation, the authors proposed that:

**H4**: YouTube Usability (YU) positively influences YouTube's perceived usefulness (PU) of people with disabilities.

## 2.5 Perceived Usefulness (PU) and Learning Experience (LE)

Perceived usefulness is a determinant of ensuring the benefit regarding adopting a system that would enhance one's work performance [46]. Perceived usefulness and perceived learning assistance have positive effects on learning performance and satisfaction of learners. Designers create eLearning opportunities and platforms that ensure an effective learning environment for the students [44]. Different aspects of blended learning increase the perceived usefulness of students concerning their learning process. As eLearning provides brisk communication, easy learning opportunities, effective study material, and examinations, it has a strongly perceived usefulness [37, 47]. Therefore, the researchers assumed that:

**H5**: Perceived usefulness (PU) positively influences the learning experience (LE) of people with disabilities.

From the proposed hypotheses, we build the following research model (Fig. 1).

# 3 Methodology

The study intended to describe the characteristics of the pure or a specific problem or a specific group and an accurate description without entering into judgment or causes. It also built its procedures on the curriculum field survey, which is one of the most curricula used in the influence of technologies, aimed to gather data and explore the objectives of this study [48, 49]. The respondents for this study were $n = 60$ disability specialists from the ministry of social development centers in Irbid, Amman and Aqaba, Jordan. The researchers randomly distributed questionnaires among $n = 30$ male and $n = 30$ female disability specialists (At a rate of 20 respondents per region north-central south). As far as the PLS-SEM (Partial Least Squares-Structural Equation Modeling) is concerned, the Smart PLS is well-known for performing in-depth data analysis [50]. The researchers applied PLS-SEM to evaluate the structural models and measurements in this study. The latent constructs' connection is connotative of the structural model, while the correlation among the indicators themselves is attributed to the computation model (outer model). The Structural Equation Model-PLS, together with the most distinguished probability method, was used to estimate the conceptual model. The researchers used different assessments to authenticate convergent validity and reliability, including Average Variance, Composite Reliability, and others.

## 3.1 *Discriminant Validity*

Three criteria should be measured while examining the discriminate validity, including the cross-loadings, Fornell-Larker criterion, and the Heterotrait-Monotrait ratio (HTMT). Table 1 shows that the Fornell-Larker is according to the designated criterion because all AVEs have a square root that is more than its correlation with the other study constructs [51]. Table 2 also exhibits that values associated with each construct are less than 0.85, which is the threshold value [51]. This further affirms the Heterotrait-Monotrait Ratio. These findings also confirm the presence of discriminant validity. Analytical results provided proof that the measurement model's assessment did not face any problems for its reliability and validity concerns. Therefore, the structural model can also be assessed using the collected data.

**Table 1** Fornell-Larcker scale

| | VT | PU | YT | VQ | YU | LE |
|---|---|---|---|---|---|---|
| VT | *0.911* | | | | | |
| PU | 0.222 | *0.868* | | | | |
| YT | 0.551 | 0.447 | *0.865* | | | |
| VQ | 0.202 | 0.544 | 0.301 | *0.896* | | |
| YU | 0.500 | 0.336 | 0.500 | 0.490 | *0.897* | |
| LE | 0.322 | 0.306 | 0.490 | 0.494 | 0.557 | *0.910* |

**Table 2** Heterotrait-Monotrait ratio (HTMT)

| | VT | PU | YT | VQ | YU | LE |
|---|---|---|---|---|---|---|
| VT | | | | | | |
| PU | 0.251 | | | | | |
| YT | 0.466 | 0.335 | | | | |
| VQ | 0.336 | 0.649 | 0.115 | | | |
| YU | 0.444 | 0.669 | 0.165 | 0.420 | | |
| LE | 0.554 | 0.701 | 0.298 | 0.467 | 0.696 | |

## 3.2 *Coefficient of Determination—$R^2$*

The researchers examined the study model by using the coefficient of determination ($R^2$ value) standard. This coefficient is the squared association between a particular endogenous construct's actual and prophesied values. Moreover, the predictive accuracy of the model can be further assessed through it. The coefficient is to indicate the combined influence of exogenous latent variables on the other endogenous latent variable. The squared relationship between the exact and prognosticated values of the variables is the coefficient. It also includes the level of variance in the endogenous constructs preserved by every recognized exogenous construct [51], considered that the value above 0.67 is high. However, the weak deals were found in the scope of 0.19–0.33, whereas the direct values were reported in the scope of 0.33–0.67. Besides, when the estimation is below 0.19, it is inadmissible. According to Fig. 2, the model reportedly contains high predictive power. According to Table 3, the $R^2$ values for Perceived Usefulness and Learning were found above 0.67. Thus, the potential of these constructs is found high.

## 3.3 *Hypotheses Testing—Path Coefficient*

The researchers utilized the Structural Equation Model technique to validate the research hypotheses. Table 4 affirms that all the mentioned values are within the

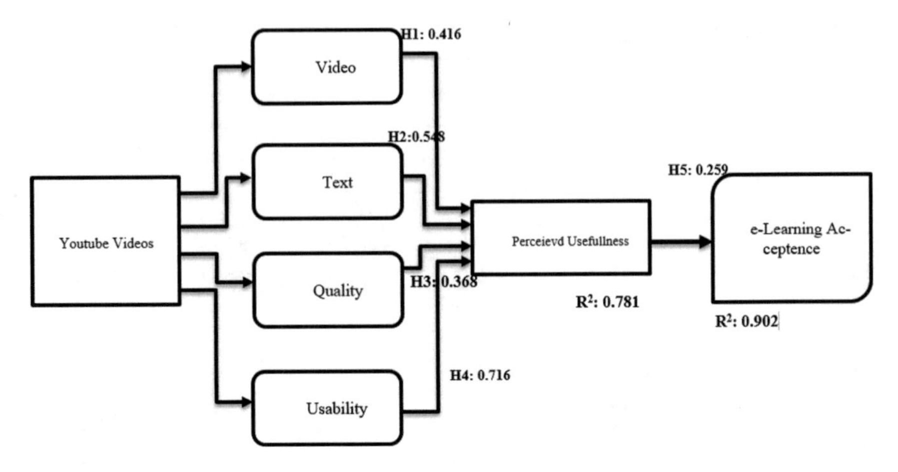


**Fig. 2** Model validation

**Table 3** $R^2$ of the endogenous latent variables

| Constructs | $R^2$ | Results |
|---|---|---|
| Perceived usefulness | 0.781 | High |
| Learning | 0.802 | High |

**Table 4** Results of hypotheses testing

| Hyp | Relationship | Path | *t*-value | *Sig* | Status |
|---|---|---|---|---|---|
| H1 | Video → Perceived Usefulness | 0.416 | 4.546 | 0.016 | Moderately Significant |
| H2 | Text → Perceived Usefulness | 0.548 | 2.435 | 0.021 | Moderately Significant |
| H3 | Quality → Perceived Usefulness | 0.368 | 1.224 | 0.010 | Moderately Significant |
| H4 | Usability → Perceived Usefulness | 0.716 | 12.118 | 0.000 | Strongly Significant |
| H5 | Perceived Usefulness → Learning | 0.259 | 25.119 | 0.000 | Strongly Significant |

given range. The obtained data validated all the research hypotheses; generally, the data supported all hypotheses. Based on the data analysis, hypotheses H1, H2, H3, H3, H4, and H5 were supported by the empirical data. The results revealed that Perceived Usefulness is positively correlated with Video ($\beta = 0.416$, $P < 0.05$), Text ($\beta = 0.548$, $P < 0.05$), Quality ($\beta = 0.368$, $P < 0.05$), and Usability ($\beta = 0.716$, $P < 0.001$), supporting the proposed hypotheses (H1, H2, H3, and H4 respectively). The relationships between Perceived Usefulness and Learning ($\beta = 0.259$, $P < 0.001$) were statistically significant, and the hypothesis H5 is generally supported. Table 4 shows a summary of hypotheses assessments.

## 4 Discussion and Conclusion

For both developed and developing regions, Covid-19 greatly challenged everyday life activities. With each coming day, Covid-19 is raising concerns for the educational policymakers and stakeholders. According to Khalid and Ali [10], the brisk spread of Covid-19 during the past few months led local governments to consider lockdown, social distancing, and even curfew as primary preventive measures. In such a situation, closing the educational institution was also important to hamper the virus transmission. However, shifting to online system, was a great challenge for the educational system as both students and instructors were not prepared for this major shift. Likewise, shifting from formal education to online education was only one challenge, but several technological barriers were another main concern. This shift impacted educational institutions on the levels, leading to serious e-learning [52]. In this context, the primary focus of the current chapter was to address the Impact of YouTube Videos on the Learning Experience of People with Disabilities. In the current chapter, the researchers utilized the quantitative method for data collection and analysis procedures. The Structural Equation Modeling (SEM) provided a strong conceptual background to this chapter. The researchers scrutinized the effect of YouTube videos directly on learning for people with disabilities according to specialists' opinions. Researchers found a positive association between the Perceived Usefulness of using YouTube videos and Learning about people with disabilities. The quality, ease of use, and texts in the video contribute significantly to people with disabilities' learning experience. Inferential statistics also validated the study hypotheses, showing a strong, significant relationship between proposed variables. Moreover, the current phenomena are unique and distinct as disable students are comparatively more facing challenges due to the closure of educational institutions. Besides disable students, educational challenges are intensified for their parents as well. Along-with the other perpetual challenges, relying on online media and resorting to e-learning are two primary concerns for disabled students and their parents [53]. This paper also ensured the potential of YouTube to provide quality learning services to disabled students. As during the current outbreak, they require accessible and usable educational services with better quality video and text [52]. The results indicated that Perceived Usefulness is significantly associated with Video, Text, Quality, and Usability, supporting the research hypotheses. This result also affirmed the previous studies that examined the influence of YouTube Videos on the Learning Experience of People with Disabilities [54]. Deduced responses manipulated, measured, and displayed appropriately also, the respondents fully address research questions. All the hypothetical postulations discussed our research queries and authenticated the relationship between stated variables. Therefore, we recommend more research on the respective phenomenon concerning the particular reasons (text, quality, usability) behind YouTube Videos usage, especially in Jordan. As validated by Daruka and Nagavci [53], blending the different techniques especially using sign-language and captions, supports e-learning, and facilitates the disabled students. Increased YouTube video for educational purposes also indicates increased social media dependency among disabled students in general.

## 5 Study Contributions, Limitations and Recommendations

The current investigation has several limitations, such as the researchers used a limited sample size, which could be further extended for large data-gathering purposes. Similarly, the chapter could also examine the other dimensions of e-learning for the disabled people during the Covid-19 outbreak. However, this chapter also contributed to the existing literature regarding the Technology Acceptance phenomenon. The researchers suggested a conceptual framework in terms of technology acceptance and disability during the healthcare crisis, which will help future researchers further consider technology, especially during crises. This study also suggests that legislators and production operators of YouTube Videos should consider the features that are crucial in accelerating eLearning [6]. Future work will contribute to a more in-depth discussion regarding the influence of YouTube Videos on People with Disabilities. Also, it will determine the most influential factors on the perception of learning.

## References


1. Al-Maroof, R.S., Salloum, S.A., Hassanien, A.E., Shaalan, K.: Fear from COVID-19 and technology adoption: the impact of Google Meet during Coronavirus pandemic. Interact. Learn. Environ. 1–16 (2020)
2. Al-Maroof, R.S., Alfaisal, A.M., Salloum, S.A. Google glass adoption in the educational environment: a case study in the Gulf area. Educ. Inf. Technol. (2020)
3. Alhawamdeh, A.K., Alghizzawi, M., Habes, M.: The Relationship Between Media Marketing Advertising and Encouraging Jordanian Women to Conduct Early Detection of Breast Cancer The Relationship Between Media Marketing Advertising and Encouraging Jordanian Women to Conduct Early Detection of Breast Canc (2020)
4. Zarocostas, J.: How to fight an infodemic (January), pp. 19–20 (2020)
5. Waris, A., Atta, U.K., Ali, M., Asmat, A., Baset, A.: COVID-19 outbreak: current scenario of Pakistan. New Microbes New Infect. **35**(20), p. 100681 (2020)
6. Alhumaid, K., Ali, S., Waheed, A., Zahid, E., Habes, M.: COVID-19 & Elearning: Perceptions & Attitudes of Teachers Towards E- Learning Acceptancein the Developing Countries, vol. 6, no. 2, pp. 100–115 (2020)
7. United Nations: Policy Brief : Education during COVID-19 and beyond (2020)
8. Alnaser, A.S., Habes, M., Alghizzawi, M., Ali, S.: The Relation among Marketing ads, via Digital Media and mitigate (COVID-19) pandemic in Jordan The Relationship between Social Media and Academic Performance: Facebook Perspective View project Healthcare challenges during COVID-19 pandemic View project, Dspace Urbe University (July) (2020)
9. Almansoori, A., AlShamsi, M., Salloum, S.A., Shaalan, K.: Critical review of knowledge management in healthcare. Stud. Syst. Decis. Control **295**(January), 99–119 (2021)

10. Khalid, A., Ali, S.: COVID-19 and its challenges for the healthcare system in Pakistan. Asian Bioeth. Rev. **12**(4), 551–564 (2020)
11. Ali, S.: Combatting against Covid-19 & misinformation: a systematic review. Hum. Arenas (0123456789) (2020)
12. Habes, M., Alghizzawi, M., Khalaf, R., Salloum, S.A., Ghani, M.A.: The relationship between social media and academic performance: facebook perspective. Int. J. Inf. Technol. Lang. Stud. **2**(1) (2018)
13. Chen, C.-M., Wu, C.-H.: Effects of different video lecture types on sustained attention, emotion, cognitive load, and learning performance. Comput. Educ. **80**, 108–121 (2015)
14. Shyu, H.C.: Using video-based anchored instruction to enhance learning: Taiwan's experience. Br. J. Educ. Technol. **31**(1), 57–69 (2000)
15. Choi, H.J., Johnson, S.D.: The effect of problem-based video instruction on learner satisfaction, comprehension and retention in college courses. Br. J. Educ. Technol. **38**(5), 885–895 (2007)
16. Nikopoulou-Smyrni, P., Nikopoulos, C.: Evaluating the Impact of Video-Based Versus Traditional Lectures on Student Learning (2010)
17. Khan, H.U.: Possible effect of video lecture capture technology on the cognitive Empowerment of higher education students: a case study of gulf-based university. Int. J. Innov. Learn. **20**(1), 68–84 (2016)
18. Habes, M.: The influence of personal motivation on using social TV: a uses and gratifications approach. Int. J. Inf. Technol. Lang. Stud. **3**(1) (2019)
19. Alghizzawi, M., Salloum, S.A., Habes, M.: The role of social media in tourism marketing in Jordan. Int. J. Inf. Technol. Lang. Stud. **2**(3) (2018)
20. Habes, M., Salloum, S.A., Alghizzawi, M., Mhamdi, C.: The relation between social media and students' academic performance in Jordan: youtube perspective. In: International Conference on Advanced Intelligent Systems and Informatics, pp. 382–392 (2019)
21. Alias, N., DeWitt, D., Siraj, S.: Development of Science Pedagogical Module: Based on Learning Styles and Technology. Pearson Malaysia (2013)
22. Alexa: Statistic report: Top sites in Spain (2015)
23. ONTSI: Las Redes Sociales en Internet (2011)
24. Clifton, A., Mann, C.: Can YouTube enhance student nurse learning? Nurse Educ. Today **31**(4), 311–313 (2011)
25. Duncan, I., Yarwood-Ross, L., Haigh, C.: YouTube as a source of clinical skills education. Nurse Educ. Today **33**(12), 1576–1580 (2013)
26. Kelsen, B.: Teaching EFL to the iGeneration: a survey of using YouTube as supplementary material with college EFL students in Taiwan. Call-EJ Online **10**(2), 1–18 (2009)
27. Alon, I., Herath, R.K.: Teaching international business via social media projects. J. Teach. Int. Bus. **25**(1), 44–59 (2014)
28. Habes, M., Salloum, S.A., Alghizzawi, M., Alshibly, M.S.: The role of modern media technology in improving collaborative learning of students in Jordanian universities. Int. J. Inf. Technol. Lang. Stud. **2**(3) (2018)
29. Condie, R., Munro, R.K.: The impact of ICT in schools-a landscape review (2007)
30. Rocha, T., Martins, J., Branco, F., Gonçalves, R.: Evaluating youtube platform usability by people with intellectual disabilities (a user experience case study performed in a six-month period). J. Inf. Syst. Eng. Manag. **2**(1), 5 (2017)
31. Rocha, T., et al.: The recognition of web pages' hyperlinks by people with intellectual disabilities: an evaluation study. J. Appl. Res. Intellect. Disabil. **25**(6), 542–552 (2012)
32. Chapman, G.: YouTube serving up two billion videos daily. AFP, vol. 17. http://www.google.com/hostednews/afp/article/ALeqM5jK4sI9GfUTCKAkVGhDzpJ1ACZm9Q. Accessed May, 2010
33. Libin, A., et al.: YouTube as an on-line disability self-management tool in persons with spinal cord injury. Top. Spinal Cord Inj. Rehabil. **16**(3), 84–92 (2010)
34. Alghizzawi, M., et al.: The impact of smartphone adoption on marketing therapeutic tourist sites in Jordan. Int. J. Eng. Technol. **7**(4.34), 91–96 (2018)

35. Brame, C.J.: Effective educational videos: principles and guidelines for maximizing student learning from video content. CBE—Life Sci. Educ. **15**(4), es6 (2016)
36. Hove, P.E.: Characteristics of instructional videos for conceptual knowledge development. University of Twente (2014)
37. Welker, B.: How important are visuals and videos in e-learning? Tesseract Learning. Available at 2019
38. The Rapid E-Learning Blog: Why E-Learning is So Effective|The Rapid E-Learning Blog (2010)
39. Backlinko: YouTube Video Description. Online Information Review (2018)
40. Pratama, Y., Hartanto, R., Kusumawardani, S.S.: Validating youtube factors affecting learning performance. In: IOP Conference Series: Materials Science and Engineering, vol. 325, no. 1, p. 12003 (2018)
41. Gernsbacher, M.A.: Video captions benefit everyone. Policy Insights Behav. Brain Sci. **2**(1), 195–202 (2015)
42. Bevan, M.: Why videos are important in education. Online Information Review (2015)
43. Brecht, H.D.: Learning from online video lectures. J. Inf. Technol. Educ. **11**(1), 227–250 (2012)
44. Islam, A.K.M.N.: Conceptualizing Perceived Usefulness In E-Learning Context And Investigating Its Role In Improving Students' Academic Performance. In: ECIS, p. 8 (2013)
45. Ahmed, I., Qazi, T.F.: Deciphering the social costs of Social Networking Sites (SNSs) for university students. Afr. J. Bus. Manag. **5**(14), 5664–5674 (2011)
46. Alsabawy, A.Y., Cater-Steel, A., Soar, J.: Determinants of perceived usefulness of e-learning systems. Comput. Hum. Behav. **64**, 843–858 (2016)
47. Ciel.viu.ca.: Types of Educational Video. Centre for Innovation and Excellence in Learning (2019)
48. Habes, M., Alghizzawi, M., Ali, S., SalihAlnaser, A., Salloum, S.A.: The relation among marketing ads, via digital media and mitigate (COVID-19) pandemic in Jordan. Int. J. Adv. Sci. **29**(7), 2326–12348 (2020)
49. Ali, S.: Social media usage among teenage girls in Rawalpindi and Islamabad. Glob. Media J. **16**(31), 1 (2018)
50. Xiaohui, Y.: Chapter 4—Data Analysis 87 (2016)
51. Zait, A., Alexandru, U., Cuza, I.: Methods for testing discriminant validity. Manag. Mark. **9** (2), 217–224 (2011)
52. Nikdel Teymori, A., Fardin, M.A.: COVID-19 and educational challenges: a review of the benefits of online education. Ann. Mil. Heal. Sci. Res. **18**(3), 19–22 (2020)
53. Daruka, Z.H., Nagavci, N.: The impact of the COVID-19 pandemic on the Education of Children with Disabilities. StatCan COVID-19 Data to insights a better Canada (45280001) (2020)
54. Hunt, P.F., Belegu-Caka, V.: Situation Analysis: Children with Disabilities in Kosovo (2017)

# IoT-Based Wearable Body Sensor Network for COVID-19 Pandemic

**Joseph Bamidele Awotunde, Rasheed Gbenga Jimoh, Muyideen AbdulRaheem, Idowu Dauda Oladipo, Sakinat Oluwabukonla Folorunso, and Gbemisola Janet Ajamu**

**Abstract** The novel severe contagious respiratory syndrome coronavirus (COVID-19) has caused the greatest global challenge and public health, after the pandemic of the influenza outbreak of 1918. According to the World Health Organization, more than 19,687,156 people have been infected by the virus, with at least 727,435 deaths globally as of 10:33 am CEST, 10 August 2020. Globally, people spend much of their time indoor to contain or avoid people infected with the virus. Until now, there has been a rapid increase in diverts research works to find a lasting solution to this worldwide threat. In the past few years, IoT has drawn convincing ground in research fields range from academic and industrial fields, especially in healthcare. The IoT revolution reshapes contemporary healthcare systems by incorporate economic, social, and technological prospects. It progresses from conventional healthcare systems to more personalized healthcare systems, where patients can be monitored, diagnosed, and treated effortlessly. Wearable body sensor network has transformed the power to change our lifestyle with abundant technologies in areas of healthcare, entertainment, transportation, retail, business, and emergency services control. The integration of wireless sensors and

J. B. Awotunde (✉) · R. G. Jimoh · M. AbdulRaheem · I. D. Oladipo
Department of Computer Science, University of Ilorin, Ilorin, Nigeria
e-mail: awotunde.jb@unilorin.edu.ng

R. G. Jimoh
e-mail: jimoh_rasheed@unilorin.edu.ng

M. AbdulRaheem
e-mail: muyideen@unilorin.edu.ng

I. D. Oladipo
e-mail: odidowu@unilorin.edu.ng

S. O. Folorunso
Department of Mathematical Science, Olabisi Onabanjo University, Ago-Iwoye, Nigeria
e-mail: sakinat.folorunso@oouagoiwoye.edu.ng

G. J. Ajamu
Department of Agricultural Extension and Rural Development, Landmark University, Omu Aran, Nigeria
e-mail: ajamu.gbemisola@lmu.edu.ng



A.-E. Hassanien et al. (eds.), *Advances in Data Science and Intelligent Data Communication Technologies for COVID-19*, Studies in Systems, Decision and Control 378, https://doi.org/10.1007/978-3-030-77302-1_14

sensor networks with simulation and intelligent systems research has developed an interdisciplinary definition of ambient intelligence to address the obstacles faced in our everyday lives. It is essential to build a reliable and efficient wearable system for monitoring during the COVID-19 outbreak. In the situation of COVID-19, an IoT-based wearable body sensor can be utilized to lower the possible spread of the pandemic using enabled/linked devices aimed at people for early diagnosis, monitoring during social distance, quarantine time, and after recovery. Therefore, this chapter reviews the role of IoT and wearable body sensor technologies in fighting COVID-19 and presents an IoT-based wearable body sensor architecture to combat the COVID-19 outbreak. IoT-based wearable body sensor can be used widely to control and track patient conditions in both towns and cities using an internal network, thus minimize pressure and tension on healthcare professionals, eliminating medical faults, reducing workload and medical staff productivity, reducing long-term healthcare costs, and enhancing patient satisfaction during COVID-19 pandemic.

**Keywords** Internet of things · Coronavirus · Wearable body sensor network · Cloud server · Healthcare · COVID-19 pandemic

## 1 Introduction

The spread of Coronavirus Disease (COVID-19) has caused substantial changes in the lifestyle of communities around the world. The first case of the pandemic was discovered in Wuhan, Chana in December 2019 and was triggered due to the Severe Acute Respiratory Syndrome Coronavirus 2 (SARS-CoV-2) virus [1–3]. As reported to World Health Organization (WHO) there have been 30, 949,804 confirmed cases and 959,116 deaths of COVID-19 Globally, as of 4:30 pm CEST, 21 September 2020, and affecting more than 180 countries. Owing to its alarming level of global spread, WHO asserted COVID-19 as a pandemic. These make governments of these countries take immediate control actions, since the healthcare sectors of the affected countries, such as isolating extremely affected areas, closing the cross border traffic amid countries, closure of schools, workplaces, and common places, regulating the activities of the general public by urging them to stay at home as much as possible. The social life and economics had significantly affected due to these measures put in place.

The societies face different challenges ranging from education, healthcare, manufacturing, supply chain management travel, tourism, and service delivery under the prevailing circumstances and in a post-COVID-19 world. As an example, the overcrowded hospitals and other healthcare facilities due to the rapid increase of COVID-19 patients and the inability to provide medical aid to normal patients due to limited movement are significant barriers to the fight against COVID-19 in the healthcare sector. Likewise, delays and rise in resource demand for manual contact tracing, and unavailability of effective and automated contact tracing software impede the actions for controlling the spread. Hence, different parties have to act within their capacity to control the prevailing COVID-19 situation, such parties as students, healthcare workers, researchers, government authorities, technology managers, engineers, and the general public. To not only safeguard but also handle the post-COVID-19 environment, digitalization and the implementation of information and communication technologies will be imperative. Technologies like Big data, 5G communication, Internet of Things (IoT), cloud computing, Artificial Intelligence (AI), and blockchain play a vital role to assist the environment in adapting to different protection and improvement of people and economies. The technologist and engineers will have to address important challenges to implement these promising solutions and realize their benefits, and carry out elaborate findings regard to risk management, resources, cost, scope, and quality.

Globally, the cases of coronavirus cases keep on increasing regardless of the practice of all measures put in place to fight the outbreak. According to a report from WHO the most affected twelve (12) countries in the world are shown in Table 1. Interestingly the United State of America has been the leading country since the outbreak early this year, and these twelve countries take 73.82% of the entire confirmed cases globally.

In an attempt to prevent the circulation of the sickness, several countries were extending the lockdown and affecting millions of people. Initially, experts raised questions about the feasibility of this initiative and cautioned that certain countries were at risk of repeating a SARS-like epidemic [4, 5]. By this time, the pandemic appears to be under control in most countries, although there is still criticism of the utilization of what others have called "draconian" steps to stifle its circulation. The globe is presently trying to monitor the exceptional virus circulation that involves the greatest sum of indispositions plus deaths. As there is no such thing as definite medical care for coronaviruses and attempts to control the circulation have yet remained unsuccessful [6, 7], there's a crucial necessity for worldwide investigation of persons with intense COVID-19 contagion.

**Table 1** The most twelve affected countries of the world by Coronavirus

| S/no | Country | Total cases |
|---|---|---|
| 1 | United States of Africa | 6,703,698 |
| 2 | India | 5,487,580 |
| 3 | Brazil | 4,528,240 |
| 4 | Russian Federation | 1,109,595 |
| 5 | Peru | 762,865 |
| 6 | Colombia | 758,398 |
| 7 | Mexico | 694,121 |
| 8 | South Africa | 661,211 |
| 9 | Spain | 640,040 |
| 10 | Argentina | 622,934 |
| 11 | Chile | 446,214 |
| 12 | Iran | 432,424 |

The emerging of new trends in technologies contributed to the commencement of the Internet of Things (IoT) and is acquiring worldwide concentration as well as becoming obtainable for monitoring, diagnosing, forecasting, and preventing arising communicable ailments. IoT in the medical organization is advantageous and enabled suitable controlling of COVID-19 persons by using interrelated wearable sensors and networks. IoT is an evolving area of investigation within infectious disease epidemiology. However, the augmented dangers of communicable ailment transmitted over worldwide integration and the pervasive obtainability of smart types of machinery, including interrelatedness of the world requires its utilization for monitoring, averting, predicting, and managing transmittable viruses.

IoT is an innovative way of combining healthcare gadgets and their applications to interact with the systems of human resources and data innovation. An inquiry on the possible outcomes of defying progressive COVID-19 pandemic by adopting the IoT strategy when providing care to all groups of the patient without any partiality in poor and wealthy. The various cloud-based IoT administrations are the exchange of knowledge, report verification, investigation, patient monitoring, data social affair, cleanliness clinical consideration, and so on. It can change the working format of the medical services while rewarding the huge volume of patients with a prevailing degree of treatment and more fulfillment, especially during this pandemic of COVID-19 outbreak, Health workers will easily concentrate on the patient and identify anyone who has come into contact with the infected person and move these people to quarantine/isolation. By functioning as an early warning system, IoT devices such as the geographic information system may be used as an important tool to curb the spread of this pandemics. Sensors like temperature and other signs might be used at airports around the world to detect people infected with the COVID-19 pandemic.

Wireless sensor network (WSN) applications are considered as one of the major fields of research for improving quality of life in the computer science and healthcare applications industries [8]. The WSN has transformed the power to change our lifestyle with abundant technologies in areas of healthcare, entertainment, transportation, retail, business, and emergency services control in addition to many other areas. The integration of wireless sensors and sensor networks with simulation and intelligent systems research has developed an interdisciplinary definition of ambient intelligence to address the obstacles we face in our everyday lives [9]. The WSNs with smart sensor nodes have become a substantial technology enabling a wide spectrum of uses. New technologies in the incorporation and mass production of single-chip sensing devices, computer chips, and radio interfaces have allowed WSN to be appropriate for several applications [8, 10–12]. They could be used, for instance, for emergency preparedness, factory equipment, surveillance systems, seismic surveillance, environmental control, agricultural practices, and health monitoring. Among the most hopeful usage of WSN is monitoring healthcare [13, 14].

Body sensor node system will help users by providing public healthcare services such as health tracking, memory improvement, home appliance control, access to health data, and emergency communication especially during this period of the

COVID-19 pandemic [15]. Constant monitoring with wearable devices will significantly raise the early diagnosis of a case of emergency conditions and diseases in patients at risk of COVID-19 outbreak, as well as provide a wide range of healthcare for people with different levels of cognitive and emotional illnesses. These systems will benefit not just the elderly and chronically ill patients, but also the households where both parents are working to provide their babies and children with good quality care services. This chapter, therefore, presents the roles of IoT and wearable body sensor technologies in combating the COVID-19 pandemic and also proposes an IoT-based body sensor framework to fight the COVID-19 outbreak.

The main contributions of the chapter are: (i) the prospects, amazing and applicability captivating IoT-based Wearable Body Sensors in fighting COVID-19 pandemic was discussed; (ii) proposed an IoT-based WBS system architecture, and (iii) the reliable practical applicability of the proposed system were presented.

The chapter is organized as follows. Section 2 explains the applications of IoT to combat the COVID-19 pandemic. Section 3 discusses WBSN in Healthcare and how it helps to fight COVID-19. Section 4 presents the IoT-Based wearable body sensors network framework for the fighting COVID-19 outbreak, while Sect. 5 discusses the practical applicability of the proposed framework. Finally, Sect. 6 concludes the chapter and discusses future works for the realization of efficient uses of IoT-based WBSN in fighting the COVID-19 pandemic.

## 2 Internet of Things Technology in Combating COVID-19

In developing countries, the healthcare sector is shifting rapidly as life expectancy grew dramatically during the 1990s [16]. Infectious illnesses are also placing increasing pressure on health-care systems in these countries [17]. The life expectancy in advanced nations during the 20th century has been raised by about 30 years. As a result, the number of elderly people has risen rapidly [16]. Also, chronic disease proliferation has put pressure on health-care systems in other countries due to a lack of funding [17]. Increasing infectious illnesses and aging populations present significant challenges, as health systems have to deal with a multitude of ailments and treatment options, but similarly a rising patient number. To avoid overloading health infrastructure and reducing the costs of healthcare, successful approaches have been shown in-house telemedicine services [18].

Telemedicine platforms are extremely diversified and typically designed to respond to a single therapeutic purpose, like mobile cardiac monitoring, and stroke recovery [19]. This attribute of telehealth systems makes them cost-effective and overloading health systems, but reflects a weakness as patient numbers and disease variety increase. The IoT can handle the need for stronger genericity and reliability. Admittedly, the IoT integrates both the efficiency and security of traditional medical equipment and the traditional capacity for dynamics, genericity, and scalability of IoT. It devises the potential to fix the aging problem and terminal illnesses by managing various sensors deployed for millions of patients, as well as being broad enough to deal with various illnesses requiring precise diverse checking and action specifications.

IoT in transmittable sickness epidemiology is an evolving area, but the omnipresent proliferation of smart technology and amplified threats of transmittable sickness conveyed via global integration and worldwide interconnectivity demand its use to anticipate, deter and monitor evolving COVID-19 pandemics [20, 21]. Web-based monitoring platforms and strategies for disease intelligence have recently appeared in many countries [20, 21] to promote risk management and prompt identification of outbreaks, but there is a shortage of systematic use of available technologies. IoT-Implemented Medical Care Surveillance in a worldwide health care system would offer local health authorities with the ability to strengthen efforts to identify, control, and avoid infectious diseases [22].

It can help diagnose infectious patients rapidly and forecast accurately the potential circulation of an ailment to alternative places using transit data. Essentially, an IoT-established observation network may assist to rebuild an epidemic and restore the economies of the source nation, rather than locking up major cities, borders, and businesses. Mobile connectivity in the context of mobile health (m-health) will as well improve the productivity of a medical care network by including multiple facilities, apps, third-party APIs, and non-health-related mobile sensors [23]. Security and Medical Care Observing Operations for instance wearable IoT allow safety monitoring in real-time and will gain developments in global health.

Due to an utter impossibility to track these large geographical regions or communities [24], such innovations may eliminate holes in monitoring current systems. Such methods were introduced in the areas of computer science and healthcare analysis but are fairly new in the area of epidemiology of transmittable diseases [25]. Despite the current global scenario, IoT-established smart sickness detection schemes can be a significant advancement in current pandemic response exertions.

For a great deal of the technology already in position (i.e. Android phones, wearable devices, net access) the function that IoT can play in reducing the circulation of the disease includes simply gathering and reviewing the already collected data. The collective responsibility of IoT and connected emerging types of machinery could influence the first detection of outbursts and deter the circulation of COVID-19 if the data was enhanced and used. Smart IoT-based disease detection systems will include continuous communicating and recording, end-to-end networking and availability, data variety and review, tracing, and warnings, including choices for inaccessible healthcare support in China and other impacted countries to diagnose and control COVID-19 outbreaks.

The heart of the IoT is an Internet-based network that grows and extends; the user side can be extended, thus enhancing the sharing of knowledge and contact between "people." IoT is the science of intelligent detection, location, tracking, and AI services for COVID-19 patients using RFID, Global Positioning Systems, various sensing instruments, information exchange, and community services if we can go deeper into its work, it will certainly yield unpredictable results.

RFID readers can be mounted on robots in the therapeutic work of stopping SARIs, and UHF stickers can be read when a drug is inserted into the device to validate the delivery of drugs. This clinical program has gained valuable expertise. When a robot triggers its RFID reader, it can collect the necessary information and quantities of all the medications in a cabinet, and a prescription can be delivered to the medical units correctly by precisely matching the drug information. In Danville, Pennsylvania, USA, the Geisinger Medical Center has implemented integrated RFID robotics to guarantee that a drug is reliably administered to all units in operation, with images sent immediately. RFID technology can also be used to create a medical waste management network that can effectively control and track all COVID-19 medical waste processes including generation, recycling, transportation, and treatment.

Managing suspected patients is a difficult problem. The explanation is that it is only possible to recognize those with fever or those that have fallen sick. The other potential virus carriers aren't isolated and are the next wave in infection. Tell them to live in their own homes or to remain in the neighborhood to minimize personal transition. The recommendations recommend self-protection by ensuring good hand and mouth health, keeping healthy eating habits, and avoiding direct contact with those with respiratory illness symptoms (such as coughing and sneezing).

Nevertheless, the impact of this policy being followed is not understood, so there is little quality assurance. Therefore, there is an immediate need that IoT be implemented to better handle this group of patients, compared to conventional medicine, IoT will track the clinical and medical status of the suspicious patients during the process, illustrating the management benefits of offering customized treatment strategies for various classes of individuals. By the utilization of wireless radar gadgets and modern IT, patients will benefit from medical services, thereby guaranteeing the health of suspicious patients and preventing their relatives' infection. The Medical Things Network (IoT), the expanded health-specific edition of the IoT [8] applied to the present situation, it should be used to build a digital forum to help people access sufficient treatment at home and to establish a robust network for policy and community organizations.

Persons with minor symptoms may receive supplies for treatment and healthcare (protective gloves, thermometers, drugs, POC COVID-19 supplies for treatment and control of infections). Inmates may upload their medical position to the IoT (medical cloud storing) portal online daily and pass their records to regional clinics, the Center for Disease Control (CDC), state and resident medical offices. Infirmaries might then suggest operational wellness sessions built on the medical status of each patient, and the régime (the CDC local as well as national medical offices) may perhaps distribute resources and establish isolation places (guesthouses or consolidated isolation amenities) as appropriate. This will minimize national health expenses, alleviate the pressures of medical equipment shortages, and deliver a centralized framework that would tolerate the régime to track infection transmission effectively, administer materials efficiently, and enforce response strategies.

IoT an extension of IoT aims to connect patients to health care facilities to monitor and control human body vital signs using communication infrastructure

[26]. Telemedicine is getting popular in remote areas where accessibility to a quality physician is limited due to different factors [27]. For example, measurement of heart rate, electrocardiography, diabetes, and signs of the vital body may be tracked remotely without patients being physically present. Examples are sensors and actuators that can be used to receive data from the patient and send the information to the cloud using a local gateway. The doctor examines the data using any mobile or desktop application provided to them and notifies the patient or medical staff taking care of the patient about the report [28].

Digital telehealth plays a very important role during the COVID-19 outbreak. A portal is created where patients interact with the doctors and the treatment is provided remotely. The benefit of employing a secured IoT system in COVID-19 is that the physicians do not directly examine the patients, hence, avoiding the spread of the virus [27]. Many countries have started operating the digital telehealth in this time of crisis. Health Arc [29] provides IoT based health care devices to the patients whose data is continuously monitored by the medical staff. The data is analyzed and the suggestions and prescriptions are provided to the patients on their mobiles or tablets. Continuous Care [30], Health net connect [31] are among the leading telehealth service providers. A person with COVID-19 symptoms can use the assessment tool provided on the digital platform such as the "COVID-19 Gov PK mobile app" [32], which is accessed by the physicians remotely. Using this tool, patients are timely guided and many precious lives can be saved. Furthermore, it also serves to reduce the number of hospitalizations, readmissions, and the bulk of patients in hospitals, all of which help in improving the quality of life and provides timely treatment to COVID-19 patients.

Medical staff associated with ambulances are usually dealing with very high pressure and error-prone situations [33]. During the current pandemic of COVID-19, the situations have become even more tensed and pressurized for medical staff dealing with COVID-19 patients. The IoT-based ambulances offer an effective solution in which remote medical experts suggest necessary actions to the medical staff dealing with the patient in the ambulance. This leads to the timely response and effective handling of the patient. The radio-frequency identification (RFID) based equipment is connected to the wireless local area network (WLAN). The information of the patient is remotely accessible by the concerned medical staff. Figure 1 displayed prospective applications of IoT to fight the COVID-19 pandemic.

IoT-aided equipment is classified as personal and clinical [33]. The user tracks the heartbeat, exercise, sleep, nutrition, and weight using these gadgets. These are useful in fighting against COVID-19 as well because the rest and sleep become very important factors for the patients suffering from this disease. Sleeping well increases the immunity of the body to fight against the virus [34]. The patient can see his reports on the portals provided by these gadget makers and provide information to the related physicians if required. IoT-based wearable gadgets can help in reducing the spread of coronavirus if certain algorithms are implemented to the existing devices. The wearable devices notify in real-time if:

**Fig. 1** Prospective applications of IoT to combat COVID-19

- the social distancing protocol is violated,
- any COVID-19 patient is in the locality, and
- the area was declared as a danger zone by the government from the perspective of coronavirus outbreak.

The automatic human body temperature sensing machines are installed in many countries where the camera is integrated with the sensor and sends real-time information to the server. The system also uses AI to recognize a face and matches it with the centralized database [35]. The use of these devices helps in tracking the patient of COVID-19. Social distancing is enforced by enabling smart infrastructure in which the environment is sensed and reports are generated in real-time for law enforcement agencies [36]. The data from sensors is constantly recorded on an online database for constant monitoring [37]. The content of hazardous gases or carbon content is communicated to the environmental protection agency and is updated on the server to be checked online [38]. Researchers are working on sensing the coronavirus which can also be operated in the same manner. IBM,

Microsoft, Huawei, and Cisco are among the top companies that provide smart infrastructure solutions [39].

Ideally, engineers and technicians would avoid warehouses, isolated locations, or busy places traveling and visiting because they fear being infected with the virus. Nevertheless, their physical appearance on site is needed because they can't get away with manually operating the equipment available. By the use of special sensors mounted in machinery and enhanced knowledge, real-world overlays, and remote expert feedback, IoT may help solve their repair problems, conduct machine operation remotely, and AI may also anticipate when machines need repair (predictive maintenance) or when they might be faced with a challenge. In this way, physical visits will be significantly decreased, helping to safeguard workers' health and welfare, thus enhancing efficiencies at the level of service of the facilities.

When supermarkets continue imposing limits on sales of goods per customer, future alternatives have also been produced by IoT and AI. Smart shelves, smart fridges, video analytics, and an end-to-end integrated supply chain will help retailers deal with planning challenges and even reduce the extreme behavior of customers due to hysteria.

A couple of years ago, a major US retail chain put IoT trackers in its trolleys to prevent daily theft. Perhaps it is time to start applying this to the store shelves of important homes, sanitizers, and everything else that is already over-selling, in an attempt to properly control supplies and prevent hoarding behavior. Smart-watches and fitness trackers will be readily available in the not-too-distant future and people with chronic conditions will be able to record temperature, asthma, heartbeat without the use of intrusive instruments. For this fact, patients will be able to opt to transmit real-time and past data to public or private hospitals anytime they feel unwell so that medical health IT departments will interact with the wearables and mobile devices of patients. In this way, the treatment of coronavirus or other diseases may be prepared even more effectively and with minimal resources expended optimally. Smart connected medical devices, such as smart home ventilators, together with video and wearables, may support patients monitoring at home, give updates to those in distress, or even alert when paramedics are required to come and move them to the hospital.

Ventilators are important in treating people that have become contaminated. The health services have not been prepared to deal with this magnitude of a pandemic, and a resulting shortage is now widespread in only the most developed hospitals. An IoT-3D printing may be a lifesaver in the face of coronavirus-induced supply shortages. With an IoT-3D printer that offers critical medical equipment, for example, replacement valves. This is what an Italian company called Isinnova does, taking a 3D printer to a Milan hospital and manufacturing incomplete valves to be shipped to a hospital in Brescia, Northern Italy. Touch screens have been the chosen user interface (UI) until recently: among them are tablets, computers, and even doors. Nonetheless, the fact that coronavirus is more quickly transmitted from a contaminated surface than by air has made direct contact sound risky.

Many UIs that don't need any physical contact are also usable. Voice has already triumphed over tactile user interfaces, especially through smart speakers and digital helpers. Despite people confining themselves inside the building, there will be increased interest in smart home apps, and the voice apps of smart speakers. Another functionality that gains ground in smartphones is beyond speech, biometrics, and their use for eye/face identification, such as the use of facial picture identification to open phones or make payments. The largest penetration is in China but the opportunity for the remaining part of the biosphere is enormous. Wearables, for instance, smart payment watches and other use cases (enabled by voice or close contact) can allow us to escape physical surface contact.

IoT is the medical-care-precise variant of IoT which could be introduced to deliver remedy or cure to health care professionals, guarantee isolation compliance which tracks disease sources [40]. By the assistance of radars embedded in smart headsets, drones, robotics, and COVID-19 self-sampling experiments, as well as data collection may be performed. The data obtained by these techniques would be forwarded for processing to a central cloud repository. The data created by such a system would provide health-care providers and government agencies better equipment to answer the COVID-19 disaster.

By these results, care professionals will be able to offer more tailor-made electronic wellness appointments for patients. Such electronic facilities will also allow patients to seek more effective treatment while reducing their access and further spread of the virus at the same time. Agency departments, together with resident public medical offices and Centers for Disease Control and Prevention (CDC), will be well prepared to distribute resources, assess quarantine needs, track outbreaks, and use this information to enforce emergency plans [40].

These technologies are now being implemented in major cities to reduce the pandemic. For example, for simultaneous tracking of COVID-19 inmates, the Shanghai Public Health Clinical Center utilizes physique fever monitors laterally utilizing facts transfer directly to the nurse post, thus minimizing possible sensitivity to healthcare workers [41, 42]. Similarly, a device now used for medical interviews with sensors was installed in Boston to assess the breathing degree and physique fever of patients. In Singapore, a contact-tracing mobile program utilizes Bluetooth wireless technologies to identify individuals who were similar to COVID-19 patients [43]. Apple and Google are working on touch monitoring and detection software that will be made available to many countries around the world that are intended to significantly step up the recognition that warns users who have been naively near to COVID-19 inmates. IoT won't solitary combat the present pandemic but could also be used to deter potential outbreaks.

## 3 The Wearable Body Sensors Network in Fighting COVID-19 Pandemic

Billions of smart devices and applications, such as WSNs, are not expected to be segregated but linked to and incorporated with communications networks in the future IoT [44]. To manage these sensor devices well, systems also need to be designed to operate properly by allowing system management organizations to track and control systems remotely without the need for substantial resources.

The IoT is known as a network that comprises of collecting medical devices and sensors that are intimately connected via digital technologies. Furthermore, IoT uses embedded sensors, cameras, thermometers, air quality sensors, ECG/EEG/EMG sensors, pressure gauge, gyroscope sensors, sensors for saturation of blood oxygen, temperature and humidity sensors, heart rate, and respiratory sensors to monitor and supervise the health of the patient continuously. The IoT senses the health status of COVID-19 patients and then transfers the clinical data to doctors and caregivers using remote cloud data centers [15].

Patient information as body temperature, heart-beat rate, blood pressure, ECG pulse, blood oxygen saturation levels (SpO2) can be obtained by a wearable sensor attached to the patient body. One of the critical layers on the Internet of Medical of Things (IoMT) a subset of IoT architecture is the connectivity of sensors and the network. This component is an essential part of the IoT ecological system which offers access to other layers in the network. IoMT environment sensors serve as frontend. These devices may be closely relevant to IoT networks whenever the transformation and processing of the signal have been identified. Nevertheless, not all sensors are the same and nowadays there are several types available in the market. For instance, two to three days of continual physicochemical measures based on sensor devices were monitored by scholars and used to collect related parameters to update specific healthcare records [13, 45]. Sensors allowed the collection of personalized health data and behaviors of patients and these data can be moved to the cloud to further the analysis.

Healthcare services focused on smartphones have a clear propensity to have a cost-effective long-term healthcare management alternative [46]. These systems can enable health care physicians to monitor their patients' health status remotely without invading into their daily routines [47]. Smartphones also have numerous embedded sensors including an image sensor, an accelerometer, an ambient light sensor, a GPS sensor, a gyroscope, a microphone, and a fingerprint [48]. These sensors help to evaluate numerous patients' health parameters including heart rate, variability in heart rate, respiratory rate, body glucose, and blood pressure only to consider a few. This makes the communication device a continuous and long-term monitoring tool for health care. These sensors allow health personnel including a doctor to monitor the progress and health parameters of the patient in real-time. The benefits of wearable sensors are shown in Fig. 2.

The WSNs with smart sensor nodes have become a substantial technology enabling a wide spectrum of uses. New technologies in the incorporation and mass production of single-chip sensing devices, computer chips, and radio interfaces have allowed WSN to be appropriate for several applications [13, 49, 50]. They could be used, for instance, for emergency preparedness, factory equipment, surveillance systems, seismic surveillance, environmental control, agricultural practices, and health monitoring. Among the most hopeful usage of WSN is monitoring healthcare [51, 52].

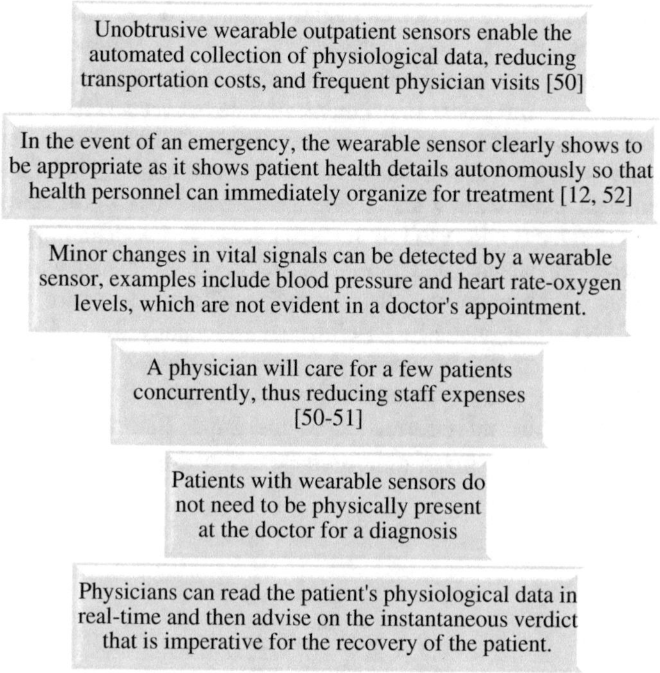

**Fig. 2** The benefits of wearable sensors

A sensor network is a network of several nodes furnished through a sensor module, a memory chip, a microprocessor, a wireless connectivity interface, and a multi-hop power source [53, 54]. With routing responsibilities, the nodes in the same proximity can interact with one another [54]. Opposing to the popular sensor nodes that are meticulously crafted and installed in the defined locations, WSNs could install an ad-hoc, making them resilient, fault-tolerant, and increasing coverage area [55, 56]. They can be used widely to control and track patient conditions in both towns and cities using an internal network. Therefore, these methods can minimize pressure and tension on healthcare professionals, eliminating medical faults, reducing workload and medical staff productivity, reducing long-term healthcare costs, and enhancing patient satisfaction [12, 57, 58].

These inaccuracies happen as a result of a lack of comprehensive and accurate information at the time and place required, leading to an erroneous treatment and issues with particular medication [58, 59]. The risk of death can be lowered if appropriate measures are provided to patients at the proper moment [12, 60]. To ensure the safety of patients and to save lives, hospital personnel must have the right to information about patients at the right and within a short period. Therefore, it is necessary to provide a safe and low transmission latency for patients with life-threatening diseases like diabetes mellitus, heart diseases, and blood pressure. To build a cluster WBAN, sensor networks can be well-positioned on the human body, and therefore, used to retrieve symptoms from patients [61, 62].

For effective and convenient transmitting of data between WBAN and personal server a battery cell sensor networks are required, and their energy consumption during transmission should be relatively low. The examples of IoT devices using for communication technology are smartphones, which use the Internet, General Packet Radio Service (GPRS), 2G to 5G technology. These technologies enable us to keep patients, physicians, and caregivers informed while also setting a pattern and identifying health variations. This is called biomedical sensor wireless networks (BWBSN) when applied to biomedical technology [63, 64]. The WBSN enables low-power, remotely operated, smart pervasive sensor nodes to be embedded to monitor body systems and physical vicinity. Every node will sense, monitor, and then forward information to the Super Sensor. Figure 3 presents selected sensors used in IoT for information gathering during the COVID-19 pandemic.

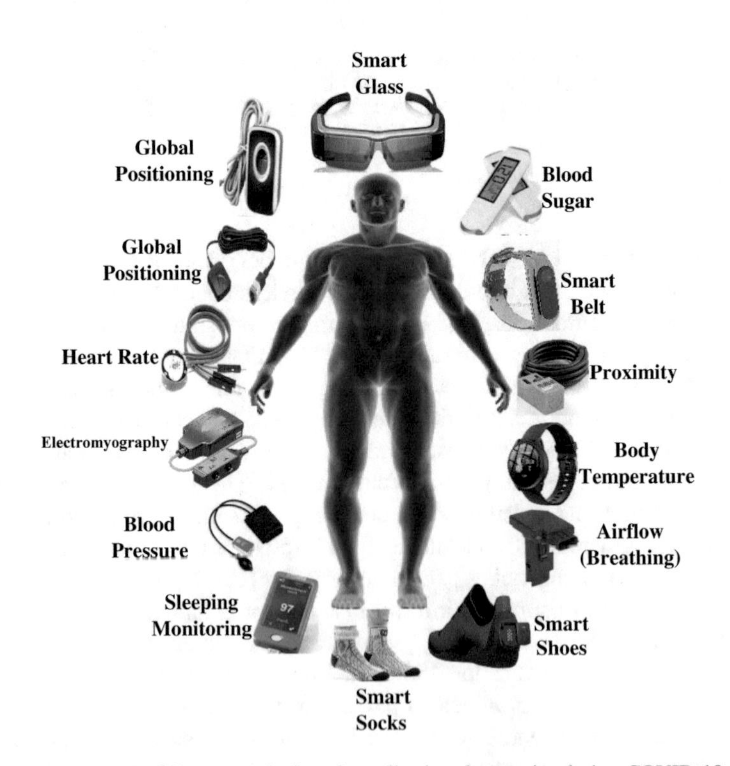


**Fig. 3** Selected worthwhile sensor devices for collection data on iot during COVID-19

## 4 IoT-Based Wearable Sensors Network Framework for Combating COVID-19 Pandemic

IoT-based WBSN Healthcare Monitoring System Architecture is a complex task, and a variety of systems were examined. Data collection and combination from multiple IoT devices is one of the major challenges facing IoT's realization in designing smart customized healthcare systems. In most cases, data is not always available in real-time so there is a challenge in evaluating or integrating a wide range of data since IoT devices obtain dynamic and complex data on medical assessment, tracking, treatment plan, and prediction in healthcare. The aggregation of data from specific sensor data sources is a major problem that needs to be addressed critically.

Therefore, it would be encouraging to investigate which of the IoT sensors enhances the efficiency of the intelligent systems with a collection of biomarkers of COVID-19 pandemic in kind. So it would be crucial to investigate whether any other background data existed which could improve the efficiency of the model. Besides, further studies are required to determine the quality of the attributes selected from each biomarker. Other relevant matters in IoT implementation for monitoring of patients, evaluation, treatment plan, and prediction is the need to design a system that can accurately switch between cloud and local classification methods with minimal processing time to ensure real-time and up-to-date provision for the patient.

This chapter proposes the framework for the IoT-based Wearable Body Sensors Network Healthcare System (IoTWBSN), which applies a range of wearable sensors interconnected to track a person's health condition. The use of body temperature and pulse, for instance, helps to collect physiological signals. The sensor data collected from these wearable sensors will be transferred directly to the cloud server due to the small computing power of the sensor nodes and the storage, as well as to avoid the use of a smartphone as a processing device. The design process of the IoT-based wearable body sensors System for COVID-19 is depicted in Fig. 4.

The proposed IoTWBSN architecture that can be used to manage WSNs is based on the perceptions of IoT devices, wearable body sensors or embedded WSNs, and Artificial Intelligence (AI). In the proposed model, WBSNs are used to collect information, send them through their programmable devices, which can enable examining the received data. Each device can be considered to be a diagnostic system due to the way they are programmed using different machine learning techniques. The sensors or devices can be connected to the Internet through the IoT layer. The devices have no energy limitations when compared with sensor networks, hence, they are connected directly to the power. The AI methods become resourceful and paramount to achieve a suitable diagnosis system, and physicians monitor the systems. The systems can report any suspicious patient or person with COVID-19 symptoms to the related experts. As displayed in Fig. 3, the framework has three layers.

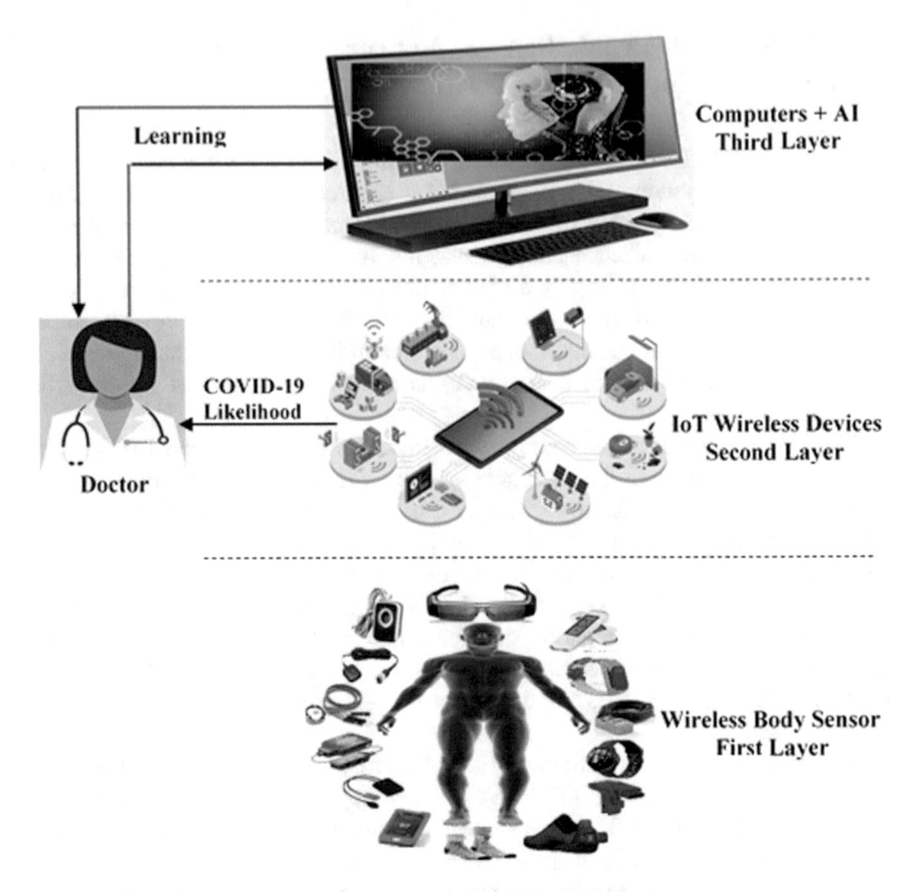


**Fig. 4** The proposed architecture IoTWBSN for COVID-19 pandemic

The first layer consists of the mobile wireless sensors that can be a move within a determined environment. The sensors are divided into embedded [65] and body wearable sensors (WSs) [66]. In any situation, a person can have several wearable sensors, and very possible for a patient to have more than one WSs. But experts determine the total number of WSs and this also depends on the required information. The second layer consists of the IoT-based devices that cover the WBSNs. In the third layer, the IoT devices link to the Internet and transfer the obtained data from the first layer to cloud computers.

The computer and AI model generate a real-time prediction, diagnosis, and monitoring events and produce the required reports needed by experts. This provides a high computing ability to enhance the IoT devices that have a lower computing skill. With the different predictor models that report COVID-19 suspicious events, the doctors and experts can promptly make a decision and advise accordingly. The third layer has two main functions (i) Generate real-time models to predict, diagnose, and monitor undesired events or signs related to COVID-19 pandemics such as heartbeat, temperature and measuring a reduced amount of insulin in a patient body and (ii) help to determine the positions of IoT-based devices in the 3D environment and the total number of IoT-based devices.

## 5 The Practical Applicability of the Proposed Framework

The integration of computer and biomedical technologies in medical systems has supported healthcare events, for instance, real-time disease analysis, remote monitoring of patients, and real-time drug prescriptions, among others. The methods have helped greatly to store both patients' personal information and their symptoms on the cloud, which can help during the COVID-19 pandemic. This aids the quality of services provided by the physicians thereby improve patients' satisfaction. IoT and wearable sensors and devices can aid early diagnosis of COVID-19 pandemics and create an efficient type of measurement in response to the outbreak. These devices have a remarkable impact on the prompt detection of the COVID-19 outbreak. For instance, IoT-wearable sensors have the power to show any part of them that is not functioning well, by capturing every single health data from the user. The results from these device users can notice any change in their health condition frequently and book an appointment with a physician before it is generated to real disease or any symptoms appear [67]. The COVID-19 pandemic might be easier to fight with the implementation of IoT and smart wearables in the healthcare system. Also, monitoring COVID-19 patients remotely would be more convenient and reduce the number of a patient admitted into hospital or isolation centers. The following are practical applications of the proposed framework.

For tracking purposes, the smartphone applications are empowered with IoT-based devices providing real-time information using Global Positioning System (GPS), Geographic Information System(GIS), etc. These have been extensively useful to intensification the chance of monitoring and detecting infected people [68, 69]. The implementation of smartphone applications with IoT-based platforms duringCOVID-19 an outbreak might have the benefits of having a complete cloud database that healthcare workers and government can monitor the COVID-19 pandemic and allow the infected person receiving treatment from home. The online hospital and real-time health information are possible by people to upload their health-related information to the cloud using an IoT-based cloud database and get health advice from physicians online within physically present at the clinic. Treatment is being administered within this platform, and without expanding the contamination patient will be cured at home. The system is cost-effective when compared with having a physical appointment at hospitals and the report obtain can be used by the government to make better decisions and action in future pandemics and will be able to manage the outbreak effectively [40].

The application of wearable smart glasses in IoT-based devices has also proved useful in detecting the COVID-19 outbreak among people. This can also be replaced with thermometer guns with its characteristic of less human interaction.

The smart glass with the functionality of optical and thermal cameras have been used to monitor people in a crowd [70]. The face recognition technology within the sensor captures the user body temperature, and tracking of an infected person becomes easier. The data from smart glasses become more reliable with help of Google Location History, which can be used to track the infected patient contact within a given time. The report of the captured data can be handy from the IoT-cloud database or sent to the physicians' smartphone for further actions [70]. For example, a Chinese company Rokid smart glasses with infrared devices to fight the COVID-19 outbreak, which can monitor up to 200 people [71]. The combination of Vuzix smart glasses with Onsight Cube camera thermal is another good example of these devices, the device helps to detect high temperatures and provide a real-time report for medical centers.

Different IoT smart thermometers that can be used to record constant measurement of body temperatures have been developed. They are developed in different types such as radiometer, touch and patch, very accurate devices, and cost-effective [72]. The devices are extremely helpful in early diagnosis, treatment, and monitoring of COVID-19 patients and any suspicious cases that can arise from the outbreak. Also, the use of infrared thermometers for taking body temperature can spread the COVID-19 virus due to the closeness of the infected person and physician, thus smart thermometers will be of help in such cases [73]. Different wearable thermometers can be used during this period such as IFiever, ISense, Ran's Night, Tempdrop among others, which report the body temperature in real-time on smartphones and can be used as sensors on IoT-based devices. These touchable devices can be a stick or worn on the skin under clothing [72]. For example, Kinsa's thermometers are been used in the USA to predict the most suspicious areas in each state based on the recorded temperature with a high possibility of being infected with the COVID-19 virus [74–76]. The proposed system can improve the chance of diagnosing new COVID-19 patients with any stress and increase people's daily lives.

A smart helmet with a thermal camera is another acceptable sensor results when compared with the infrared thermometer gun due to its lower human interactions [73]. The image and location of the user's face are taken when the high temperature is detected by an optical camera on the smart helmet, and then sent to the assigned IoT-cloud database with an alarm, then experts can differentiate the infected person, and take necessary action immediately. The use of the devices allows physicians and other related officers have access to facial recognition, temperature dark spot visioning, and personal information of the user in the crowd. The smart helmet has the storage capacity to keep all of the captured data within the helmet, and thus serve as a backup for the IoT-cloud database [77]. Moreover, the smart helmet integrated Google Location and the history of this can be used to find the places visited by the infected person after discovery [78]. This wearable device has been used successfully in countries like Italy, the United Arab Emirates (UAE), and China to monitor crowds within two meters and has shown good results [79].

For instance, a Chinese company produced a smart helmet called KC N901, which has an accuracy of 96% for high body temperature discovery and has been used by the aforementioned countries [79].

# 6 Conclusion and Future Research Directions

The global epidemic of COVID-19 has become the major hub of scientific research. The new digital technologies will act as a perfect solution to this global crisis. IoT can address the problem of detecting, surveillance, mapping contacts, and controlling this viral infection. The Introduction of IoT-based technologies to the present COVID-19 pandemic situation can be used to build a social forum to help individuals access appropriate treatment at home and to develop a robust repository on disease control for government and healthcare organizations. This helps in the use of IoT for the diagnosis and healthcare devices to be obtained from a person with minor symptoms (Preventive disguises, thermometers, medicines, personalized COVID-19 infection diagnosis, and control kits). Patients were able to upload their overall health to the IoT (clinical data storage) site online regularly and exchange their relevant data to area hospitals, the Center for Disease Control (CDC), and state and local health offices. Therefore, this chapter proposed an IoTWBSN framework to fight the COVID-19 pandemic by first presents the roles of IoT and wearable body sensor technologies in combating the COVID-19 outbreak. The framework can be used to obtain real-time data and information, which can be used by physicians and of relevant experts in clinical sciences. The results show that the proposed IoTWBSN architecture helps track and spread chronic diseases such as COVID-19. For future research purposes, it should be noted that progress has been recorded in the study on the IoT-based healthcare system in customized e-healthcare, but it is necessary to address a series of research questions and operational problems. In previous studies, various challenges concerning information safety and confidentiality, mobility control, and applications are identified during design processes. Thus, there is a need to concentrate on such challenges, particularly as they relate to IoT and prediction, diagnosis, exams, and monitoring of disease. In this context, an intelligent security model should be thoroughly studied to minimize the risks identified. We agree that conventional healthcare is certain to see a major conceptual change in the nearest future, as the technological revolution will put sophisticated software and its related services in the patients' fingertips and offer greater connection to reliable healthcare services for both patients and doctors in different areas. The full implementation of the proposed framework will be carry out as future direction. The lack of awareness amongst the healthcare stakeholders has a major hurdle in the adoption of IoT-based Wearable Body Sensors needs to be addressed in future for adequate implantation of the system in healthcare system especially in fighting future pandemic.

## References

1. Ogundokun, R.O., Lukman, A.F., Kibria, G.B., Awotunde, J.B., Aladeitan, B.B.: Predictive modelling of COVID-19 confirmed cases in Nigeria. Infect. Dis. Modelling **5**, 543–548 (2020)
2. Asai, A., Konno, M., Ozaki, M., Otsuka, C., Vecchione, A., Arai, T., Taniguchi, M.: COVID-19 drug discovery using intensive approaches. Int. J. Mol. Sci. **21**(8), 2839 (2020)
3. Helmy, Y.A., Fawzy, M., Elaswad, A., Sobieh, A., Kenney, S.P., Shehata, A.A.: The COVID-19 pandemic: a comprehensive review of taxonomy, genetics, epidemiology, diagnosis, treatment, and control. J. Clin. Med. **9**(4), 1225 (2020)
4. Rahman, M.S., Peeri, N.C., Shrestha, N., Zaki, R., Haque, U., Ab Hamid, S.H.: Defending against the novel coronavirus (COVID-19) outbreak: how can the internet of things (IoT) help to save the world? Health Policy Technol. **9**(2), 136–138 (2020)
5. Allam, Z., Jones, D.S.: Pandemic stricken cities on lockdown. Where are our planning and design professionals [now, then, and into the future]? Land Use Policy 104805 (2020)
6. Pullano, G., Pinotti, F., Valdano, E., Boëlle, P.Y., Poletto, C., Colizza, V.: Novel coronavirus (2019-nCoV) early-stage importation risk to Europe, January 2020. Eurosurveillance **25**(4), 2000057 (2020)
7. Zhao, S., Lin, Q., Ran, J., Musa, S.S., Yang, G., Wang, W., Wang, M.H.: Preliminary estimation of the basic reproduction number of novel coronavirus (2019-nCoV) in China, from 2019 to 2020: a data-driven analysis in the early phase of the outbreak. Int. J. Infect. Dis. **92**, 214–217 (2020)
8. Joyia, G.J., Liaqat, R.M., Farooq, A., Rehman, S.: Internet of medical things (IOMT): applications, benefits, and future challenges in the healthcare domain. J. Commun. **12**(4), 240–247 (2017)
9. Cook, D.J., Augusto, J.C., Jakkula, V.R.: Ambient intelligence: technologies, applications, and opportunities. Pervasive Mob. Comput. **5**(4), 277–298 (2009)
10. Magsi, H., Sodhro, A.H., Chachar, F.A., Abro, S.A.K., Sodhro, G.H., Pirbhulal, S.: Evolution of 5G on the internet of medical things. In: International Conferences on Computing, Mathematics, and Engineering Technologies (iCoMET), March 2018, pp. 1–7. IEEE (2018)
11. Sodhro, A.H., Sangaiah, A.K., Pirphulal, S., Sekhari, A., Ouzrout, Y.: Green media-aware medical IoT system. Multimedia Tools Appl. **78**(3), 3045–3064 (2019)
12. Awotunde, J.B., Adeniyi, A.E., Ogundokun, R.O., Ajamu, G.J., Adebayo, P.O.: MIoT-based big data analytics architecture, opportunities and challenges for enhanced telemedicine systems. Stud. Fuzziness and Soft Comput. **410**, 199–220 (2021)
13. Darwish, A., Hassanien, A.E.: Wearable and implantable wireless sensor network solutions for healthcare monitoring. Sensors **11**(6), 5561–5595 (2011)
14. Bibri, S.E.: The IoT for smart sustainable cities of the future: an analytical framework for sensor-based big data applications for environmental sustainability. Sustain. Urban Areas **38**, 230–253 (2018)
15. Manogaran, G., Chilamkurti, N., Hsu, C.H.: Emerging trends, issues, and challenges on the internet of medical things and wireless networks. Pers. Ubiquit. Comput. **22**(5–6), 879–882 (2018)
16. Christensen, K., Doblhammer, G., Rau, R., Vaupel, J.W.: Ageing populations: the challenges ahead. Lancet **374**(9696), 1196–1208 (2009)
17. Yach, D., Hawkes, C., Gould, C.L., Hofman, K.J.: The global burden of chronic diseases: overcoming impediments to prevention and control. JAMA **291**(21), 2616–2622 (2004)
18. Darkins, A., Ryan, P., Kobb, R., Foster, L., Edmonson, E., Wakefield, B., Lancaster, A.E.: Care coordination/home telehealth: the systematic implementation of health informatics, home telehealth, and disease management to support the care of veteran patients with chronic conditions. Telemed. e-Health **14**(10), 1118–1126 (2008)
19. Ekeland, A.G., Bowes, A., Flottorp, S.: Effectiveness of telemedicine: a systematic review of reviews. Int. J. Med. Inform. **79**(11), 736–771 (2010)

20. Christaki, E.: New technologies in predicting, preventing, and controlling emerging infectious diseases. Virulence **6**(6), 558–565 (2015)
21. Udgata, S.K., Suryadevara, N.K.: COVID-19: challenges and advisory. In: The Internet of Things and Sensor Network for COVID-19, pp. 1–17. Springer, Singapore (2020)
22. Ahmadi, H., Arji, G., Shahmoradi, L., Safdari, R., Nilashi, M., Alizadeh, M.: The application of the internet of things in healthcare: a systematic literature review and classification. Universal Access Inf. Soc. 1–33 (2019)
23. Abiodun, M.K., Awotunde, J.B., Ogundokun, R.O., Adeniyi, E.A., & Arowolo, M.O: Security and information assurance for IoT-Based big data. Stud. Comput. Intell. **972**,189–211 (2021)
24. Wu, F., Wu, T., Yuce, M.R.: An internet-of-things (IoT) network system for connected safety and health monitoring applications. Sensors **19**(1), 21 (2019)
25. Hammad, T.A., Abdel-Wahab, M.F., DeClaris, N., El-Sahly, A., El-Kady, N., Strickland, G. T.: Comparative evaluation of the use of artificial neural networks for modeling the epidemiology of schistosomiasis mansoni. Trans. R. Soc. Trop. Med. Hyg. **90**(4), 372–376 (1996)
26. Rodrigues, J.J., Segundo, D.B.D.R., Junqueira, H.A., Sabino, M.H., Prince, R.M., Al-Muhtadi, J., De Albuquerque, V.H.C.: Enabling technologies for the internet of health things. IEEE Access **6**, 13129–13141 (2018)
27. Chen, S.C.I., Hu, R., McAdam, R.: Smart, remote, and targeted health care facilitation through connected health: qualitative study. J. Med. Internet Res. **22**(4), e14201 (2020)
28. Poppas, A., Rumsfeld, J.., Wessler, J.D.: Telehealth is having a moment: will it last? (2020)
29. Olsen, G.A.: U.S. Patent application no. 15/339,639 (2017)
30. Crowley, R., Daniel, H., Cooney, T.G., Engel, L.S.: Envisioning a better US health care system for all: coverage and cost of care. Ann. Internal Med. **172**(2_Supplement), S7-S32 (2020)
31. HealthnetConnect.: Healthcare delivery, remimagined. https://healthnetconnect.com/. Accessed 12 Sept 2020
32. Ohannessian, R., Duong, T.A., Odone, A.: Global telemedicine implementation and integration within health systems to fight the COVID-19 pandemic: a call to action. JMIR Public Health Surveill. **6**(2), e18810 (2020)
33. Habibzadeh, H., Dinesh, K., Shishvan, O.R., Boggio-Dandry, A., Sharma, G., Soyata, T.: A survey of healthcare internet of things (HIoT): a clinical perspective. IEEE Internet Things J. **7**(1), 53–71 (2019)
34. Bai, Y., Yao, L., Wei, T., Tian, F., Jin, D.Y., Chen, L., Wang, M.: Presumed asymptomatic carrier transmission of COVID-19. JAMA **323**(14), 1406–1407 (2020)
35. Konstantakopoulos, I.C., Barkan, A.R., He, S., Veeravalli, T., Liu, H., Spanos, C.: A deep learning and gamification approach to improving human-building interaction and energy efficiency in smart infrastructure. Appl. Energy **237**, 810–821 (2019)
36. Gupta, M., Abdelsalam, M., Mittal, S.: Enabling and enforcing social distancing measures using smart city and its infrastructures: a COVID-19 use case. ArXiv preprint arXiv:2004.09246 (2020)
37. Mehmood, R., Katib, S.S.I., Chlamtac, I.: Smart Infrastructure and Applications. Springer International Publishing, Berlin (2020)
38. Ullo, S.L., Sinha, G.R.: Advances in smart environment monitoring systems using IoT and sensors. Sensors **20**(11), 3113 (2020)
39. Kamal, M., Aljohani, A., Alanazi, E.: IoT meets COVID-19: status, challenges, and opportunities. ArXiv preprint arXiv:2007.12268 (2020)
40. Abiodun, M.K., Awotunde, J.B., Ogundokun, R.O., Misra, S., Adeniyi, E.A., Arowolo, M.O., & Jaglan, V.: Cloud and big data: A mutual benefit for organization development. J. Phys.: Conf. Ser. **1767**(1), 012020 (2021). IOP Publishing
41. Koh, D.: SPHCC employs IoT tech and wearable sensors to monitor COVID-19 patients. Mobi Health News. https://www.mobihealthnews.com/news/asia-pacific/sphcc-employs-iot-tech-and-wearable-sensors-monitor-covid-19-patients (2020). Accessed 12 Sept 2020

42. Ogundokun, R.O., Awotunde, J.B.: Machine learning prediction for COVID-19 pandemic in India. medRxiv (2020)
43. Baharudin, H., Wong, L.: Coronavirus: Singapore develops a smartphone app for efficient contact tracing. https://www.straitstimes.com/singapore/coronavirus-singapore-develops-smartphone-app-for-efficient-contact-tracing
44. Sheng, Z., Wang, H., Yin, C., Hu, X., Yang, S., Leung, V.C.: Lightweight management of resource-constrained sensor devices in the internet of things. IEEE Internet Things J. **2**(5), 402–411 (2015)
45. Nichols, S.P., Koh, A., Storm, W.L., Shin, J.H., Schoenfisch, M.H.: Biocompatible materials for continuous glucose monitoring devices. Chem. Rev. **113**(4), 2528–2549 (2013)
46. Zhang, Y., Sun, L., Song, H., Cao, X.: Ubiquitous WSN for healthcare: recent advances and prospects. IEEE Internet Things J. **1**(4), 311–318 (2014)
47. You, I., Choo, K.K.R., Ho, C.L.: A smartphone-based wearable sensor for monitoring real-time physiological data. Comput. Electr. Eng. **65**, 376–392 (2018)
48. Nemati, E., Batteate, C., Jerrett, M.: Opportunistic environmental sensing with smartphones: a critical review of current literature and applications. Curr. Environ. Health Reports **4**(3), 306–318 (2017)
49. Benini, L., Farella, E., Guiducci, C.: Wireless sensor networks: enabling technology for ambient intelligence. Microelectron. J. **37**(12), 1639–1649 (2006)
50. Deng, Z., Wu, Q., Lv, X., Zhu, B., Xu, S., Wang, X.: Application analysis of wireless sensor networks in nuclear power plant. In: International Symposium on Software Reliability, Industrial Safety, Cyber Security, and Physical Protection for Nuclear Power Plant, August 2019, pp. 135–148. Springer, Singapore (2019)
51. Belfkih, A., Duvallet, C., Sadeg, B.: A survey on wireless sensor network databases. Wireless Netw. **25**(8), 4921–4946 (2019)
52. Farsi, M., Elhosseini, M.A., Badawy, M., Ali, H.A., Eldin, H.Z.: Deployment techniques in wireless sensor networks, coverage, and connectivity: a survey. IEEE Access **7**, 28940–28954 (2019)
53. Yousefi, M.H.N., Kavian, Y.S., Mahmoudi, A.: On the processing architecture in wireless video sensor networks: node and network-level performance evaluation. Multimedia Tools Appl. **78**(17), 24789–24807 (2019)
54. Venugopal, K.R., Kumaraswamy, M.: An introduction to QoS in wireless sensor networks. In: QoS Routing Algorithms for Wireless Sensor Networks, pp. 1–21. Springer, Singapore (2020)
55. Yick, J., Mukherjee, B., Ghosal, D.: Wireless sensor network survey. Comput. Netw. **52**(12), 2292–2330 (2008)
56. Akyildiz, I.F., Vuran, M.C.: Wireless Sensor Networks, vol. 4. Wiley, New York (2010)
57. Govinda, K.: Body fitness monitoring using IoT devices. In: Contemporary Applications of Mobile Computing in Healthcare Settings, pp. 154–169. IGI Global (2018)
58. Varshney, U.: Pervasive healthcare: applications, challenges, and wireless solutions. Commun. Assoc. Inf. Syst. **16**(1), 3 (2005)
59. Varshney, U.: Mobile health: four emerging themes of research. Decis. Support Syst. **66**, 20–35 (2014)
60. Benjamin, D.M.: Reducing medication errors and increasing patient safety: case studies in clinical pharmacology. J. Clin. Pharmacol. **43**(7), 768–783 (2003)
61. Alumona, T.L., Idigo, V.E., Nnoli, K.P.: Remote monitoring of patients' health using wireless sensor networks (WSNs). IPASJ Int. J. Electron. Commun. (IIJEC) **2**(9) (2014)
62. Vijendra, S.: Efficient clustering for high dimensional data: subspace based clustering and density-based clustering. Inf. Technol. J. **10**(6), 1092–1105 (2011)
63. Panigrahy, S.K., Dash, B.P., Korra, S.B., Turuk, A.K., Jena, S.K.: Comparative study of ECG-based key agreement schemes in wireless body sensor networks. In: Recent Findings in Intelligent Computing Techniques, pp. 151–161. Springer, Singapore (2019)

64. Velez, F.J., Chávez-Santiago, R., Borges, L.M., Barroca, N., Balasingham, I., Derogarian, F.: Scenarios and applications for wearable technologies and WBSNs with energy harvesting. Wearable Technol. Wireless Body Sensor Netw. Healthc. **11**, 31 (2019)
65. Rabby, M.K.M., Alam, M.S., Shawkat, M.S.A.: A priority-based energy harvesting scheme for charging embedded sensor nodes in wireless body area networks. PloS One **14**(4), e0214716 (2019)
66. Chen, C.M., Xiang, B., Wu, T.Y., Wang, K.H.: An anonymous mutual authenticated key agreement scheme for wearable sensors in wireless body area networks. Appl. Sci. **8**(7), 1074 (2018)
67. Haghi, M., Thurow, K., Stoll, R.: Wearable devices in the medical internet of things: scientific research and commercially available devices. Healthc. Inf. Res. **23**(1), 4–15 (2017)
68. El Khaddar, M.A., Boulmalf, M.: Smartphone: the ultimate IoT and IoE device. Smartphones Appl. Res. Perspect. **137** (2017)
69. Adeniyi, E.A., Ogundokun, R.O., & Awotunde, J.B.: IoMT-based wearable body sensors network healthcare monitoring system. Stud. Comput. Intell. **933**, 103–121 (2021)
70. Mohammed, M.N., Hazairin, N.A., Syamsudin, H., Al-Zubaidi, S., Sairah, A.K., Mustapha, S., Yusuf, E.: 2019 novel coronavirus disease (Covid-19): detection and diagnosis system using IoT based smart glasses. Int. J. Adv. Sci. Technol **29**(7 Special Issue) (2020)
71. Bright, J., Liao, R.: Chinese startup Rokid pitches COVID-19 detection glasses in the US. China Publishing Company (2020)
72. Tamura, T., Huang, M., Togawa, T.: Current developments in wearable thermometers. Adv. Biomed. Eng. **7**, 88–99 (2018)
73. Mohammed, M.N., Hazairin, N.A., Al-Zubaidi, S., AK, S., Mustapha, S., Yusuf, E.: Toward a novel design for coronavirus detection and diagnosis system using IoT based drone technology. Int. J. Psychosoc. Rehabil. **24**(7), 2287–2295 (2020)
74. Chamberlain, S.D., Singh, I., Ariza, C.A., Daitch, A.L., Philips, P.B., Dalziel, B.D.: Real-time detection of COVID-19 epicenters within the United States using a network of smart thermometers. medRxiv (2020)
75. Dubov, A., Shoptaw, S.: The value and ethics of using technology to contain the COVID-19 epidemic. Am. J. Bioethics 1–5 (2020)
76. McNeil, D.G.: Can smart thermometers track the spread of the coronavirus? The New York Times (2020)
77. Mohammed, M.N., Syamsudin, H., Al-Zubaidi, S., AKS, R.R., Yusuf, E.: Novel COVID-19 detection and diagnosis system using IOT based smart helmet. Int. J. Psychosoc. Rehabil. **24** (7) (2020)
78. Ruktanonchai, N.W., Ruktanonchai, C.W., Floyd, J.R., Tatem, A.J.: Using Google location history data to quantify fine-scale human mobility. Int. J. Health Geogr. **17**(1), 28 (2018)
79. Ghosh, S.: Police in China, Dubai, and Italy are using these surveillance helmets to scan people for COVID-19 fever as they walk past and it may be our future normal. Bus. Insider (2020)

# The Dark Side of Social Media: Spreading Misleading Information During COVID-19 Crisis

**Noor Aamer Al Shehab**

**Abstract** The Internet has heralded a new era of information technology and facilitated communication through different platforms. Several governments, organizations, and individuals recognize that social media has yielded significant advantages to their practices and easily delivers their messages. For the first time in their history, some governments and businesses offer jobs related to the social media landscape, such as Social Media Specialist, due to the technological and information revolution. Certainly, social media enables its users to distribute information and reach a mass audience irrespective of their geographic and demographic boundaries. Therefore, social media could breed new challenges when misused and exploited in dispensing rumors and misleading facts to the public, especially in crises. This chapter aims to underscore social media's existence and how it paves the way to smooth the flow of information and communication worldwide. By employing social media tools in their practices, both governments and businesses witness exceptional outcomes and acknowledge that social media laid the foundation for more effective communication. This chapter will also highlight social media's disadvantages in increasing false information and inadequate facts that drive more uncertainty, sadness, anger, and lack of confidence among the public. It will also address the concept of crisis in general and focus on crisis management and crisis communication. Moreover, some countries will clarify the actions relevant to information filtering via prohibiting some popular websites and social media platforms. Furthermore, it will identify how governments and businesses deal with rumors in crisis, for example, COVID-19 outbreak, wars, financial scandals, business interruptions, and the like. Finally, the chapter will deliver the best practices that manage and control the fake information in social media and determine the best ways to spread accurate, reliable, and sufficient facts.

**Keywords** Social media · Twitter · Facebook · Instagram · Crisis · Crisis management · Crisis communication · Information · Rumors · COVID-19 pandemic

---

N. A. Al Shehab (✉)
Ahlia University, Manama, Bahrain



A.-E. Hassanien et al. (eds.), *Advances in Data Science and Intelligent Data Communication Technologies for COVID-19*, Studies in Systems, Decision and Control 378, https://doi.org/10.1007/978-3-030-77302-1_15

## 1 Introduction

No doubt, Social Media or Web 2.0 yielded a remarkable digital renaissance that has changed the entire world. Huge investments and efforts are allocated to boost the information technology infrastructure by multiplying the Internet's speed and delivering novel solutions that innovatively collapse the geographic boundaries and embrace communication and knowledge sharing. In recent years, the high penetration of smartphones by all ages has led to a dramatic increase in surfing the Internet browser and becoming actively engaged in social media platforms. Indeed, social media is a wide umbrella that covers Facebook, Twitter, Instagram, Google, LinkedIn, Snapchat, WhatsApp, TikTok, Pinterest, and furthermore. The number of computers connected to the Internet would reach 50 billion devices by 2020 [46], and the number of users of social media around the world is beyond 3.8 billion in January 2020. There are more than 5.19 billion people using cell phones [120, 121].

Social media tools provide key benefits to their users and entail some challenges and drawbacks. Ultimately, selecting the most appropriate platform depends on the user's needs, the convenience of use, information technology literacy skills, and the country's culture and regulations. Studies showed that Internet use is primarily impacted by the digital skills and efficacy acquired by individuals [87]. Because of globalization, competitive markets, and advanced technologies, several educational authorities worldwide started embedding information technology skills into their pedagogies to prepare skillful graduates who can build resilient communities in front of the ongoing threats [103].

While Facebook is mostly preferable by Americans, as stated in 2019, statistics with more than 169 million users followed by Instagram with nearly 121 million. The line is the most popular in Japan, according to a 2019 report, and approximately 80 million users [120, 121]. Besides, China has its own rules and regulations regarding social media platforms where all of which are strictly banned except its national social media sites. WeChat has been introduced and improved by the Chinese government to facilitate formal information release and increase public participation [15]. It is noticeable that governments and organizations nowadays employ social media tools to communicate with their stakeholders interactively and to spread meaningful information to them.

To put it differently, governments tend to pay public attention to critical aspects. To point out: weather forecasting, traffic status, upcoming events, the announcement of public holidays, job vacancies, latest politics and economic facts, national and international news, recent crimes, and the updates of disasters are obtainable via social media approaches. Besides this, governments of developing countries in Africa and other parts of the world reach massive students through social media sites to run their classes online to avoid disruptions in the educational process under certain circumstances. As well recognized, crises can take different forms and shapes. It could be a natural disaster, E-coli outbreak, terrorism, financial scandals, and much more [28].

Likewise, social media expands the business's connectivity with global markets, supply chain networks, and customers with superior impact. Parveen et al. [104] argued that with lower costs and efforts, organizations now enjoy better management of their brand by maintaining high social capital, enhancing stock performance, and increasing the amount of online transactions [44]. Advertising in social media cannot be overlooked since it accelerates individuals' awareness about the available products and services, the affordable prices, the features, and extra further information. Here, social media platforms play a major role in communication between different stakeholders to upturn the certainty and downturn the blurred areas [96].

Referring to several types of research, ordinary individuals use free diverse social media platforms for varied reasons such as: "communication and collaboration with others, meeting friends, sharing knowledge, identification of news and events, job hunting, acceptance from others, passing the time and escapism from reality" [145]. Studies show that some people use different accounts in the same application, while others use diverse social media platforms. However, some individuals post their real personal photos. The rest do not do so. Moreover, nicknames and true names are both used in the world of social media. Researchers try to capture the time that people spend on each social media site to gain profounder understanding of what they are looking for and what the future trends would look like.

In light of the persistent spread of the COVID-19 pandemic, which destructively hit the entire globe during 2020, most governments declared stay at home precautions for several months and encouraged remote works [143]. Moreover, students from all levels enrolled in distance learning during the last semesters, where schools and universities altered their strategies to cope with the dynamic changes. Correspondingly, the number of internets and social media users has been considerably increased. In this regard, countries supply their optimum financial and technical support to handle this crisis and motivate public to process their services through e-Government portals. Furthermore, various countries launched hotlines to deal with their public and respond to all queries to keep them on the right track and disconnect them from confusing information.

It is important to realize that social media is a vehicle to share personal thoughts and opinions in a responsible way freely, unlike the traditional media Web 1.0 where individuals cannot click like, post, subscribe, re-tweet, or comment to the content. Crijns et al. [32] stated that organizations should deal appropriately with the negative comments in their social media sites since this impair their reputation and could affect their profitability. Despite of the returns of social media, effective communication during crises verifies how governments and organizations are certainly good as they adequately represent themselves [40].

In bad times, organizations have crisis responsibility to minimize the anger; sadness and ambiguity among stakeholders by explaining the real situation for them and protecting their rights and privileges from additional threats [112]. As individuals and businesses are both responsible for what they release in social media, governments are also requested to apply transparency and credibility in all of its functions. It is found that individuals tend to indicate information relevant to their existing beliefs and avoid irrelevant information [85]. Hence, researchers initiate new horizons in social media landscape that should be explored as this is a vibrant and crucial domain that attracts most governments, organizations, and individuals alike. Here in this chapter, the focus will be applicable to social media's heavy use in communication, especially in crises.

## 2 Literature Review

### *2.1 The Rise of Social Media*

As per Forbes [52], the World Wide Web has been turned out for public access in 1991. The recent two decades have witnessed excellent developments in the information technology sphere. Web 2.0 has created a big jump in this field even though the traditional Web 1.0 possessed read-only content [58]. Internet and smartphones spread widely over the world. We reached a stage where everybody is a media specialist, as advanced technologies have intensively and extensively enlarged. The phenomenon of social media affects the behavior of governments, businesses, and individuals. Statistics reflect that mostly 79% of the world population access social media. So, the tremendous academic investigations on the growth and use of social media platforms have become apparent [59].

Kaplan and Haenlein [77] defined that Social Media is "a collection of Internet applications that lay ideological and technological foundations for Web 2.0 which embrace the creation, edit, and share of User-Generated Content (UGC)." This is advantageous when users express themselves to others within an interactive environment that allows for a simple exchange of information, thoughts, resources, experiences, interests, and practices.

In 2011, Bergh et al. [13] categorized social media into Content Community Sites (CCSs), Social Networking Sites (SNSs), and Social Media Platforms. To clarify, content communities permit the exchange of multimedia content, such as YouTube. At the same time, Facebook is an example of a social networking site that allows the users to effectively share their thoughts, photos, videos, posts, and the like. A media platform refers to any service, site, or method which delivers media to a targeted audience and might allow for discussion and sharing.

Subject to its coverage, social media also can be classified to global, regional, and local [101]. Those platforms are characterized by the users' languages, cultural backgrounds, education and digital literacy levels. However users are made of diverse cultural structures. They formulate a global network-based community within the complex social media ecosystems. Social media impacts each and every country and affects the entire global consumer segments. The international social media platforms are not similar in terms of their formats, personalities and features. Thus, this could affect the degree of users' interactions and perceptions.

In reality, social media varies in its ability to enable its users to effectively communicate and resolve the uncertainties. Media Richness Theory introduced by Daft and Lengel [36] explained that media methods are not operating equally. Therefore, communication and understanding levels differ from e-mail communication to Facebook communication. The first one recorded the lowest media richness as it provides a text-based message only. However, Facebook is richer in media since pictures and videos are possible to be shared accompanied by texts. The concept of Media Richness encompasses the information amount that a specific medium permits to be transferred in a specific time interval. Tied with this, social media embraces the “Social Presence” that was introduced by Short et al. [119] which is defined as “the degree of salience of the other person in the interaction and the consequent salience of the interpersonal relationships”. There are two pillars that construct the theory of Social Presence which are: Intimacy which involves physical presence, physical space, eye contact, body language, etc. and Immediacy which refer to psychological distance between communicator and the receivers. When it comes to social media, social presence is indicated also by the “privacy, settings, purpose of communication and interactivity” [133]. It is observed that literature review has not offered any “Theory of Social Media” although there are many authors who considered and studied various social phenomena within the context of social media.

It is interesting to remember that Facebook was founded in 2004 by four students from Harvard University in the United States. After one year in 2005, the American platform YouTube was instituted by three PayPal employees. In 2006, a group of four friends launched Twitter as an American microblog with only 140 characters in every tweet. Although this may be true that Instagram was created in 2010 after the existence of the previously mentioned social media platforms, there is an extreme increase in the number of its users on a daily basis [102]. According to Mansfield’s study [91], an average of 118 min daily is spent on social media by more than 1.8 billion people. It mentioned that 77% of them are actively enrolled in generating content through their likes, comments, posts, shares, re-tweet, and link clicks. Another phenomenon has been emerged as a result of the social media which is the “Social Media Influencer” who create and build their personalities through different social media platforms such as: personal blogs, Instagram accounts and YouTube channels. Actually, they consistently construct and manage their contents in order to attract a massive number of followers [2].

It must be remembered that social media has generated enormous benefits. It seems that it has transformed the world into a truly global village. By the same token, all social media platforms have a global reach. To illustrate, a new nation which is larger in population than both China and India has appeared lately which is called “Facebook Nation” [117]. Beside this, OMNICORE website in 2020 showed that there are beyond 500 video hours are posted on YouTube in every minute and more than 5 billion uploaded videos are watched and shared every day.

Social media produces significant network impacts where real-time connectivity enables texts, images, sounds, and videos to go viral. Identically, Twitter has been considered the most effective tool in monitoring and evaluating risks and concerns for specific audience needs [83]. Similarly, Instagram has become socially popular, especially among youths, as it is a visual-oriented platform that encourages people to communicate and collaborate more. It is the fifth social media application preferred by one billion monthly active users worldwide [100]. Social media is not only designed for individual uses but also organizational activities. Kim and Liu [80] recognized that organizations have not fully consumed the available rewards of social media to stay in touch with its stakeholders.

Seemingly, social media platforms have pulled out the rug from the traditional media. Seltzer and Mitrook [113] explained that social media tools have greater dialogic, interactive and faster features than traditional media devices. Unlike traditional media, social media offers its users access, share, participate, and spread information and resources through private, semi-private and publicly networks. Yet, traditional mass media such as newspaper and radio broadcasting run in one way direction.

In addition, traditional media in general is subjected to the control of governments where "gatekeeping" practices have not been modified yet. White [142] admitted that all news, articles, and images are carefully selected by "gatekeepers" before their release in the traditional media. Consequently, any news and evidence outside the scope of newspaper or TV channel would not be published. It is acknowledged that newspaper websites habitually employ social media applications to broadcast the content to their audience rapidly. In the same fashion, studies demonstrated that some journalists have appreciated the social media tools to post their articles and opinions related to their own beliefs and values instead of the limitations drawn by the press or channel. They agreed to Facebook and Twitter terms and mentioned that they are implementing "same values, but new tools." Shang et al. [115] found that press articles expressively focused upon "corruption, diplomacy, and elections" while Twitter deals greatly with the personal messages of congratulations, condolences, thank-you, and relationships.

On the other hand, Austin et al. [8] claimed that people find that traditional media is more trustworthy than social media. Information supplied by television and newspaper is more credible and reliable as they are revealed. People prefer traditional media to create information and receive knowledge. Conversely, they use social media to obtain exclusive and deeper information and stay in connection with their favorite networks. Studies elaborated that people would mix between the traditional mass media, particularly television live shows and social media platforms, generating a "Multi-Screen" phenomenon [139]. Although mass media such as television and radio broadcast news on regularly scheduled timings, social media is superior in delivering instant, bottomless, and rationalized facts.

## 2.2 *The Benefits of Social Media*

The Fourth Industrial Revolution today boosts social media employment and allows for a continuous increase in the number of users that hits billions. Social media is a real Tsunami where information and media literacy become the core to handling revolutionary data selection and management. Numerous papers addressed the importance of social media in communication and knowledge sharing. Currently, communication channels through social media are more affordable and reliable. The inherent and immediate interaction through social media networks shapes closer and mutual relationships beyond geographical borders. Moreover, social media serves as a vehicle to exchange explicit and tacit knowledge [20]. Small and medium-sized enterprises (SMEs) revealed that social media methods improve their innovation practices. With this viewpoint, there are major shifts in the global mindsets that prompt governments and business organizations to promptly knock on social media doors [64].

### 2.2.1 Governments in Social Media

During the contemporary digital era, governments are executing new media technologies to enhance their governance. The government of China realized the tremendous power of social media and accelerated the infrastructure of its applications, microblogs, and WeChat. In 2015, the Cyberspace Administration Office of China [34] stated that there must be a centralized Social Media Strategy that articulates specific online tasks for the government: leveraging formal information, responding to citizens and public queries, and offering some civic services as well. Along with this, studies have correlated the positive experience of Internet use with high participation in online governmental affairs [126]. Thomas and Streib [130] mentioned that individuals who regularly accessed governmental websites are more familiar with public matters than others.

Likewise, politicians worldwide are eagerly relying on social media since it provides golden opportunities to reach massive voters, audiences, and even rivals. Twitter has been considered the favorite social media tool in diplomatic communication through interactive dialogues among countries' spokesperson. Back to the Iran Swift Release event in 2015, it was relatively new that Kerry and Zarif communicated freely and positively on Twitter in front of the global audience even though their countries are not friends [40]. As diplomacy is the "art of communication", countries should carefully select the words in their social media accounts to represent themselves appropriately. Coupled with this, electoral candidates adopt social media platforms to communicate with their people throughout the election campaigns by texting, posting videos, tweeting, and much more [21].

It is equally important to spot that governments are keenly increasing public awareness via social media. Governments used to spread information that is pertinent to public interests such as: "economy, education, health concerns, diplomacy, parliamentary issues, corruption, international relationships, terrorism, philanthropic crisis, elections, national holidays, congratulations and condolences messages." Social media encourages people's engagement to directly dialogue with legislators, civic officials, and the whole government agencies. Sometimes, the governments empower nominated public figures to deliver a message on their behalf through segregated social media methods [105].

As a matter of the fact that the novel coronavirus has spread everywhere; governments employ social media platforms to interactively deal with the public to inform them about the number of positive cases, the number of recovered patients, the number of death cases, the daily locations of swab tests around the country, the updates of vaccine, the working hours of hospitals and clinics, and the timing of national COVID-19 periodic meetings. Indeed, people seeking directions from the government to set the tone of coping with this pandemic in a clear, calm, and professional manner. Social media saves the government's time and expenditures since traditional public outreach are costly [131]. The ongoing lockdown of schools and universities enforces governments to communicate with students through social media as well. Virtual classes are running via varied applications such as Zoom and Microsoft Teams, for which online learning assessments and assignments are also piloted.

Before everything, all governments need a holistic strategy for the implementation of social media. This is an energetic field that has to be well monitored to donate fruitful outcomes. This kind of strategy has to assign what, when, why, where, and how to deliver precise, reliable, and sufficient information. Transparency and openness are important to minimize the uncertainty and misleading information principally at the time of crises.

### 2.2.2 Businesses in Social Media

There are dozens of main changes brought by social media to all types of businesses and industries. Thus, it is extremely required to review and modify business strategies, planning methods and push for more creative works [81]. Social media is considered to become the leader in advertisement, marketing, and e-commerce. However, there is a strong debate about the most effective method that measures the exact return of investment (ROI) of social media [122]. Due to the wild proliferation of digital media, business organizations set social media on the top of their agendas to enhance their position within the marketplace [25]. Interestingly, Okazaki and Taylor [101] confirmed that global social media would bring both: global reach and personalization at the same time.

To enumerate, Microsoft, Amazon, Facebook, and many multinational companies can establish virtual communities and attract followers from all parts of the world to interact and exchange information, interests, and experiences. These big behemoths have reached dramatic growth in market during the pandemic [45]. The research projected that further than 75% of consumers make online comparisons between products and prices before their physical visit to the automobile dealer. A concept of "Webrooming" has arisen in the opposite of the term "Showrooming" [75]. The prime goal is to promote the products or services to reach a maximum considerable audience. Digital advertising generates around 30% of total platform accounts' revenue and acts as the fastest growing income source. To illustrate, most

U.S. News agencies admitted that social media is the foremost driver of increasing their digital advertising revenue and website traffic [76].

Social media such as Facebook, YouTube, and Instagram are welcoming the free posts of advertisements. This allows organizations to entice consumers' attention to what they do in terms of their brands, services, activities, job vacancies, events, donations, etc. Kamboj et al. [74] found that there is a direct link between social media adverts and purchase intentions. A lot of firms employ social media in their value creation procedures where certain new products are posted through different platforms before bringing them to the retail shops. This yields a great experience to customers, especially when prices are lower. In 2014, Jin et al. [70] commented that "organizations have no chance about whether to adopt social media into their strategies; the only option is to do so." Consequently, SMEs confront the potential of failure because they have to remain competitive by overcoming their technological barriers, as Van Scheers supposed in 2011 [110]. SMEs which are more aware of the latest trends in social media are less exposed to risky situations [61].

In comparison to the traditional tools of knowledge sharing, social media involves employees from different territories to become more active in discussion and faster in knowledge sharing practices [82]. Social media certainly provides new effective communication and distributions that shortened time and facilitated interactive communities [96]. Similarly, businesses virtually contact their consumers and stakeholders to clarify ambiguities and justify some actions like massive layoffs and frequent product recalls. If there is a disconnection between parties, false and inaccurate information would take place, which ultimately hurt the organization's reputation.

### 2.2.3 People in Social Media

As mentioned earlier, the development of smartphones and the unceasing connectivity to the Internet manifest all human activities with no exception. People can create their webpages and invite their families, friends, colleagues, and others to share miscellaneous information and resources. Moreover, social media eases communication among people through texting, recording voice notes, making video calls, sending emojis, notifications, photos, and much more. Coupled with this, social media encourages people to take photos of themselves "Selfie" while engaged in multiple events to post them on Facebook, Instagram, Snapchat, and other mediums. Additionally, posting "stories" allows social media audiences to be cultivated due to the increasing number of impressions. Statistics prove that there are approximately 93 million selfies presented on social media each day [114]. Social media yields a lasting digital legacy where users capture, manage and archive their major life occasions to re-experiencing their memories [6].

Researchers originated that individuals lean towards adopting social media platforms to acquire acceptance from others, passing time, escapism from reality, entertainment, job hunting, and staying informed about news, events, brands, shopping offers, travel deals, and latest trends [41]. Individuals on social media share their shopping experiences with multiple brands and recommend products and services to others. Lee et al. in 2015 implemented "Use and Gratifications Theory" to discover the motives behind using Instagram by individuals and found their four gratifications, which are: "Informativeness, Community support, Status-seeking and Self-representation" [7].

Social media indeed provides full support to the pedagogic functions for which students can utilize various sites, networks, and platforms to communicate, collaborate, and exchange knowledge. Lovejoy et al. in 2012 identified the role of Twitter in public relations campaigns and highlighted that it is treasured for educational and professional uses as well. By taking NASA's social media accounts, there is valuable information about physics and space exploration in each of its posts. Users find it beneficial and credible as a source of knowledge. Social media also becomes the backbone of the mobile learning "m-learning," especially in Africa, as stated by UNESCO [132]. Studies show that mobile penetration in Africa would reach 79% by 2020, which means the access to use of technology will multiply [57].

Ramo and Prochaska [107] claimed that Americans employ social media platforms to team up with others who are interested in well-being, having common health issues, or may want to take part in a research study. In the same fashion, millions of users consume social media to learn more about their work, hobbies, interests, and more. YouTube and Instagram possess personalized accounts that fill specific hobbies or interests with video tutorials about cooking, makeup, designing, and the like, mostly provided by social media influencers.

Along with this, it is documented that social media platforms are the getaways for protesters and activists to broadcast the violations against their human rights immediately. According to Taki, in 2013, social media has been attached to activists to be more demanding. Specifically, in the Pacific region, social media has been utilized for communication, disaster awareness, political disputes, activism, and rejuvenation identification [141]. Similarly, the recent Arab rebellions in Egypt and Libya in 2011 have been ignited via social media.

Because individuals are more engaged in social media, they are likely to have mutual comments and feelings. Online comments drive for both positive and negative impacts. Na and Rhee [97] explained that the public goes through the online comments to understand different opinions, analyze the present issues, and criticize news. Positive comments enable the community to blossom and prosper, whereas negative ones push for more anger, conflicts, and harm their reputation. Consequently, ordinary users should participate in the social media world in a responsible and liable way as their words are accountable. Most governments established strict laws and regulations to combat cybercrimes, including the misuse of social media. It seems that some people's sole mission is to distribute fake and misleading information or criticize the governments' policies that adversely influence public opinion, mainly at the time of crises. This invites us to deliberate: Is social media bad?

## 2.3 The Dark Side of Social Media

However, social media's sphere offers several benefits and merits to its users; it breeds some challenges and threats [28]. There are several digital dangers facing governments, including cyber-attacks and the spread of false information among the public. In 2020, the Norwegian Parliament declared that it had been targeted for a major cyber-attack that leads to a breach in several members' email accounts [11, 9, 10]. Moreover, the Stock Exchange of New Zealand confronted multiple days of disruptions after an unknown cyber-attack [108]. Since information is publicly disseminated in social media in real-time regardless of its accuracy, this would affect the reputational judgments about specific governments, organizations, or individuals. Henceforth, scholars presented the term "Media Reputation," which stated that the overall image is directly influenced and evaluated by media [38].

Governments improve their extents of responsiveness to address their public's concerns to shed light on the grey areas [109]. Governments worldwide provide venues for public expressions through interactive communication in social media [22]. In 2012, Magro [90] debated that the development in social media strategy would empower the transparency measures of governments that could enhance the public image and promote a transparent government.

For the public welfare, governments undertake strict actions to prevent the delivery of forged information by executing cautious surveillance to social media accounts, eliminate improper ones, banned any misleading websites and networks, and advise the public to extract information from the formal and authorized channels [51]. Governments started to create, implement, and review their social media laws and regulations to chase rumors and increase the trust between governments and their people. Occasionally, guilty persons pay financial penalties or even spend time in jail due to their illegal activities of abusing social media.

In 2020, one of the most surprising events is the case of money laundry in Kuwait. This had impacted the reputation of Kuwait and threatened its socio-economic conditions. The key people who are mainly enrolled in money laundry activities are social media influencers. Kuwait struggles to take the necessary actions to incubate the money laundry propaganda tied with other corruption cases in governments. Ministry of Interior in Kuwait requested the public to avoid engagement in fabricating and distributing false information in this regard. As authoritatively circulated, several social media celebrities had their bank accounts frozen and faced a travel bank until the end of their investigations [3]. Officials pointed out that facts and evidence should only be taken from the trustworthy channels; however, all details are considered confidential since several political leaders in the government and national army are expected to be guilty.

Another example of the dark side of social media is during the uprising of Libya in 2011 when Gaddafi's regime decided to block all kinds of communication, including mobiles, television, and the Internet, to isolate Libyans from the world protestors' actions cannot be shared through Twitter or YouTube [136]. In 2009,

Freedom House [55] listed Libya in the top 20 countries with greater media boundaries.

From governments to businesses, it is agreed that commercial organizations are seeking beneficial methods to maintain long-lasting success. Hajli et al. [62] approved that social media and e-commerce provide significant marketing opportunities that upturn sales and people's purchasing actions. Therefore, while individuals communicate and exchange their personal experiences toward a certain product or service, there is a potential for bad comments to occur accompanied by "dislike" clicks. Again, the business reputation would be adversely affected, and this causes more serious losses. It is not easy to monitor user-generated content on all social media platforms even when bad comments are deleted or reported as spam. In more advanced cases, companies sue people in court who intentionally harm the company's reputation [98].

On the other hand, employees inside the organizations might also harm their workplace's reputation via sharing confidential and sensitive information through their social media networks where there is no stopping point. Employees might possess valuable information related to internal budgeting and planning. The disclosure of such material information affects the investor's and stakeholders' decisions [67]. One of the illustrations is disseminating the new product features on social media, so competitors take advantage to stay ahead in the market. Shoemaker and Reese [118] claimed that it is important to explore the internal and external factors that impact the corporate media. Also, an aviation organization (United Airlines) lost a market value of around 800 million dollars in one day only due to reputational loss when passengers upload a video on social media showing how the aircraft was offensive toward travelers [84].

It is observed that companies such as IKEA used to recall products from time to time due to quality deficiency, which in few cases lead to child deaths. Usually, stakeholders blame the company for such accidents and push it to review its crisis responsibility. Quality failure strikes the reputation and drives to increase the claims against the company. Although most governments and organizations have a habit of hiding bad news, reputation must be considered a long-term game [24].

Companies that carry out actions that contribute directly to society cannot be forgotten and stay longer in people's memory. During the spread of the COVID-19 outbreak, the owner of a Brazilian brand posted a video on his Instagram account promising people that he will never cut jobs; however, his company laid off 600 employees after his speech. As per an Edelman survey [42], 35% of Brazilians mentioned that they would boycott this brand and invite their friends.

Correspondingly, social media stimulates risks and crises. To protect the social community, India in 2018 allocated 5 billion dollars to standardize the corporate social responsibility by training employees to deal with multiple situations. Both governments and business firms should implement the art of communication where wise words are used along with emotional support to minimize the general anger, sadness, and fears [117].

Remarkably, social media caused harmful impacts on people's lives. Researchers found that social media is a modern kind of addiction [60]. Studies confirmed that the increased use of digital screens had caused vision-related problems associated with neck and back pain. The nonstop use of social media allows for the immediate existence of "Time Out" centers in order to halt children and youth from the constant texting and posting [1]. Moreover, trying to get a perfect photo shot could lead to accidental deaths. Furthermore, being unhappy with your selfies or the quality of your life due to the comparisons might also drive suicides and attempted suicides. Furthermore, receiving bad comments from followers could lead to the same conclusion. Besides this, some social media users prefer themselves after applying digital filters, which enhance their overall look and provide a better appearance. This has eventually led to "Snapchat Dysmorphia" phenomenon [106].

Lately, the reflective ramifications of social media on typical individuals have been widely spread. The increased tendencies of stolen accounts, blackmailing kids and women, death threats, and drug promotions push societies to reach a critical juncture. Tied with this, youths' heavy use of Internet and social media dissuade them from serious study and allow them to spend their entire day in conversations, entrainments, and browsing. Twitter and Instagram received countless spam reports as a result of fake, stolen and inappropriate accounts [123]. Shen and Guangyan [116] debated that the credibility of information in social media is questionable because its virtual nature does not always permit to assess of the source's reliability. For this, biased and fabricated information might be transferred, and here, the focal question is: How governments deal with distorted information?

## 2.4 Dealing with Misleading Information

It is recommended that governments should prepare an effective roadmap to provide the foundations for the right communication methods to encourage honesty and reliability in knowledge circulation. In fact, developing countries lack for proper preparation and resilience at the times of crises due to their limited capacity [4]. There should be authorized websites, networks, and individuals' accounts that yield accurate facts and reflect the government's vision. Social media allow users to modify posts and correct errors and mistakes in the same time. Deviations from the assigned media regulations and mobile phones' abuse are exposed to the country's laws. Depending on media policies, some websites and social media platforms are banned in certain countries. According to their openness, transparency, objectives, and the degree of risk associated to the occurred event; governments react differently [127]. Remarkably, it can be seen that the Governor of Dubai Sheikh Mohamed Bin Rashed Al Maktoum deals with national and regional affairs through his social media accounts. Even locals and general audiences communicate directly with his highness to share their thoughts and criticisms as well.

On the other hand, the Indonesian government subverted online communication in West Papuan. It slowed the Internet to the minimum to filter information and inhibit rumors during the political uprising [51]. Also, Chinese authorities aggressively fight the distribution of information and try to bury the facts. Referring to the SARS epidemic in 1990, the Chinese government did not deal with transparency to its public. The local officials were lying about the number of infected people as per Tai and Sun [125]. This was not the only time that China was covered up and manipulated with evidence. The whole Chinese dairy industry was affected by the containment crisis of Sanlu Milk Company when leaders in the government delete all negative online exposures against the company [137].

Historically, it is well known that traditional media in the Arab region have been under government control, which continues to practice a strong role in gatekeeping news in both mass and social media contents [88]. Stirred by the uprisings in neighboring counties in Tunisia and Libya, Facebook gave credit to start the revolution in Egypt in 2011. Protesters exaggeratedly use social media requesting for their basic privileges, uploading raw footage, videos, and the overall chaos. Even though Mubarak attempt to block the Internet and disrupted Blackberry Messenger and mobile phone services, Egyptians preferred the online fight. They carried on the proliferation of information through social media [5].

Identically, social media was a part of the uprising in Iran in 2009. The same scenario has been repeated that government disconnected digital activists who found it difficult to uncensored news. Along with this, it blocked websites to stop the stream of information [19]. Protestors in Iran recognized that they could not upload textual and visual content, although they subsequently documented their struggle on social networks producing a "Green Revolution," which did not thrive. The Washington Times [128] indicated that Twitter was heavily used during the uprising to reach all audiences beyond Iran. The Human Rights Watch of Iran [69] claimed that social media tolerates the actual birth of "Citizen Journalism".

In somewhere else in the world, Twitter users who stay in South East Queensland in Australia in 2011 dedicated themselves to broadcasting and re-tweeting the flood updated news that assists the government in arranging emergency backups [18]. In a similar case, the Kenyan government, during its electoral crisis in 2007, encouraged the citizen journalists to gear up the elections through social media since it has a powerful impact upon Kenyans [146].

Sometimes, silence is not golden. The Sanador Hospital in Romania in 2018 has no online engagement and lack of communication on social media. After a child died in the surgery department, the hospital issued press release twice and instantly closed the case claiming that they strictly followed all the medical procedures, and this happened accidentally. People started accusing the hospital and the surgery team and push for an urgent and honest investigation. After a while, the hospital decided to terminate four doctors since they caused the child's death [12]. Therefore, organizations must exploit social media to create a dialogue with the public, show compassion, and clear any misinterpretations at the time of disasters. To put it differently, the Lebanese government in August 2020 shows its sympathy and support to the Beirut Port Explosion victims and injured people. The spokespersons promised to inspect the causes of such catastrophe and ensure that those responsible are held accountable and ultimately brought to justice [11, 9, 10].

The Chinese popular social media application TikTok acquires a large number of users across the world. It gives users 15 s to share their videos, which are soundtracked by music. According to the website, TikTok was downloaded 738 million times in 2019 [92]. China and the United States are not in the same direction as there is an outstanding trade competition. With this in mind, the President of U.S. A. Donald Trump has no mind to tweet several times a day, announced within the electoral season that he has the intention to halt TikTok and WeChat in the States in the current month of September 2020 as per New York Times [37]. Social media continues to open the world up and multiplies the potential of rumors and fake information to be rapidly scattered. Therefore, the message's delivery should be performed adequately in a professional manner to achieve desirable crisis communication goals.

## *2.5 Who Should Deliver Messages to Stakeholders?*

The term "Stakeholders" here refers to any groups or persons who can influence or are influenced by realizing a corporation's purpose [56]. During uprisings and disasters, social media has been notably praised for bonding people. The heavy exchange of information within social media has vividly increased leaders' role to embrace their communication strategy by speaking up at early stages and implementing the truth in all fast-moving crises. There is no point in hiding bad news as it would be eventually discovered. Indeed, fragile crisis communication can make the situation worse and drive for "Double Crisis" and reputational impairment [54]. Thoughtful leaders recognize that effective crisis communication could turn every disaster into an opportunity, for instance: a marketing campaign [140].

It is vigorous to spot that swift media communication during disasters is highly recommended because, in unattainable representatives or information shortage, news agencies will seek alternative sources. When the source is not accredited, fake and inaccurate information will be disseminated. With this, the quick reply mainly within the first 24 h enhances the organization's image in front of the public and helps reduce negative emotions. Representative or spokesperson should provide a comprehensive and true explanation to the media to minimize the uncertainty [65]. Liu et al. [86] advised crisis communication practitioners to spread disaster information through all platforms and in all types of formats to produce better outcomes. Conversely, Formentin et al. [53] argued that traditional media possess a stronger impact on decreasing adverse responses from social media.

Kavanaugh et al. [79] claimed that the U.S. government is in an attempt to employ social media during crises. However, it is still unconfident about utilizing the platforms to respond to thousands of incoming questions and comments. Ironically, the Netherlands' crisis professionals stayed at a "premature level" of using Twitter when the major fire of Moerdijk happened. Dabner [35], stated that governments should possess a strong presence in social media and form their target groups before any crisis occurs.

According to the study of Lovejoy et al. in 2012, Twitter has the greatest potential to influence stakeholders' excessive number during crises among all social media platforms. The feature of real-time interaction is a superior power on Twitter [144]. Ebrahim and Seo [43] argued that Twitter is being used for "propaganda and governmental agenda" in the Middle East area. In a crisis, communicators usually fail to control the unlimited comments of users in Facebook accounts. In times of catastrophe, the public appreciates social media platforms' ability to correct inaccurate information, whereas traditional television news takes a longer time. The official information initiated from public authorities and media organizations has a relatively significant influence over the spread of information as per the investigation of Hughes and Palen [68].

Since the attributions about the organization's crisis responsibility have major consequences on stakeholders' reactions, the importance of delivering proper messages during disasters becomes essential. In essence, the accountability and credibility of information releasing and sharing is not the responsibility of a single person but the full team. The senior management takes the lead within disasters by providing a convenient environment that allows for more understanding and resilience. Hence, communicators should apply "Coping Mechanism" which improves the cognitive disagreements and replace them with consonant perceptions [49]. Most social media users are not passive as they read, share, and seek emotional support. Depending on Coombs [27], emotions should be tied with attribution as many people try to attribute firms to the crisis. Moreover, organizations must adhere to show sympathy toward stakeholders throughout crisis and never overlooked the emotional side.

The amplified debates on who bears the responsibility in crisis communication mandate more investigations in this field. Most organizations responded to crises via their Public Relation department (PR) or the Communication department. A study conducted by Theaker [129] exhibited that nearly 78% of PR practitioners believe that social media is to be the foundation for their functions. 48% of the surveyed PR employees agreed that social media facts are completely reliable and do not necessitate extra counter checking. Communication specialists utilize social media platforms and networks to deal with public concerns by re-tweeting, creating hashtags, and providing hyperlinks to supply information [89].

While big firms assign their Managing Director or CEO to respond to catastrophic events, many organizations hire external professionals who shape the framework of both crisis management and crisis communication. Outsourcing experts should be before crisis to become familiar with the history, organizational culture, previous threats and problems, the nature of employees, and the information's sensitivity. Additionally, those professionals provide training for internal employees to implement the best practices to keep data secured and effectively communicate with the public when the crisis hits their firms [73]. Furthermore, organizations should be aware of what their employees know since certain material and sensitive information are private and cannot be disclosed. Some information is likely considered non-material in itself, but valuable in further information. For that reason, Brown and Ki [17] demonstrated that the stakeholders could assess the level of reputational hazard during a crisis because obtaining a piece of confidential information could cost the organization a fortune.

On the other hand, studies express that there is "no standalone department or position with a sole responsibility for crisis communication." It mentioned that the employees should involve crisis communication, and no peripheral experts are needed at all. Different employees who are nominated on a case-to-case basis require intuitive communication without training or using relevant instruments [124]. Responding successfully to a crisis mandates two important things: the message itself and the medium used for delivery. In 2011, Schultz et al. disputed that the medium is more important than the message, whether it is a traditional or a new media type. The adequate and wise way to deliver the message yields optimistic attitudes among recipients.

## 2.6 Crisis Management in Literature

Coombs [29] explained the term "Crisis" as the stakeholders' perception of a rare event that utterly threatens the organization's performance and generates undesirable outcomes. In this volatile world, organizations should recognize that they are not protected from the crisis. Based on different criteria, researchers classify crises in relevance to timing, cause, intensity, and crisis nature. Bernstein and Bonafede [14] categorized crisis to: "creeping crisis, slow-burn crisis, and sudden." Creeping crisis are "foreshadowed by a serious of events that decision makers do not view as part of pattern." The event of 800 migrants who had drowned in the Mediterranean Sea in 2015 because of a capsized boat was an example of a creeping crisis.

Along with this, the COVID-19 outbreak is considered by some researchers as a creeping crisis that is often difficult to recognize in earlier stages. The slow-burn crisis is the sort that is gradually developed over time and provides advanced signals and warning before the actual strike happened, such as famine resulted from drought and harvest failure. Ultimately, crises are already occurring and need an immediate response before things get worse, such as earthquakes or workplace violence.

Most crises damage the organization's functions and the brand's image if they are not managed properly [134]. For this, organizations should quickly respond to recover the situation and diminish the effects of catastrophe. The prudent method to cope with disasters and crises is to set a well-planned crisis management strategy [63].

Fink [50] defined Crisis Management as the art of getting rid of the risk and ambiguity to improve the control over your destiny. Author Johnson [71] prefers to use the term "Crisis Leadership" instead of crisis management. They acknowledged that employing the same leadership competencies under significant pressures that generate a crisis is the art and ability to employ the same leadership competencies. Leadership in crisis times could extremely affect the organization's goals and outlines both the internal and external atmosphere. It also influences the degree of stakeholders' engagement [93].

Crisis leaders play a vital role in the execution of internal and external communication and collaboration. It is important to develop the ability to craft a good message received by the public. Communication researchers can assist crisis managers in learning how to communicate during such circumstances [31]. Leaders enhance their connections with the public and cultivate deeper and meaningful relationships by utilizing social media platforms [94]. It is equally essential to be completely aware of what people are stating online since their opinions and biases might affect others' thoughts and attitudes [66]. Jin et al. [70] argued that the primary negative emotions arising during crises are: anger, sadness, and fright. Without a doubt, the most dominant emotion after recalling products is anger, as per Choi and Lin's study in 2009 [26]. Coombs and Holladay [30] cited that emotional support can directly impact the public's perceptions when organizations employ apology, sympathy, and compensation. During the pandemic of H1N1 in 2009, organizations incorporated sympathy to respond toward public anxieties, and it was the only affirmative emotion used in diverse media to regain stakeholders' trust. Dezenhall and Weber [39] mentioned that emotions are the driver to run the world instead of plans.

Indeed, social media sites and platforms facilitate the flow of information and ease stakeholders' access to media. Emergency managers and communication practitioners appreciate social media's role in speeding up the communication that allows for providing proper explanations to interested stakeholders [16]. Henceforth, regular updates about the crisis and showing compassion are the keys to better resilience and understanding by the public. Organizations contribute to providing adjusted information when necessary to help stakeholders psychologically cope with stress and suspicions. Ulmer et al. [134] identified that crises should not always be viewed as dangers and negative forces but also as new openings for learning, progress, and growth. To illustrate, aviation disasters are extremely destructive. However, top management should implement a strong post-crisis communication strategy to restore their stakeholder trust and their airline legitimacy [63]. It is worth mentioning that the fundamental concern in all crises should be the public safety that their physical and psychological protection is on the topmost of their priorities. Other important concerns come subsequently, such as financial retrieval and reputation.

To attain an excellent crisis management strategy, leaders should introduce different tools and methods during the lifecycle of a crisis: pre-crisis, crisis, and post-crisis. Meyer [95] preferred the short plans that emphasize vulnerability assessments. Nevertheless, Bernstein and Bonafede [14] disagreed with using template crisis plans since they are unworkable in their point of view. On the other hand, Chandler in 2010 advocated that there are six main phases within each crisis: (1) Warning which highlights the importance of preparedness, (2) Risk Assessment whereby the degree of damage is calculated and addressed, (3) Response that entails the communication strategy to media, public and stakeholders (4) Management phase in which organizations decide their convenient way to handle the crisis (5) Resolution of crisis when company compensates affected stakeholders and finally (6) Recovery phase to regain the reputation and trust of stakeholders.

## 2.7 The Concept of Crisis Communication

Fearn-Banks [47] identified that crisis communication is the mutual dialogue between the organization and its stakeholders before, during, and after the crisis to redeem the organizational image and reputation. Moreover, Coombs in 2010 mentioned that Crisis Communication is "The collection, processing, and dissemination of information required to address a crisis". He underscored that organizations should make extensive efforts to control the severity of losses in the three levels of crisis. Firstly in the pre-crisis phase, relevant information about potential threats and risks is gathered, and expected decisions should be made in case of emergencies. Also, employees should be well trained, especially those who are more likely to be enrolled in crisis management. Secondly, the crisis team should create and distribute information during the crisis depending upon the available and updated facts. Making decisions here is greatly linked to actual newsfeed and evidence that reflects the high commitment and responsibility of the organization. Thirdly, it is important to amend changes and re-evaluate the situation regularly to provide ad hoc messages during the post-crisis phase. Kader Ali et al. [72], claimed that poor communication leads to weighty negative outcomes.

Coombs' theory of Situational Crisis Communication (SCCT) introduced in 2010, has been widely held in researching crises. Situational crises involved an unpredictable or unprepared event that is often beyond individuals' control. For instance, a situational crisis could be a natural disaster or a key person's sudden death. The mentioned theory includes four stages of (1) Bolstering: by expressing thanks to stakeholders for their loyalty and commitment and remind them about the positive results of the past, (2) Denying: which is used when a company is not guilty; hence they could state that the crisis does not occur or somebody else is to be blamed, (3) Diminishing: that is applicable for companies with no or rare similar cases in their history by justifying their actions to diminish their responsibility and (4) Rebuilding: through being responsible, apologizing and offering suitable compensations to those affected. However, internal crisis communication and "stealing thunder" are added to the outlined strategies that help resolve problems. Internal dialogue mitigates the stress inside the organizations and motivates employees to become brand ambassadors during crisis time. Stealing thunder means the organization takes the lead to spread the information and facts before another media party protects its reputation.

It is found by Kim and Liu [80] that social media is most often used in crisis management. Heavy users believe that social media possess credibility during crises and claim that Twitter minimizes the undesired crisis reactions compared to blogs and newspapers [111]. With this perspective, governments have begun to exploit social media in crisis communication, yet researchers appealed that they are not entirely utilized. During the COVID-19 outbreak, Google and Twitter have employed their channels to support small businesses by offering them to communicate with their audiences to generate more sales returns and increase their brands' value.

It is essential to select the best timing to communicate with the public during emergencies through social media. Responding too late could lead to a loss of present or potential followers and allow for more rumors to a blowout. Studies prove that social media platforms should be used only in "filling the silence". At the same time, crises occur by permitting new information to be logically developed from the investigations without any aimlessly reporting. It is alright to say that there is nothing new to be stated, so audience's perceptions about your credibility would stay conserved [138].

Studies found that "Key Publics" respond quicker than others during the crisis and enjoy greater communication influence over the inactive public. Ni and Kim [99] admitted that the active public did not have common characteristics even though they are playing an essential role in increasing or decrease issues and conflicts. Utz et al. [135] stated that those key publics share and discuss crisis news through various media methods, including traditional and modern social media. Although this may be true that there is a lack of information at the first hours of crisis, an emergency manager or crisis communication representative should assign a spokesperson to clarify the situation or at least share initial information over several media channels. The weight given to social media motivates organizations to capitalize on two-way communication by sharing their side of the story and allowing queries from stakeholders to flow. Businesses should also provide a quick response to the government and vice versa to ensure that no contradictions of information are disclosed to the public.

In reality, bulks of organizations dedicate a significant amount of time and resources to focus on their operational side of the crisis. Yet, few of them concern about the readiness of their crisis communication. Moreover, they spend huge amounts on building their brands' images; however, the poor crisis communication strategy destroys their efforts when the crisis hit their companies. Organizations are not immune to crises and disasters. Therefore, it is vital to establish a proactive plan that boosts the organization's reputation, productivity, revenue, compliance, and technical support. Furthermore, organizations should bear in mind that rebuilding their image after a crisis exceeds the earlier planning of crisis management.

It is better to acquire educational messages and materials to help organizations cope with such threats and dangers. Along with this, they need to dispense resources and time for the pre-crisis situation because these are undersized during emergencies. Crook et al. [33] advised organizations to do some pre-crisis work to fill the potential silence that allows for the birth of "Information Vacuums" and "Media Hijacks". Stakeholders demand answers, and people are hungry for information. On the occasion of muteness, a second crisis would occur [78]. Social media is a fertilized ground for hearsays and hoaxes, particularly when credible information is not timely filed. Coombs [28] claimed that silence drives to uncertainty and permit others to take control. He agreed that silence reflects that organizations are trying to hide something or they might be guilty. However, some others justified that they tend to stay away from communication with the public because of the potential of exploiting their words against them in court. Thus, organizations should master the art of both crisis management and crisis communication to meet their intended goals.

### 2.8 Best Practices in Crisis Communication

A short quote stated that "90% of crisis response is communication." In 2012, Holladay claimed that media relations are the core of crisis communication since they enable organizations to recognize how their replies would look like. Public crisis communication entails oral, visual, and written interactions [48]. Here below are ten suggested best practices in crisis communication provided by literature [23] and developed by the author:

1. Plan for success: organizations from the beginning should acquire appropriate vision, mission, goals, strategies, plans, and notification protocols that embrace their success and lead to desirable outcomes at all times.
2. Prepare messages in advance: Management has to craft various communication messages aligned with their commitment to different potential emergencies. Both training in crisis communication and the readiness of communication messages should be well-prepared in advanced stages.
3. Monitor the chatter: It is advisable to pay attention to tools and practices that can enhance the organization's prosperity and alleviate threats and pressures that would greatly influence the safety side and the operations' continuity.
4. Keep messages simple: Thoughtful leaders have to bear in mind that employees and stakeholders during disasters possess negative emotions. Therefore, they have to keep their messages simple, direct, short, honest, updated, and emotionally supportive.
5. Words matter: Words are essential to direct the public reaction since social media contents are rapidly shared in front of the entire world. Convenient words would motivate the audience to act properly by accepting the new situation and coping with challenging surroundings. Yet, random and unperceptive words drive for more fears, anger, and confusion.
6. Stay in Connection with Audience; victims first: When a crisis strikes, organizations have to put the victims and their families on their priority and lessen their pain by employing compassion, sympathy, or maybe compensation. Being in touch with audience's needs and concerns during and after a crisis can restore the organization's image and reputation.
7. Assign a spokesperson: Between verbal and printed media, it is more favored to choose a spokesperson who provides clear and sufficient explanations and facts about the event. That is why governments and leading companies implement this procedure to mitigate the worries and ambiguities.
8. Coordinate with governments and media: To avoid information inconsistency during the crisis, all organizations should harmonize with the available media agencies and the government's instructions.

9. Design your own social media strategy: The weight given to social media imposes organizations to craft their social media strategy to utilize its benefits in information exchange, interactive dialogue, marketing campaigns, and furthermore.
10. Evaluate your response to the crisis: After each crisis, organizations' leadership has to assess their plans and actions that have been executed during the crisis. Good performance means the organization possesses adequate succession and resilient plans, whereas failing to cope with vibrant threats indicates that they need to revise and improve their plans once again.

# 3 Conclusion and Limitations

This section of the chapter comprises the conclusion and limitations of the previous literature review about employing social media in crisis communication. However, recommendations were provided earlier under the "Best Practices" headline.

## *3.1 Conclusion*

To sum up, this chapter started with the need for fast, updated, timely, and interactive exchange of communication and information approaches within this vibrant and dynamic world where digital proliferation extensively embraces the utilization of social media websites and platforms. To keep up with the increased number of social media users worldwide, governments and business organizations recognize the importance of crafting their social media strategies that maintain their image, reputation, public safety, and functions. In addition, they noticed the benefits of social media to reach a massive audience to achieve their anticipated goals. Furthermore, they paid attention to the risks accompanied by social media exclusively when fabricated facts and rumors are distributed everywhere and eventually drive for harmful consequences, especially during crises. This chapter also highlighted how some governments and organizations respond to crises through social media applications over the last two decades. Tied with this, the concept of crisis management and crisis communication were discussed, and then finally, certain best practices to attain effective crisis communication were delivered.

## *3.2 Limitations*

In the research field, it is deemed necessary shortly to learn more about social media functions and implementations in all life aspects. It is also vital to develop social media theories and go through deeper insights into how to deal with distorted information during crises, essentially the information vacuum. Governments must accelerate the awareness about cybercrimes to students at schools and universities and the public due to the extremely growing number of Internet and social media users. Putting physical and psychological effects aside, many court cases reflect how digitalization destructs societies in terms of breaching others' privacy and security, stealing bank and social media accounts, trying to promote unacceptable and unethical practices, and broadcasting misleading information during crises. As discussed above, exploiting social media platforms to dispense false information could harm the government, business firm, and individual reputation. It needs a long time and great effort to be restored, especially during crises and bad times when fears and uncertainties are widely spread.

## References

1. Acharjee, S.: Look up: social media and the addiction no one is talking about. Hay House Publishers, Alexandria, NSW (2016)
2. Agrawal, A.J.: Why influencer marketing will explode in 2017. Forbes, 27 Dec 2016. https://www.forbes.com/sites/ajagrawal/2016/12/27/why-influencer-marketing-will-explode-in-2017/#4f159a0020a9 (2016)
3. Al-Khuwaildi, M.: Social media influencers accused of money laundering in Kuwait. Available at: https://english.aawsat.com/home/article/2421936/social-media-influencers-accused-money-laundering-kuwait (2020). Accessed 22 Sept 2020
4. Al Kurdi, O.: A critical comparative review of emergency and disaster management in the Arab World. J. Bus. Soc. Econ.Dev. **1**(1), 24–41 (2021)
5. Ali, S., Fahmy, S.: The icon of the Egyptian revolution: using social media in the toppling of a Mideast government. In: Berenger, R. (ed.) Social Media Go to War: Civil Unrest, Rebellion and Revolution in the Age of Twitter. Marquette Books, Washington, DC (2012)
6. Areni, C.: Ontological security as an unconscious motive of social media users. J. Mark. Manag. **35**(1), 75–96 (2019)
7. Arti, S., AlDaihani, A., Mustafa, A., AlHashemi, M.: Instagram your life: why do female Kuwait University students use instagram? Media Watch **10**(2), 251–261 (2019)
8. Austin, L., Liu, B.F., Jin, Y.: How audiences seek out crisis information: exploring the social-mediated crisis communication model. J. Appl. Commun. Res. **40**(2), 188–207 (2012)
9. BBC News: Hackers attack Norwegian parliament. Available at: https://www.bbc.com/news/technology-53985422 (2020). Accessed 22 Sept 2020
10. BBC News: Ikea to pay family $46m after child killed by falling drawers. Available at: https://www.bbc.com/news/world-us-canada-51017438 (2020). Accessed 22 Sept 2020
11. BBC News: Beirut explosion: what we know so far. Available at: https://www.bbc.com/news/world-middle-east-53668493 (2020). Accessed 22 Sept 2020
12. Barbu, R., Cmeciu, C.: Crisis communication in Romania and social media influencers and followers. Case study: patient's death at the Sanador Hospital. J. Media Res. **12**(3(35)), 5–17
13. Bergh, B.V., Lee, M., Quilliam, E., Hove, T.: The multidimensional nature and brand impact of user-generated Ad parodies in social media. Int. J. Advertising Res. **30**(1), 103–131 (2011)
14. Bernstein, J., Bonafede, B.: Manager's guide to crisis management. McGraw Hill Profressional, New York (2011)

15. Bertot, J.C., Jaeger, P.T., Grimes, J.M.: Using ICTs to create a culture of transparency: e-government and social media as openness and anti-corruption tools for societies. Gov. Inf. Q. **27**(3), 264–271 (2010)
16. Bridgeman, R.: Crisis communication and the net. In: Anthonissen, P.F. (ed.) Crisis communication: Practical PR Strategies for Reputation Management and Company Survival, pp. 169–177. Kogan Page, UK (2008)
17. Brown, K.A., Ki, E.J.: Developing a valid and reliable measure of organizational crisis responsibility. Journalism Mass Commun. Q. **90**(2), 363–384 (2013)
18. Bruns, A., Burgess, J.E., Crawford, K., Shaw, F.: #qldfloods and @QPSMedia: crisis communication on Twitter in the 2011 South East Queensland floods. research Report, Media Ecologies Project ARC Centre of Excellence for Creative Industries & Innovation (CCI), Queensland University of Technology, pp. 1–58 (2012)
19. CNN World: Ahmadinejad hails election as protests grow, 13 June. Available at: https://edition.cnn.com/2009/WORLD/meast/06/13/iran.election/ (2009). Accessed 22 Sept 2020
20. Cevik, A.A.A., Akoglu, H., Eroglu, S.E., Dogan, N.O., Altunci, Y.A.: Review article: social media, FOAMed in medical education and knowledge sharing: local experiences with international perspective. Turkish J. Emerg. Med. **16**, 112–117 (2016). https://doi.org/10.1016/j.tjem.2016.07.001
21. Chadwick, A.: Internet Politics: States, Citizens, and New Communications Technologies. Oxford University Press, New York (2006)
22. Chadwick, A., May, C.: Interaction between states and citizens in the age of the Internet: e-government in the United States, Britain, and European Union. Governance **16**(2), 271–300 (2003)
23. Chandler, R.C.: The Six Stages of Crisis. s.l.: Everbridge (2010)
24. Chen, Y., Ganesan, S., Liu,Y.: Does a firm's product-recall strategy affect its financial value? An examination of strategic alternatives during product-harm crises. J. Market. **73**(6), 214–226 (2009)
25. Choi, S.M.: Guest editorial: advertising and social media. Int. J. Advert. **30**(1), 11–12 (2011). https://doi.org/10.2501/IJA-30-1-009
26. Choi, Y., Lin, Y.-H.: Consumer responses to Mattel product recalls posted on online bulletin boards: exploring two types of emotion. J. Pub. Relat. Res. **21**, 198–207 (2009). https://doi.org/10.1080/10627260802557506
27. Coombs, T.: Protecting organization reputations during a crisis: the development and application of situational crisis communication theory. Corp. Reput. Rev. **10**(3), 163–176 (2007)
28. Coombs, T.: Ongoing Crisis Communication: Planning, Managing and Responding, 3rd edn. Sage Publications, California (2012)
29. Coombs, W.T.: Parameters for crisis communication. In: Coombs, W.T., Holladay, S. J. (eds.) The Handbook of Crisis Communication, pp. 17–53. Blackwell Publishing Ltd., UK (2010)
30. Coombs, W.T., Holladay, S.J.: Further exploration of post-crisis communication: effects of media and response strategies on perceptions and intentions. Pub. Relat. Rev. **35**, 1–6 (2009). https://doi.org/10.1016/j.pubrev.2008.09.011
31. Coombs, W.T., Holladay, S.J.: How publics react to crisis communication efforts: comparing crisis response reactions across sub-arenas. J. Commun. Manag. **18**(1), 40–57 (2014)
32. Crijns, H., Cauberghe, V., Claeys, A.S., Hudders, L.: The use of ambiguity markers in crisis communication and the moderating role of source of information disclosure. In: EMAC (2017)
33. Crook, B., Glowacki, E.M., Suran, M., Harris, J.K., Bernhardt, J.M.: Content analysis of a live CDC twitter chat during the 2014 Ebola outbreak. Commun. Res. Rep. **33**(4), 349–355 (2016). https://doi.org/10.1080/08824096.2016.1224171
34. Cyberspace Administration Office of China: Unprecedented development of government social media [in Chinese]. Retrieved from http://www.cac.gov.cn/2015-02/07/c_1114291869.htm (2015)

35. Dabner, N.: "Breaking ground" in the use of social media: a case study of a university earthquake response to inform educational design with facebook. Internet Higher Edu. **15**(1), 69–78 (2012)
36. Daft, R.L., Lengel, R.H.: Organizational information requirements, media richness and structural design. Manage. Sci. **32**(5), 554–571 (1986)
37. David McCabe: President Trump says China must cede control of TikTok or he 'won't make the deal.' Available at: https://www.nytimes.com/2020/09/21/business/trump-tiktok-deal.html (2020). Accessed 22 Sept 2020
38. Deephouse, D.L.: Media reputation as a strategic resource: an integration of mass communication and resource-based theories. J. Manag. **26**, 1091–1112 (2000)
39. Dezenhall, E., Weber, J.: Damage control: the essential lessons of crisis management. Prospecta Press, Westport, CT (2011)
40. Duncombe, C.: Twitter and transformative diplomacy: social media and Iran–US relations. Oxford University Press, USA (2017)
41. Durukan, T., Bozacı, İ, Hamsioğlu, B.: An investigation of customer behaviors in social media. Eur. J. Econ. Financ. Adm. Sci. **44**, 148–157 (2012)
42. EDELMAN: Barometer Edelman Trust. Trust special report: Trust in brands and Coronavirus pandemic (Brazilian and global data). Available in: https://www.edelman.com.br/sites/g/files/aatuss291/files/2020-04/2020%20ETB%20Brands%20and%20the%20Coronavirus_Brasil%20com%20Global_POR.pdf (2020). Accessed 22 Sept 2020
43. Ebrahim, H., Seo, H.: Visual public relations in middle eastern higher education: content analysis of twitter images. Media Watch **10**(1), 41–53 (2019)
44. Eggers, F., Hatak, I., Kraus, S., Niemand, T.: Technologies that support marketing and market development in SMEs—evidence from social networks. J. Small Bus. Manage. **55** (2), 270–302 (2017). https://doi.org/10.1111/jsbm.12313
45. Elali, W.: The importance of strategic agility to business survival during corona crisis and beyond. Int. J. Bus. Ethics Gov. **4**(2), 1–7 (2021)
46. Evans, D.: The internet of things: how the next evolution of the internet is changing everything. Cisco IBSG, USA (2011)
47. Fearn-Banks, K.: Crisis communications: a case approach, 4th edn. Routeledge, New York (2007)
48. Ferguson, D., Wallace, J.D., Chandler, R.: Rehabilitating your organization's image: public relations professional's perceptions of the effectiveness and ethicality of image repair strategies in crisis situations. Pub. Relations J. **6**(1) (2012)
49. Festinger, L.: A Theory of Cognitive Dissonance. Stanford University Press, Stanford, CA (1957)
50. Fink, S.: Crisis Management: Planning for the Inevitable. Authors Guild Backinprint.com Edition, Lincoln, NE (1986)
51. Firdaus, F.: Indonesia deploys troops to West Papua as protests spread. Al Jazeera. Retrieved from https://www.google.com/url?sa=t&rct=j&q=&esrc=s&source=web&cd=&cad=rja&uact=8&ved=2ahUKEwiogs-i1P3rAhUbUhUIHdoRBfYQFjABegQIBxAB&url=https%3A%2F%2Fojs.aut.ac.nz%2Fpacific-journalism-review%2Farticle%2Fview%2F1079&usg=AOvVaw0cDJ53hSWGVWYOZwrFbmA, 21 Aug 2019
52. Forbes: A very short history of the internet and the web. Available at: https://www.forbes.com/sites/gilpress/2015/01/02/a-very-short-history-of-the-internet-and-the-web-2/#22729cfd7a4e (2015). Accessed 22 Sept 2020
53. Formentin, M., Bortree, D.S., Fraustino, J.D.: Navigating anger in happy valley: analyzing Penn State's facebook-based crisis responses to the Sandusky scandal. Pub. Relations Rev. **43**(4), 671–679 (2017). https://doi.org/10.1016/j.pubrev.2017.06.005
54. Frandsen, F., Johansen, W.: Crisis communication, complexity, and the cartoon affair: a case study. In: Coombs, W.T., Holladay, S.J. (eds.) The Handbook of Crisis Communication, pp. 425–448. Blackwell, Malden, MA (2010)

55. Freedom House: Freedom of the press 2009. Freedom House. Available at: https://freedomhouse.org/sites/default/files/FOTP%202009%20Final%20Full%20Report.pdf (2009). Accessed 22 Sept 2020
56. Freeman, R.E.: Strategic Management: A Stakeholder Approach. Pitman, Boston (1984)
57. Frost and Sullivan: Mobile connectivity and growing data services drive the telecoms market in Africa [Blog post]. Retrieved from http://ww2.frost.com/news/press-releases/frost-sullivanmobile-connectivity-and-growing-data-services-drive-telecoms-market-africa/ (2014)
58. Gagnon, K., Sabus, C.: Professionalism in a digital age: opportunities and considerations for social media in health care. Phys. Ther. **95**(3), 406–409 (2015)
59. Gjylbegaj, V.: Role and influence of social media on the PR industry: uses and opportunities in UAE. Media Watch **11**(2), 356–362 (2020)
60. Gokcearslan, S., Mumcu, F.K., Haslaman, T., Cevik, Y.D.: Modelling smartphone addition: the role of smartphone usage, self-regulation, general self-efficacy and cyberloafing in university students. Comput. Hum. Behav. **63**, 639–649 (2016)
61. Grazzi, M., Jung, J.: ICT, innovation and productivity: evidence from Latin American firms. IACEA (2015)
62. Hajli, N., Sims, J., Zadeh, A.H., Richard, M.-O.: A social commerce investigation of the role of trust in a social networking site on purchase intentions. J. Bus. Res. **71**, 133–141 (2017)
63. Hansson, A., Vikstrom, T.: Succesful crisis management in the airline industry: a quest for legitimacy through communication? Uppsala University, Uppsala (2011)
64. Hitchen, E.L.N., Ferras, P.A., Mussons, X.S.: Social media: open innovation in SMEs finds new support. J. Bus. Strateg. **38**(3), 21–29 (2017). https://doi.org/10.1108/JBS-02-2016-0015
65. Holladay, S.J.: Are they practicing what we are preaching? An investigation of crisis communication strategies in the media coverage of chemical accidents. In: Coombs, T., Holladay, S. (eds.) The Handbook of Crisis Communication, pp. 159–180. Wiley-Blackwell Ltd, UK (2012)
66. Hsueh, M., Yogeeswaran, K., Malinen, S.: "Leave your comment below": can biased online comments influence our own prejudicial attitudes and behaviors? Hum. Commun. Res. **41** (4), 557–576 (2015). https://doi.org/10.1111/hcre.12059
67. Huang, K., Li, M., Markov, S.: what do employees know? Evidence from a social media platforms. Account. Rev. **95**(2), 199–226 (2020)
68. Hughes, A.L., Palen, L.: Twitter adoption and use in mass convergence and emergency events. Int. J. Emergency Manage. **6**(3/4), 248–260 (2009)
69. Human Rights Watch: Iran. Available at: https://www.hrw.org/world-report/2010/country-chapters/iran (2010). Accessed 22 Sept 2020
70. Jin, Y., Liu, B.F., Austin, L.: Examining the role of social media in effective crisis management: the effects of crisis origin, information form, and source on publics' crisis responses. Commun. Res. **41**, 74–94 (2014). https://doi.org/10.1177/0093650211423918
71. Johnson, T.: Crisis leadership: how to lead in times of crisis, threat and uncertainty. Bloomsbury, London, England (2018)
72. Kader Ali, N.N., Soon, B.Y., Goh, L.S., Ahmad Razi, N.A.: Symptoms versus problems (SVP) in household high speed broadband (HSBB): regaining momentum for Unifi, Malaysia. Problems Perspect. Manage. Int. Res. J. **13**(2), 330–346 (2015–2). ISSN 1727–7051, 1810–5467 (online)
73. Kaigo, M.: Social media usage during disasters and social capital: twitter and the great East Japan earthquake. Keio Commun. Rev. **34**, 19–35 (2012)
74. Kamboj, S., Sarmah, B., Gupta, S., Dwivedi, Y.: Examining branding co-creation in brand communities on social media: applying the paradigm of Stimulus organism response. Int. J. Inf. Manage. **39**(April), 169–185 (2018)
75. Kang, J.-Y.: Showrooming, webrooming, and user-generated content creation in the omnichannel era. J. Internet Commerce **17**(2), 145–169 (2018)
76. Kanuri, V., Chen, Y., Sridhar, H.: Scheduling content on social media: theory, evidence, and application. J. Mark. **82**(6), 89–108 (2018)

77. Kaplan, M., Haenlein: Users of world unite! The challenges and opportunities of social media. Bus. Horiz. **53**(1), 59–68 (2010). https://doi.org/10.1016/j.bushor.2009.09.003
78. Karl Grebe, S.: Things can get worse: How mismanagement of a crisis response strategy can cause a secondary or double crisis: the example of the AWB corporate scandal. Corp. Commun. Int. J. **18**, 70–86 (2013)
79. Kavanaugh, A.L., Fox, E.A., Scheetz, S.D., Yang, S., Tzy Li, L., Shoemaker, D.J., Natsev, A., Xie, L.: Social media use by the government: from the routine to the critical. Gov. Inf. Q. **29**(4), 480–491 (2012)
80. Kim, S., Liu, B.F.: Are all crises opportunities? A comparison of how corporate and government organizations responded to 2009 flu pandemic. J. Pub. Relations Res. **24**, 69–85 (2012)
81. Knoll, J.: Advertising in social media: a review of empirical evidence. Int. J. Advert. **35**(2), 266–300 (2016). https://doi.org/10.1080/02650487.2015.1021898
82. Kwahk, K.Y., Park, D.H.: The effects of network sharing on knowledge-sharing activities and job performance in enterprise social media environments. Comput. Hum. Behav. **55**, 826–839 (2016). https://doi.org/10.1016/j.chb.2015.09.044
83. Lachlan, K.A., Spence, P.R., Lin, X.: Expressions of risk awareness and concern through Twitter: on the utility of using the medium as an indication of audience needs. Comput. Hum. Behav. **35**, 554–559 (2014). https://doi.org/10.1016/j.chb.2014.02.029
84. Lazo, L.: After United dragging incident, 3 major airlines change policies affecting bumped passenger. Washington Post, 17 Apr: https://www.washingtonpost.com/news/dr-gridlock/wp/2017/04/17/after-united-dragging-incident3-major-airlines-change-policies-affecting-bumped-passengers/ (2017)
85. Lee, E.-J.: That's not the way it is: how user-generated comments on the news affect perceived media bias. J. Comput. Mediated Commun. **18**, 32–45 (2012)
86. Liu, B.F., Fraustino, J.D., Jin, Y.: How disaster information form, source, type, and prior disaster exposure affect public outcomes: jumping on the social media bandwagon? J. Appl. Commun. Res. **43**(1), 44–65 (2015)
87. Livingstone, S., Helsper, E.: Gradations in digital inclusion: children, young people and the digital divide. New Media Soc. **9**(4), 671–696 (2007)
88. Loewenstein, A.: Viewpoint: why Iran's twitter revolution failed. BBC World Service, 24 Mar. Available at: http://www.bbc.co.uk/worldservice/programmes/2010/03/100324_iran_blogging.shtml (2010). Accessed 22 Sept 2020
89. Lovejoy, K., Waters, R.D., Saxton, G.D.: Engaging stakeholders through Twitter: how nonprofit organizations are getting more out of 140 characters or less. Pub. Relations Rev. **38**, 313–318 (2012). https://doi.org/10.1016/j.pubrev.2012.01.005
90. Magro, M.J.: A review of social media use in E-government. Adm. Sci. **2**(2), 148–161 (2012)
91. Mansfield, M.: Social media statistics 2016. Small Business Trends (November 27), https://smallbiztrends.com/2016/11/social-media-statistics-2016.html (2016)
92. Mansoor, I.: TikTok revenue and usage statistics (2020). Available at: https://www.businessofapps.com/data/tik-tok-statistics/ (2020). Accessed 22 Sept 2020
93. Men, R.L., Stacks, D.W.: The effects of authentic leadership on strategic internal communication and employee-organization relationships. J. Pub. Relations Res. **26**(4), 301–324 (2014). https://doi.org/10.1080/1062726X.2014.908720
94. Men, L.R., Tsai, W.S.: Public engagement with CEOs on social: motivations and relational outcomes. Pub. Relations Rev. **42**(5), 932–942 (2016). https://doi.org/10.1016/j.pubrev.2016.08.001
95. Meyer, G.: Corporate Smokejumper. Crisis Management: Tools, Tales and Techniques. Blue Blaze Books, Newark, DE (2017)
96. Mukoolwe, E., Korir, J.: Social media and entrepreneurship: tools, benefits and challenges: a case study of women online entrepreneurs on kilimani mums marketplace on facebook. Int. J. Humanit. Soc. Sci. **6**, 248–256 (2016)

97. Na, E.K., Rhee, J.W.: A Study on Readers' Comments: Changes in the Online News Usage Pattern and the Implications for Discursive Public. Korea Press Foundation, Seoul, Korea (2008)
98. NeimanLab: What happened after 7 news sites got rid of reader comments. Retrieved from http://www.niemanlab.org/2015/09/what-happened-after-7-news-sites-got-rid-of-reader-comments/ (2015)
99. Ni, L., Kim, J.N.: Classifying publics: communication behaviors and problem-solving characteristics in controversial issues. Int. J. Strateg. Commun. **3**(4), 217–241 (2009)
100. OMNICORE Website: YouTube by the numbers: stats, demographics & fun facts. Available at: https://www.omnicoreagency.com/youtube-statistics/ (2020). Accessed 22 Sept 2020
101. Okazaki, S., Taylor, C.R.: Social media and international advertising: theoretical challenges and future directions. Int. Mark. Rev. **30**(1), 56–71 (2013)
102. Ortiz-Ospina, E.: The rise of social media. Available at: https://ourworldindata.org/rise-of-social-media (2019). Accessed 22 Sept 2020
103. Osgerby, J., Rush, D.: An exploratory case study examining undergraduate accounting students' perceptions of using twitter as a learning support tool. Int. J. Manag. Educ. **13**(3), 337–348 (2015). https://doi.org/10.1016/j.ijme.2015.10.002
104. Parveen, F., Jaafar, N.I., Ainin, S.: Social media usage and organizational performance: reflections of Malaysian social media managers. Telematics Inform. **32**(1), 67–78 (2015)
105. Peres, R., Talwar, S., Alter, L., Elhanan, M., Friedmann, Y.: Narrowband influencers and global icons: universality and media compatibility in the communicationpatterns of political leaders worldwide. J. Int. Mark. **28**(1), 48–65 (2020)
106. Rajanala, S., Maymone, M.B.C., Vashi, N.A.: Selfies—living in the era of filtered photographs. JAMA Facial Plast. Surg. (2020). https://doi.org/10.1001/jamafacial.2018.0486
107. Ramo, D.E., Prochaska, J.J.: Prevalence and co-use of marijuana among young adult cigarette smokers: an anonymous online national survey. Addict. Sci. Clin. Pract. **7**(5), 1–7 (2012)
108. Reuters: New Zealand's stock exchange hit by second cyber attack. Available at: https://www.reuters.com/article/us-nzx-cyber/new-zealands-stock-exchange-hit-by-second-cyber-attack-idUSKBN25M08O (2020). Accessed 22 Sept 2020
109. Roberts, A., Kim, B.Y.: Policy responsiveness in post-communist Europe: public preferences and economic reforms. Br. J. Polit. Sci. **41**(4), 819–839 (2011)
110. Van Scheers, L.: SMEs' marketing skills challenges in South Africa. Afr. J. Bus. Manage. **5**(13), 5048–5056 (2011)
111. Schultz, F., Utz, S., G€oritz, A.: Is the medium the message? Perceptions of and reactions to crisis communication via twitter, blogs and traditional media. Pub. Relations Rev. **37**(1), 20–27 (2011)
112. Schwarz, A.: How publics use social media to respond to blame games in crisis communication: the love parade tragedy in Duisburg 2010. Pub. Relations Rev. **38**(3), 430–437 (2012)
113. Seltzer, T., Mitrook, M.A.: The dialogic potential of weblogs in relationship building. Pub. Relations Rev. **33**, 227–229 (2007). https://doi.org/10.1016/j.pubrev.2007.02.011
114. Senft, T.M., Baym, N.K.: What does the selfie say? Investigating a global phenomenon. Int. J. Commun. **9**, 1588–1606 (2015)
115. Shang, S.S.C., Ya-Ling, Wu., Li, E.Y.: Field effects of social media platforms on information-sharing continuance: do reach and richness matter? Inf. Manage. **54**(2), 241–255 (2017)
116. Shen, H., Guangyan, W.: Can dynamic knowledge-sharing activities be mirrored from the static online social network in Yahoo! Answers and how to improve its quality of service? IEEE Transactions on Systems. Man Cybern. Syst. **47**(12), 3363–3376 (2017). https://doi.org/10.1109/TSMC.2016.2580606
117. Sheth, J.N.: Borderless media: rethinking international marketing. J. Int. Mark. **28**(1), 3–12 (2020)

118. Shoemaker, P., Reese, S.: Mediating the Message. Theories of Influence on Mass Media Content. Longman Publishers, White Plains, NY (2014)
119. Short, J., Williams, E., Christie, B.: The Social Psychology of Telecommunications. Wiley, London (1976)
120. Statista: Number of monthly active LINE users in Japan from 1st quarter 2013 to 2nd quarter 2020. Available at: https://www.statista.com/statistics/560545/number-of-monthly-active-line-app-users-japan/ (2020). Accessed 22 Sept 2020
121. Statista: Number of social network users worldwide from 2017 to 2025. Available at: https://www.statista.com/statistics/278414/number-of-worldwide-social-network-users/ (2020). Accessed 22 Sept 2020
122. Stelzner, M.: Social media marketing industry report: How marketers are using social media to grow their businesses. Retrieved from https://www.academia.edu/20766156/How_Marketers_Are_Using_Social_Media_to_Grow_Their_Businesses_M_A_Y_2_0_1_5 (2013)
123. Strick, B., Syavira, F.: Papua unrest: Social media bots 'skewing the narrative'. BBC News, 11 Oct 2019. Retrieved from https://www.bbc.com/news/world-asia-49983667
124. Szabóová, V.: Crisis Communication in the Creative Industries. Marketing Identity, Slovak Republic (2019)
125. Tai, Z., Sun, T.: Media dependencies in a changing media environment: the case of the 2003 SARS epidemic in China. New Media Soc. **9**(6), 987–1009 (2007)
126. Taipale,S.: The use of E-government services and the Internet: the role of sociodemographic, economic and geographical predictors. Telecommun. Policy **37**(4–5), 413–422 (2013)
127. Taki, M.: Editorial. In: Taki, M., Coretti, L. (eds.) The Role of Social Media in the Arab Uprisings—Past and Present, vol. 9, pp. 1–2. University of Westminster, London, UK (2013)
128. The Washington Times: Editorial: Iran's Twitter revolution, 16 June. Available at: http://www.washingtontimes.com/news/2009/jun/16/irans-twitter-revolution/ (2009) Accessed 22 Sept 2020
129. Theaker, A.: The Public Relations Handbook. Abingdon-on-Thames, Routledge (2016)
130. Thomas, J.C., Streib, G.: E-democracy, E-commerce, and E-research: examining the electronic ties between citizens and governments. Adm. Soc. **37**(3), 259–280 (2005)
131. Tran, T., Bar-Tur, Y.: Social media in government: benefits, challenges, and how it's used. Available at: https://blog.hootsuite.com/social-media-government/ (2020). Accessed 22 Sept 2020
132. Traxler, J.: Potential of learning with mobiles in Africa [Blog post]. Retrieved from https://www.wise-qatar.org/mobile-learning-africa-john-traxler/ (2013)
133. Tu, C.H.: On-line learning migration: From social learning theory to social presence theory in CMC environment. J. Netw. Comput. Appl. **23**(1), 27–37 (2000)
134. Ulmer, R.R., Sellnow, T.L., Seeger, M.W.: Effective crisis communication: moving from crisis to opportunity, 2nd edn. Sage, Thousand Oaks, CA (2011)
135. Utz, S., Schultz, F., Glocka, S.: Crisis communication online: how medium, crisis type and emotions affected public reactions in the Fukushima Daiichi nuclear disaster. Pub. Relations Rev. **39**(1), 40–46 (2013)
136. Vandewalle, D.: Is this Libya's new revolution? CNN Opinion, 21 Feb. Available at http://edition.cnn.com/2011/OPINION/02/21/vandewalle.libya.uprising/index.html (2011). Accessed 22 Sept 2020
137. Veil, S.R., Yang, A.: Media Manipulation in the Sanlu milk contamination crisis. Pub. Relations Rev. **38**(5), 935–937 (2012)
138. Wang, B., Zhuang, J.: Crisis information distribution on Twitter: a content analysis of tweets during Hurricane Sandy. Nat. Hazards **89**(1), 161–181 (2017). https://doi.org/10.1007/s11069-017-2960-x
139. Wang, Y.: Brand crisis communication through social media. A dialogue between brand competitors on Sina Weibo. Corp. Commun. **21**(1), 56–72 (2016a). https://doi.org/10.1108/CCIJ-10-2014-0065

140. Wang, Y.: How do television networks use twitter? Exploring the relationship between twitter use and television ratings. South. Commun. J. **81**(3), 125–135 (2016b). https://doi.org/10.1080/1041794X.2015.1116593
141. Webb-Gannon, C., Webb, M.: 'More than a music, it's a movement': West Papua decolonization songs, social media, and the remixing of resistance. Contemp. Pac. **31**(2), 309–343 (2019). https://doi.org/10.1353/cp.2019.0025
142. White, D.M.: The gatekeeper: a case study in the selection of news. Journal. Q. **27**(4), 383–390 (1950)
143. World Health Organization: Statement on the second meeting of the international health regulations (2005) Emergency committee regarding the outbreak of novel coronavirus (2019-nCoV). Available at: https://www.who.int/news-room/detail/30-01-2020-statement-on-the-second-meeting-of-the-international-health-regulations-(2005)-emergency-committee-regarding-the-outbreak-of-novel-coronavirus-(2019-ncov) (2020). Accessed 22 Sept 2020
144. Xifra, J., Graub, F.: Nanoblogging PR: the discourse on public relations in Twitter. Pub. Relations Rev. **36**, 171–174 (2010). https://doi.org/10.1016/j.pubrev.2010.02.005
145. Ziani, A., Elareshi, M., AlRashid, M.: Social impact of digital media: growth pattern of facebook in the Arab world. Media Watch **8**(2), 177–191 (2017)
146. Zuckerman, E.: Citizen media and the Kenyan electoral crisis. In: Cottle, S., et al. (eds.) Citizen Journalism: Global Perspectives, pp. 187–196. Peter Lang, New York (2009)

Printed in the United States
by Baker & Taylor Publisher Services